14th International Conference on Metrology and Properties of Engineering Surfaces

(Met & Props 2013)

Journal of Physics: Conference Series Volume 483

Taipei, Taiwan
17-21 June 2013

ISBN: 978-1-63266-704-5
ISSN: 1742-6588

Printed from e-media with permission by:

Curran Associates, Inc.
57 Morehouse Lane
Red Hook, NY 12571

Printed by Curran Associates, Inc. (2014)

For permission requests, please contact the Institute of Physics at the address below.

Institute of Physics
Dirac House, Temple Back
Bristol BS1 6BE UK

Phone: 44 1 17 929 7481
Fax: 44 1 17 920 0979

techtracking@iop.org

Additional copies of this publication are available from:

Curran Associates, Inc.
57 Morehouse Lane
Red Hook, NY 12571 USA
Phone: 845-758-0400
Fax: 845-758-2634
Email: curran@proceedings.com
Web: www.proceedings.com

TABLE OF CONTENTS

Preface

PAPERS

Journal of Physics: Conference Series **483** (2014) 011001 doi:10.1088/1742-6596/483/1/011001

Preface

Proceedings of the 14th International Conference, Taipei, Taiwan, 17th–21st June, 2013

Organized by:

Center for Measurement Standards/Industrial Technology Research Institute
Mechanical and Systems Research Laboratories/Industrial Technology Research Institute
National Taiwan University
National Cheng Kung University
National Taiwan University of Science and Technology
National Tsing Hua University

Greetings from Chairman of International Programme Committee

When Professor Ken Stout and I founded this series of conferences in the United Kingdom more than thirty years ago, we did not anticipate its longevity or its success. Since that first meeting at Leicester, the conference has been often held in England, but also in several other European countries: France, Poland and Sweden, as well as in the United States. Ken, sadly no longer with us, would be proud of what it has achieved and has come to represent. Generations of researchers have presented their new ideas and innovations here which are now embodied in many textbooks and international standards.

But this conference in 2013 marks a new departure and perhaps a new future. For the first time it is being held in Asia, reflecting the historic rise of the economies of the Pacific Rim, adding modern technology to their long-existing traditions of ordered insight and precise craftsmanship. Many of you have travelled far to attend this meeting, and we hope you will feel your trouble has been rewarded. We have an excellent selection of papers from all over the world from many of the world's experts, embodying the consolidation of tested ideas as well as the latest advances in the subject. These will be set in context by a glittering array of keynote and invited speakers.

On behalf of the International Programme Committee, I am glad to acknowledge the hard work of the members of the Local Organising Committee in putting the programme together and making all the arrangements, and to accept their hospitality. It is my privilege and pleasure to welcome you all to the 14th International Conference on Metrology and Properties of Engineering Surfaces here in Taipei.

Tom Thomas
Halmstad, 1st June 2013

Journal of Physics: Conference Series **483** (2014) 011001 doi:10.1088/1742-6596/483/1/011001

Greetings from Chairman of Local Organizing Committee

It is the great honor of Center for Measurement Standards (CMS), metrology group of Industrial Technology Research Institute (ITRI), to host the 14th International Conference on Metrology and Properties of Engineering Surfaces (Met & Props 2013) from 17th–21st June, 2013, in Taipei, Taiwan.

In collaboration with four local universities, National Taiwan University (NTU), National Cheng-Kung University (NCKU), National Taiwan University of Science and Technology (NTST) and National Tsing-Hua University (NTHU), we have spent more than one year to prepare this Conference since the approval by the International Programme Committee (IPC). With the guidance from the IPC, we are able to go through the laborious, but important, process of paper selection and review from more than 100 submissions, and also to maintain the tradition in gathering the high quality and state-of-the-art papers. Finally, more than 65 full papers are collected in the programme (oral and poster), and over 120 surface metrologists from 17 countries (or economies) will attend the Conference.

As stated in the preface by Professor Thomas, this series of conferences were founded by Tom and late Professor Ken Stout in the United Kingdom more than thirty years ago. I was lucky to join Ken's research group in Birmingham, and to start my journey over surface metrology in 1989, under the financial support from ITRI. With the encouragement from Professor Liam Blunt and endeavors of my colleagues, we are able to hold the Conference first time in emerging Asia, and to "carry on the heritage and pave the way to the future" (a Chinese proverb) in surface metrology.

Taiwan is also known as Formosa, from Portuguese Ilha Formosa, which means "Beautiful Island". Besides the inspiring scientific arrangements, I encourage you to taste Taiwan's wonderful gourmet cuisine, and to explore the beauty of the sweet-potato-shaped island.

I wish you a joyful, fruitful and memorable stay.

Victor TY Lin, PhD
Chairman
Local Organizing Committee
Met & Props 2013

Journal of Physics: Conference Series **483** (2014) 011001 doi:10.1088/1742-6596/483/1/011001

International Programme Committee

Professor Mohamed El Mansori (Arts et Metiers ParisTech, France)
Professor H Zahouani (Ecole Centrale de Lyon, France)
Professor B-G Rosen (Halmstad University, Sweden)
Professor Tom R Thomas (Halmstad University, Sweden)
Professor Liam Blunt (University of Huddersfield, UK)
Professor Richard Leach (National Physical Laboratory, UK)
Professor Chris Brown (Worcester Polytechnic Institute, USA)
Dr Jia-Ruey Duann (Center for Measurement Standards, ITRI, Taiwan)

International Scientific Committee

Professor H Zahouani (Ecole Centrale de Lyon, France)
Dr Rolf Krüger-Sehm (Physikalisch-Technische Bundesanstalt, Germany)
Professor Pawel Pawlus (Rzeszów University of Technology, Poland)
Professor B-G Rosen (Halmstad University, Sweden)
Professor Tom R Thomas (Halmstad University, Sweden)
Professor Liam Blunt (University of Huddersfield, UK)
Professor Derek Chetwynd (University of Warwick, UK)
Professor Jane Jiang (University of Huddersfield, UK)
Professor Richard Leach (National Physical Laboratory, UK)
Professor Paul Scott (University of Huddersfield, UK)
Dr Andrew Yacoot (National Physical Laboratory, UK)
Professor Chris Brown (Worcester Polytechnic Institute, USA)
Dr Chris Evans (University of North Carolina at Charlotte, USA)
Professor Jay Raja (University of North Carolina at Charlotte, USA)
Dr Ted Vorburger (National Institute of Standards and Technology, USA)
Dr Andrew Baker (National Measurement Institute, Australia)
Professor David Lee Butler (Nanyang Technological University, Singapore)
Dr Benny Cheung (The Hong Kong Polytechnic University, China)
Professor Yetai Fei (Hefei University of Technology, China)
Dr Kazuya Naoi (National Metrology Institute of Japan, Japan)
Dr Heui-Jae Pahk (SNU Precision Co. Ltd., Korea)
Professor Jiu-Bin Tan (Harbin Institute of Technology, China)
Ms. Siew-Leng Tan (National Metrology Centre (NMC/A*STAR), Singapore)
Mr. A. Tonmueanwai (National Institute of Metrology, Thailand (NIMT), Thailand)
Professor Kazuhisa Yanagi (Nagaoka University, Japan)

Local Organizing Committee

Dr Victor Tzeng-Yow Lin (Center for Measurement Standards, ITRI, Taiwan)
Professor Kuang-Chao Fan (National Taiwan University, Taiwan)
Professor Jen-Fin Lin (ASME Fellow, National Cheng Kung University, Taiwan)
Professor Chao-Chang Chen(National Taiwan University of Science and Technology, Taiwan)
Professor Shih-Chieh Lin (National Tsing Hua University, Taiwan)
Professor Liang-Chia Chen (National Taiwan University, Taiwan)
Professor Fang-Jung Shiou (National Taiwan University of Science and Technology, Taiwan)
Professor Chun-Hui Chung (National Taiwan University of Science and Technology, Taiwan)
Professor Pin-Chuan Chen (National Taiwan University of Science and Technology, Taiwan)
Dr Wen-En Fu (Center for Measurement Standards, ITRI, Taiwan)

Peer review statement

All papers published in this volume of *Journal of Physics: Conference Series* have been peer reviewed through processes administered by the proceedings Editors. Reviews were conducted by expert referees to the professional and scientific standards expected of a proceedings journal published by IOP Publishing.

Challenges in Measurement of Surface Topography beyond Micro-scale

T Goto[1]
Mitutoyo Corp.,
SE-Sankyo Bldg.11F, Kita-ku Kita 7-jyo Nishi 1-chome 1-2, Sapporo, JAPAN

E-mail: tomonori_goto@mitutoyo.co.jp

Abstract. Precision measuring machine manufacturers have been flooded with requests for more precise measurements. Not only high-accuracy and high-throughput, but also non-destructive and non-contact sensing are demanded for the latest precise measurement machines. It is well known that a variety of fine structures beyond micro-meter scale are formed on the surface of industrial products. Thus it is natural trend that the demand of 3D surface measurement with nano-scale will be drastically increased. Reflecting such a situation, Mitutoyo puts much effort into the development of a variety of measurement systems which satisfy such complex demands as a comprehensive manufacturer. In this report, some of the challenges related to industrial manufacturing will be introduced.

1. Introduction

The advancement of non-contact devices has grown dramatically over the last few years. Consequently there will be a general trend in the surface texture measurement of using non-contact optical sensor. A variety of optical sensors are provided in the market. As a matter of course, new types of sensors will be under development. Meanwhile the customers required high precision measurement system in their production line. The key words to satisfy these demands are high-accuracy, high-throughput and non-contact sensing. Yet, for the contact sensing there remains a lots of demands, like improving productivity of the traditional evaluation system and making possible to evaluate the complex contour that has not measured in the past. Reflecting such a situation, Mitutoyo puts much effort in the development of a variety of measurement systems satisfy such complex demands as a comprehensive manufacturer.

As our challenges contributing to the industrial manufacturing, Quick Vision with non-contact sensor head, Surftest probe which is a new sensing head for CMM and allows evaluating surface roughness and CV-4500/SV-C4500 enables the measurement of pitch diameter of screw thread with the dual-sided stylus. The photographs of these instruments are shown in figure 1.

[1] To whom any correspondence should be addressed.

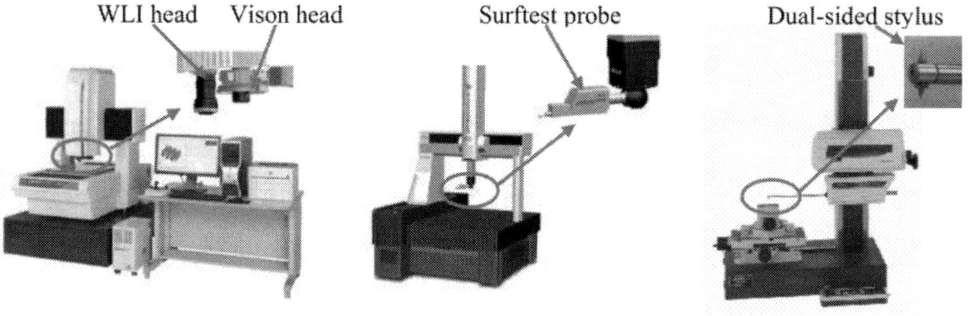

Figure 1. From left, Hyper QVWLI 404, Crysta Apex S with Surftest probe, CV-4500

2. Quick Vision with non-contact sensor head

Two optical sensing techniques, white light interferometry (WLI) and point from focus (PFF) are equipped on Mitutoyo Quick Vison (QV) series that realises both non-contact measurement and minute 3D evaluation in single machine. The QV series can be mounted a non-contact sensing probe on its main three-dimensional coordinate measuring body with a vision probe. The hybrid measurement system with built in WLI head enables in sub-micro to nano scale surface measurement. With PFF head which uses the same optical system as the image sensor, the system provides micro to sub-micro scale measurement. WLI is well known technique which enables both of accuracy and wide-range measurement. That's why this technique is widely used in the industrial measurement instruments. The accuracy of PFF is lower than WLI in general, but it has a great advantage that no additional optical system is required for the customer who has already owned the typical vision measurement system. In the table 1, these two different sensing techniques are compared. As shown in the table 2 and 3, both for WLI and PFF, we provide self-developed objective lenses which maximize the machine ability. The accuracy of QVPFF is depends on the model as you can see in the table 4.

Table 1. Comparison of characteristics of WLI and PFF

	WLI	PFF
Principle	White Light Interferomertry	Focus Position Detection
Z resolution	1～10nm	50～100nm
Application example	IC package L&S FPD photo spacer Solar cell MEMS	IC package IC socket Metallic parts Plastic parts

Table 2. Objective lens for WLI

Magnification	10×	25×
Field of view	0.32×0.24mm	0.13×0.10mm
Z-axis measuring range	170μm	
Repeatability	$2\sigma \leqq 0.01\mu m$	

Table 3. Objective lens for PFF

Magnification		2.5×	5×	10×	25×
Working distance		40.6mm	33.5mm	20.0mm	13.0mm
Field of view	PT 1×	2.49×1.86mm	1.24×0.93mm	0.62×0.46mm	0.25×0.19mm
	PT 2×	1.24×0.93mm	0.62×0.46mm	0.31×0.23mm	—

Table 4. Specification of QVPFF

Model of QV	QV ELF	QV Apex	Hyper QV	ULTRA QV
Default objective lens	2.5×			5×
Z-axis measuring range	21.0mm			5.2mm
Repeatability 2σ	≦2.0μm	≦1.5μm		≦0.7μm
E1 XY	2+3L/1000μm	1.5+3L/1000μm	0.8+2L/1000μm	0.25+L/1000μm

Some measurement examples of QVWLI and QVPFF are shown in figure 2 and 3. WLI does not depend on N.A., therefore it is possible to take high precision z-axis measurement on high aspect ratio work-piece.

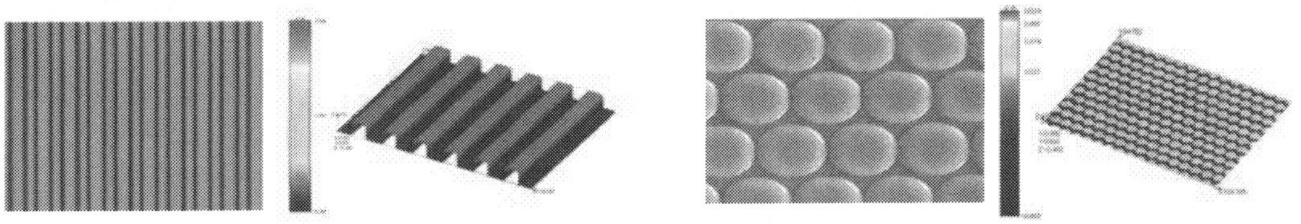

Vision image WLI measurement Vision image WLI measurement

Figure 2. Measurement examples of QVWLI
(left) nanoimprint mold (right) micro lens array

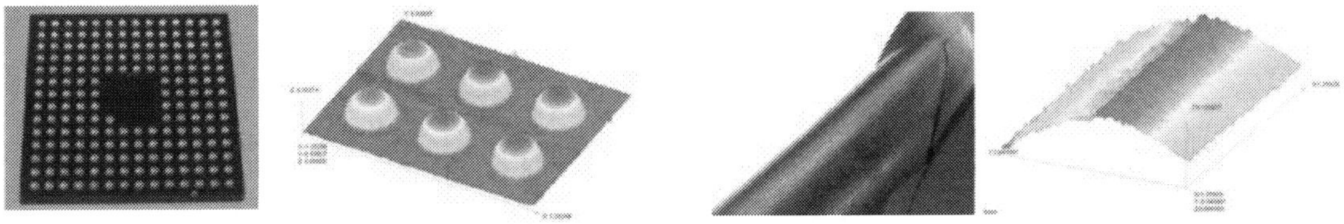

Work apearance PFF measurement Work apearance PFF measurement

Figure 3. Measurement examples of QVPFF
(left) ball grid array (right) plastic molding component

3. Surftest Probe

To satisfy the demand of higher throughput evaluation in the production line needing both of dimension and surface texture, we enables the surface texture measurement with the traditional CMM by mounting roughness sensor as a CMM head. The customer can evaluate both of three-dimensional coordinates and surface roughness only with changing its measuring probe. Moreover the probe changing is automatically operated. This means that the new CMM head "Surftest probe" contributes to the drastic improvement of productivity. In table 5, the specification of roughness sensor Surftest probe is listed. In addition, various types of roughness detectors as illustrated in figure 4 are prepared, and allow performing wide-ranging measurement.

Table 5. Specification of Surftest probe

Measuring range	Auto, 40, 100, 360μm
Traverse length	16mm
Measuring speed	0.25, 0.5mm/s
Stylus tip radius	2μm
Measuring force	0.75mN

Figure 4. Detector types

4. CV-4500/SV-C4500

CV-4500 is a high precision contour measuring instrument with the addition of a new function for continuously measuring top and bottom surfaces. The variable measuring force function has become more useful, enabling a wide variety of efficient, high precision measurements. When combined with the dual-sided stylus that is a new product with diametrically opposed contacting tips, the instrument can continuously measure in the upward and downward slopes without the need to change the arm orientation or reset the work-piece fixture. Eventually, the system expands the possibilities for the evaluation of precise work-pieces like a screw and thin metal sheet as shown in the figure 5. The specification of CV-4500 series is listed in table 6, and some measurement examples are shown in figure 6. SV-C4500 provides dual purpose measurement of surface roughness and contour combined with high accuracy by changing its stylus.

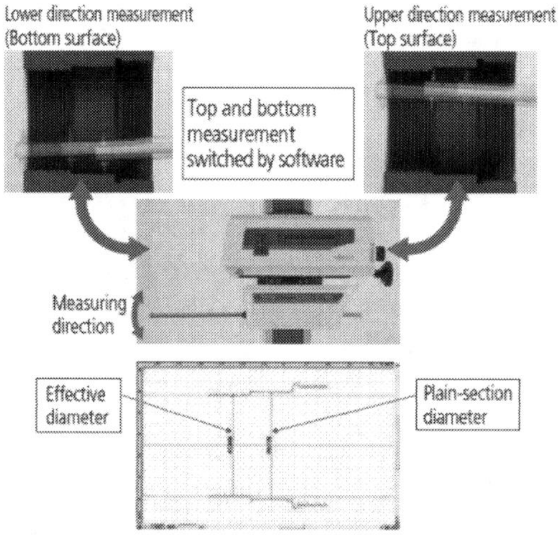

Figure 5. Top and bottom measurement on CV-4500 with the dual-sided stylus

Table 6. Specification of CV-4500 series

Model		CV-4500S4	CV-4500H4
Measuring range	X axis	100mm	
	Z1 axis	60mm	
Vertical travel		300mm	500mm
Linear displacement accuracy Z1 axis		0.8 + |2H|/100μm H: Measurement height from the horizontal position	
Measureing force(software controlled)		10-50mN	
X axis driving speed		up to 80mm/s	
Stylus tip radius		25μm	
Inclining range		±45°	

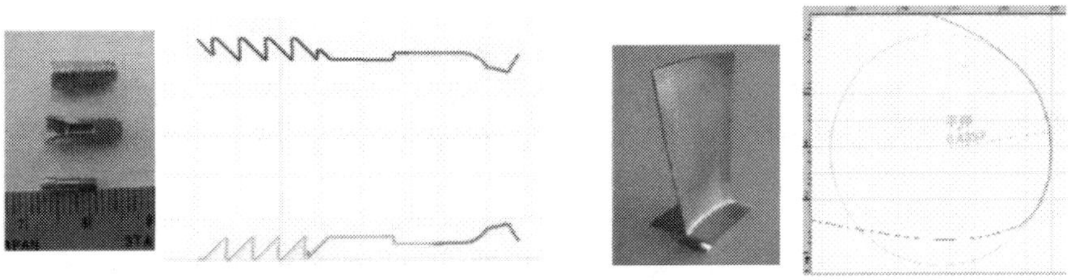

Figure 6. Measurement examples of CV-4500 with the dual-sided stylus
(left) implant screw (right) turbine blades used in jet engine

In the evaluation of a screw, which specification is strictly stated in the national standard, the angle of thread is about 60 degrees in general and for the implant placement it becomes 90 degrees. Our new system CV-4500 allows to get the pitch diameter of thread as same as the conventional three-wire method. The tip contacting angle to a thread is estimated about 63 degrees for upward slope, and 57 degrees for downward slope as shown in the figure 7. According to this simple consideration, the tip sphericity should be assured around 70 degrees for the precise measurement of a screw thread. If the poor tip was used, the measured profile would be strongly influenced by the deformed tip shape. However, if the proper calibration is performed to compensate the tip shape, it is possible to use any tip without sphericity assurance. For example, the two-dimensional outline shape of a tip was detected by tracing a very sharp edge. Once the tip shape is known, it is easy to compensate the error.

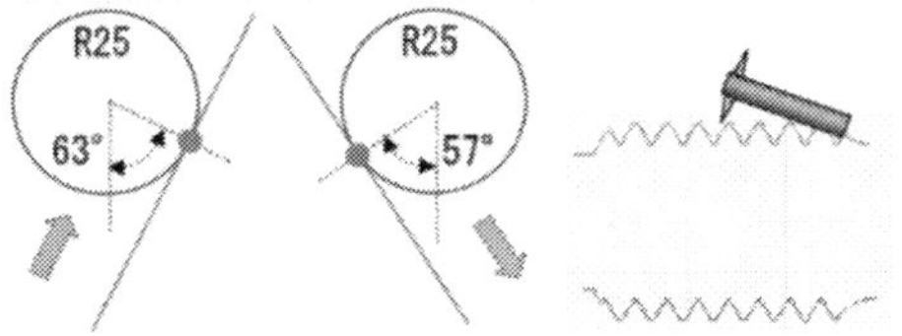

Figure 7. The tip contacting angle

5. Conclusion

Several cutting-edge products of Mitutoyo are introduced briefly. As seen in these examples, the well-balanced measurement system should be provided to satisfy complex customers' demands. As the world's leading manufacturer of metrology equipment, systems and software, Mitutoyo understand the role that accuracy plays in customers' efforts to manufacture better, faster and at lower cost. Mitutoyo continues to develop the most advanced and sophisticated metrology equipment available. If there come up with any problems in measurement and evaluation, please contact to our nearest branch offices.

6. Acknowlegement

I would like to express deepest appreciation to the committee host The Center for Measurement Standards of the Industrial Technology Research Institute (CMS/ITRI) who allows an opportunity to do a presentation.

Single-pass kelvin force microscopy and dC/dZ imaging: applications for graphene-related nanomaterials

Jing-jiang Yu and Shijie Wu
Nano Scale Sciences Division, Agilent Technologies, Inc.
4330 W. Chandler Blvd., Chandler, AZ 85226, USA

E-mail: jing-jiang_yu @agilent.com

Abstract. Development of advanced imaging strategies that could either expand the breadth of what is measurable in material science or break the current limits on detection sensitivity and spatial resolution is highly desired for atomic force microscopy (AFM)-associated metrology. We report that single-pass Kelvin force microscopy (KFM) and capacitance gradient (dC/dZ) imaging can be performed in the intermittent contact regime, and it allows simultaneous acquisition of a sample's morphology as well as its two distinctive electric properties with nanometer-scale spatial resolution. Applications of this technique for enhanced sensing of graphene will be presented.

1. Introduction

Graphene and its derivatives have emerged as a promising family of materials for nanoelectronics in a postsilicon era.[1] From an assembly point of view, these atomically thin carbon sheets of either single-layer graphene (SLG) or few-layer graphene (FLG) are well compatible with existing planar device architectures. On the other hand, one significant advantage of graphene-related nanomaterials is their highly tunable electrical properties such as carrier type or density, and rich electronic band structures. Potential applications of graphene sheets as ultrathin transistors, sensors and other nanoelectronic devices require them supported on an insulating substrate. Therefore, a quantitative understanding of charge exchange at the interface and spatial distribution of the charge carriers is critical for the device design. While the impact of the substrate and interlayer interactions on the electronic structures of FLG is discussed extensively in prior theoretical simulation,[2-7] experimental data of directly measuring the graphene-substrate interactions are lacking. We report that AFM-based technique Kelvin force microscopy as an experimental means to investigate the local electrical properties of both single-layer and few-layer graphene films on silicon dioxide. The effect of the film thickness on the surface potential is detected and quantitative measurements are obtained.

2. Methodology and Materials

2.1 Materials

The graphene samples were prepared by the mechanical exfoliation method and deposited onto Si substrates coated with a SiO2 layer.

2.2 Methodology

An Agilent 5600LS AFM system with a MAC III unit that is equipped with three lock-in amplifiers (LIAs) is used in our studies. These digitally-controlled analog LIAs have a broad bandwidth up to 6 MHz to cover the operational bandwidth of the photodetector employed in the microscope. Single-pass KFM and dC/dZ imaging is achieved by the simultaneous use of the probe flexural resonance frequency (ω_{mech} in the first LIA targeting the mechanical tip-sample interactions) for surface profiling and a much lower frequency (ω_{elec} in the second LIA for monitoring the electrostatic interactions) for sample surface potential measurements. In addition, the single-pass KFM at Agilent is operated in the intermittent regime, thus significantly enhancing the localized surface potential measurements.

3. Results and Discussion

3.1 Identify single-layer graphene on SiO_2 via high-resolution AFM imaging

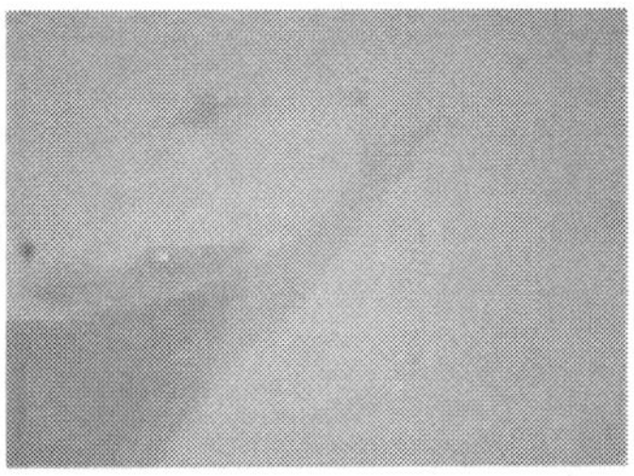

Figure 1. An optical photograph of exfoliated graphene flakes with various thickness on a silicon dioxide substrate.

Figure 1 is an optical photograph of exfoliated graphene flakes on a silica substrate, in which various colors corresponding to different film thickness (or number of layers) are noticeably captured. A darker contrast is related to a thicker film. An example of identifying single layer graphene on silica by AFM imaging is shown in the high-resolution topographic image in Figure 2A. The morphological difference between the graphene film and the substrate is observed. The graphene layer is much smoother so individual granular features within the substrate area are resolved. Such conclusion is further supported by the cross-session cursor profile (Figure 2C, corresponding to the blue line drawn in Figure 2A), in which a better view of the surface corrugation of graphene film vs. substrate is provided and quantified measurements of the surface roughness at each region can be derived. It shows that the typical peak-to-peak surface corrugation in the silica region is almost doubled compared to that of graphene layer. In addition, Figure 2C shows that the height of this graphene film is about 0.464 nm. Statistical measurement of the apparent height of a single graphene layer is performed using the "step height" function in AFM post processing software PicoImage (Figure 2B). The resulting value is 0.582 nm, similar to that derived from local cross-session measurement in Figure 2C. Although the interlayer spacing of HOPG is 0.34 nm in theory, practical AFM measurements of a single graphene layer on Si/SiO_2 always lead to higher values due to the weak sample/substrate interactions and presence of ambient species (nitrogen, oxygen, argon, or water) between SiO_2 and graphene, between the graphene sheets, and on the graphene sheet.[8] Assuming the thickness of graphene films is governed by a linear relationship equation: $h = n \times t + t_0$, where n is an integer as the number of layers, t is the approximate theoretical thickness of each graphene layer, and t_0 is a systematic offset (i.e., independent of n), the best-fit values for t and t_0 based on a combined Raman scattering and AFM studies were 0.35±0.01 nm and 0.33±0.05 nm, respectively.[9] Therefore, our experimentally measured apparent height of single-layer graphene on silica is in a good agreement with this model in the case of n=1.

3.2 Differentiate and quantify the variation in surface potential between single-layer graphene and a silica substrate

After locating a sample site containing single-layer graphene, *in situ* KFM imaging is conducted subsequently to measure the surface potential of the same area. Figure 3A is the topographic image that is almost identical to Figure 2A except that the scan size is slightly increased (from 5 um × 5 um to 6 um × 6 um). Figure 3B and 3C are simultaneously acquired phase and surface potential images, respectively. Despite the solid fact that phase contrasts have been proven to be associated with materials mechanical nature (stiffness, friction) or adhesive properties and can be utilized to distinguish different components of heterogeneous polymer or other thin film materials, it turns out the

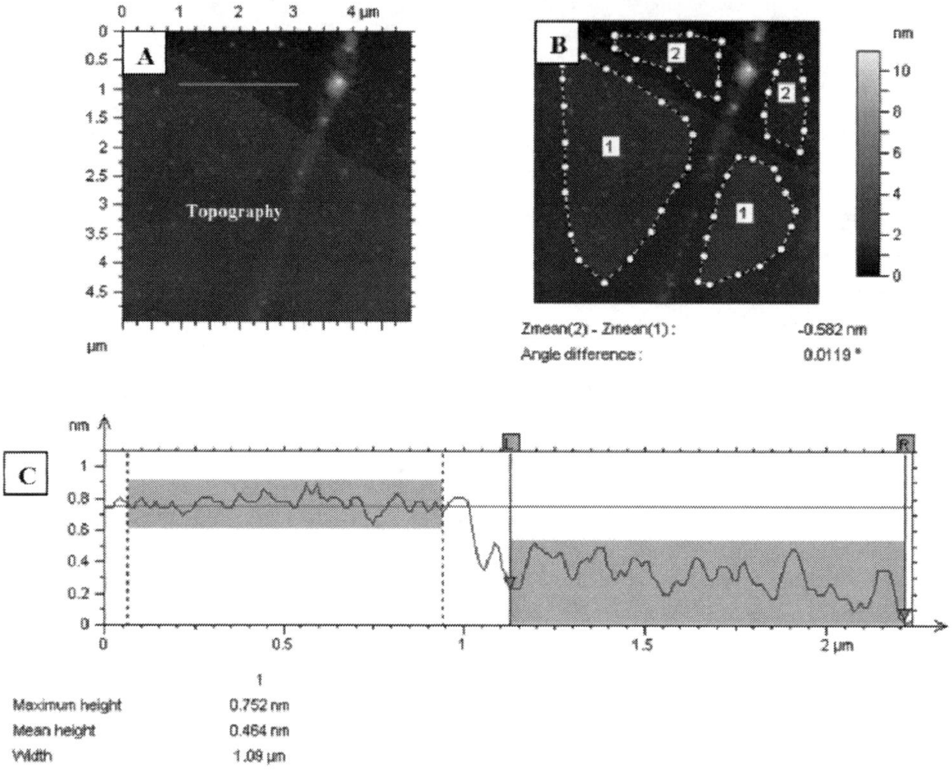

Figure 2. An example of identifying single-layer graphene on silica. [A] A high-resolution topographic image (5 um × 5 um). [B] The average apparent height of single layer graphene on silica substrate is measured using AFM data post processing software PicoImage. [C] Cross- section profile corresponding to the blue cursor line drawn in [A]. Statistic measurements of the average variation in surface potential between single-layer graphene and silica substrate using AFM data post processing software PicoImage.

sensitivity of phase imaging may not be sufficient in this extreme case dealing with only a monolayer of carbon atoms. Figure 3B is such homogenous that contrast variations between graphene layer and bare substrate is negligible. Therefore, it is very encouraging that different contrasts are observed in surface potential image Figure 3C and their boundary is matching exactly to that between the two regions (in the corresponding topographic image(Figure 3A). At this point, it is demonstrated that KPFM technique can complement phase imaging and serve as an effective means to detect graphene-related nanomaterials on a silica substrate.

In additions, Figure 3C indicates that the surface potential of single-layer graphene is slightly higher than that of silica. Quantified measurement in Figure 3D shows the surface potential variation between two different materials is about 60 mV.

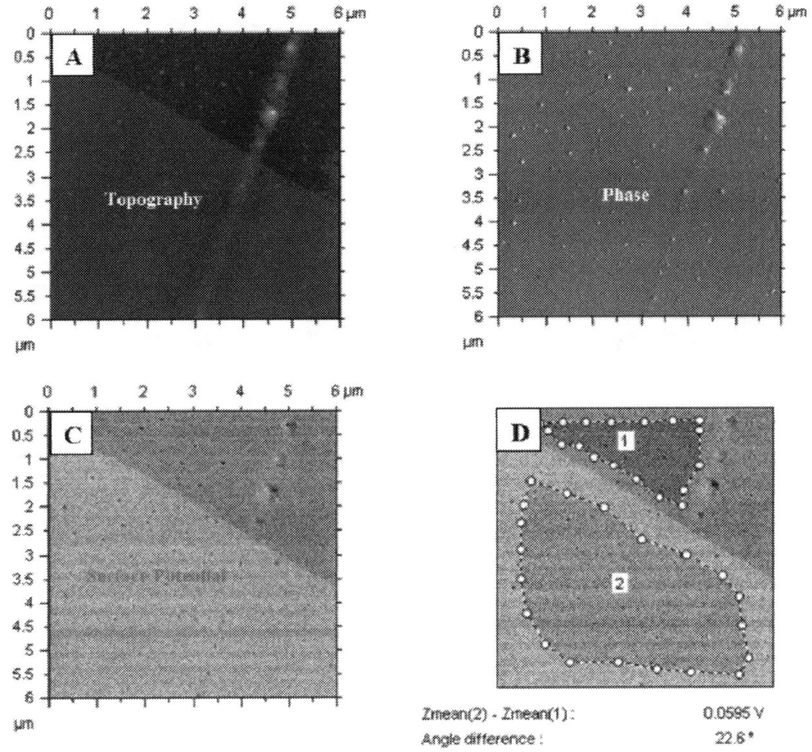

Figure 3. KFM imaging of single-layer graphene on silica. [A] A high-resolution topographic image (6 um × 6 um). [B] and[C] Corresponding phase and surface potential images of [A], respectively. [D]

3.3 Single-pass KFM and capacitance gradient (dC/dZ) measurements of few-layer graphene

More valuable data comes from our single-pass Kelvin force microscopy and capacitance gradient (dC/dZ) measurements of few-layer graphene samples. A typical example is presented in Figure 4. Multiple steps are well resolved and three main regions (labeled as I, II, and III, respectively) are distinguished in topography image (Figure 4A). Figure 4D is the cross-section profile corresponding to the purple line drawn in Figure 4A, from which the region I is indicated to be about 0.75 nm higher than region II using region III as a background baseline. This value is in good agreement with two times of the HOPG interlayer spacing (0.34 nm). Actually, this calculated bilayer change in the film thickness is further confirmed by a well-defined single atomic step resolved between region I and region II in Figure 4A. Once again, the corresponding phase image (data not shown here) is similar to Figure 3B and thus hard to be used as an evidence to differentiate few-layer graphene films from a bare silica substrate. Furthermore, although surface potential measurements prove to be sufficient to distinguish single-layer graphene from SiO_2, the dependence of surface potentials on the film thickness (illustrated in Figure 4C) brings an unavoidable complexity to the case of few-layer graphene. Additional probing based on materials' other electric properties will substantially benefit for a successful identification of region III as a substrate area. At this point, we are taking the advantage of the availability of a third LIA by connecting it directly to the photodetector and monitor the tip response at the frequency of $2\omega_{elec}$. The oscillation of the tip induced by the electrostatic force at this $2\omega_{elec}$ frequency is in proportional to the capacitance gradient (dC/dZ), which is related to the local dielectric permittivity of the sample. It is exciting to see that both region I and region II, regardless of containing different graphene layers, exhibit a homogenous contrast that is brighter than that of region III in dC/dZ image Figure 4B. Such results can be validated by the fact that the dielectric constant of graphite material (ε=10-15) is much higher than that of silica (ε=3.9). Therefore, it is demonstrated here that the combination of KFM and capacitance gradient (dC/dZ) measurements allow an unambiguous distinction between interface layer and multi-layer graphene.

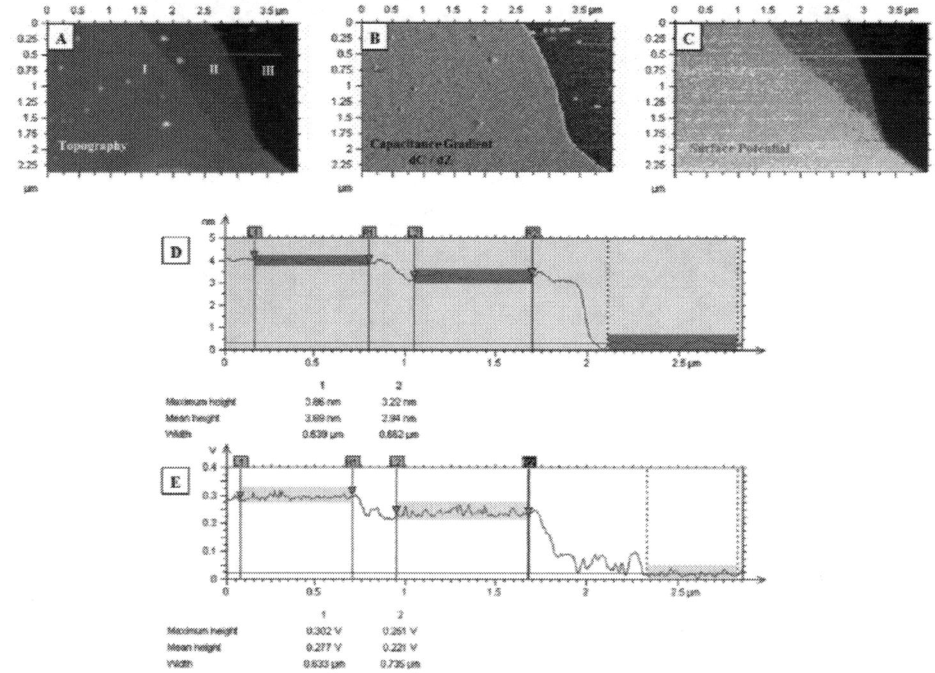

Figure 4. Single-pass KFM and capacitance gradient (dC/dZ) measurements of few-layer graphene on silica. [A] A high-resolution topographic image (4 um × 2.4 um). [B] and [C] Corresponding capacitance gradient and surface potential images of [A], respectively. [D] Cross-section cursor profile corresponding to the purple line drawn in topographic image [A]. [E] Cross-section cursor profile corresponding to the blue line drawn in surface potential image [C].

A layer-dependent behavior is detected in our surface potential measurements of FLG films, evidenced by the observation that the thicker film in region I has a brighter contrast than that of region II in Figure 4C. The cross-sectional view of the blue line drawn in surface potential image indicate that the variation of surface potential between graphene film at region II (assigned to be nine layers based on an apparent height of 2.94 nm derived in Figure 2C) vs. silica substrate is about 221 mV, which is much higher than the value of 60 mV that is revealed in Figure 3D in the case of a single-layer garphene. Continued rise of two additional layers results into a further increase of 56 mV in region I with respect to region II.

The conclusion that FLG surface potential change monotonically with graphene layers based on our KFM studies is consistent to the observations by many other groups. For example, Filleter et al found a surface potential reduction of 135 mV for single layer as compared to bilayer graphene grown

Journal of Physics: Conference Series **483** (2014) 012002 doi:10.1088/1742-6596/483/1/012002

epitaxially on a SiC substrate.[10] In contrast to a linear relationship between FLG thickness and the number of layers, the variation of surface potential between the two neighboring layers decrease exponentially when the thickness of graphene films goes up. It is reported that FLG surface potential may approach a "bulk" value for samples with five or more graphene samples.[11] To the best of our knowledge, the possible threshold has been extended to ten layers by Lee et al that a tiny surface potential difference between seven- and nine-layers of exfoliated graphene on silica has been resolved.[12] Here, we report our successful differentiation and quantitative measurement of the surface potential between nine- and eleven-layer graphene films.

Conclusions

Kelvin force microscopy is employed to investigate the local electric properties of either single-layer or few-layer graphene films on a silicon substrate. A thickness-dependent surface potential behavior of graphene films is observed and quantitative measurements of surface potentials for both single-layer graphene and few-layer graphene are reported. In addition, it proves that local electrical property-based measurements (surface potential or/and capacitance gradient) have a higher sensitivity than mechanical property-based phase imaging in both detection and differentiation of graphene sheets with either single-layer or multilayer structures.

Acknowledgements. The graphene samples that were used in studies described in this application note were kindly provided by Dr. Ryan Cooper and Prof. Jeffrey W. Kysar from Columbia University.

References

1. Berger *at al.* Electronic confinement and coherence in patterned epitaxial graphene. *Science* **2006,** *312,* 1191-1196.
2. Guinea *at al.* Charge distribution and screening in layered graphene systems. *Phys. Rev. B* **2007**, *75*, 235433.
3. McCann *at al.* Asymmetry gap in the electronic band structure of bilayer graphene. *Phys. Rev. B* **2006**, *74*, 161403.
4. Min *at al.* Ab initio theory of gate induced gaps in graphene bilayers. *Phys. Rev. B* **2007**, *75*, 155115.
5. Zhou *at al.* Substrate-induced bandgap opening in epitaxial graphene. *Nat. Mater.* **2007**, *6*, 770–775.
6. Adam *at al.* A self-consistent theory for graphene transport. *Proc. Natl. Acad. Sci.* **2007**, *104*, 18392.
7. Katsnelson *at al.* Electron scattering on microscopic corrugations in graphene. *Phil. Trans. R. Soc. A* **2008**, *366*, 195.
8. Ishigami *at al.* Atomic structure of graphene on SiO_2. *Nano Lett.* **2007,** *6,* 1643-1648.
9. Gupta *at al.* Raman scattering from high-frequency phonons in supported n-graphene layer films. *Nano Lett.* **2006,** *12,* 2667-2673.
10. Filleter *et al.* Local work function measurements of epitaxial graphene. *Appl. Phys. Lett.* **2008,** *93*, 133117.
11. Datta *et al.* Surface potentials and layer charge distributions in few-layer graphene films. *Nano Lett.* **2009,** *9,* 7-11.
12. Lee *et al.* The interlayer screening effect of graphene sheets investigated by Kelvin probe force microscopy. *Appl. Phys. Lett.* **2009,** *95,* 222107.

Reconstruction of freeform surfaces for metrology

N El-Hayek[1,4], H Nouira[1], N Anwer[2], M Damak[3,4] and O Gibaru[4]

[1] Laboratoire Commun de Métrologie (LNE-CNAM), Laboratoire National de Métrologie et d'Essais (LNE), 1 Rue Gaston Boissier, 75015 Paris, France,
[2] Ecole Normale Supérieure de Cachan, The University Research Laboratory in Automated Production, 61 avenue du Président Wilson, 94235 Cachan, France.
[3] GEOMNIA: 3D Metrology Engineering and Software Solutions, 165 Avenue de Bretagne, EuraTechnologies 59000 Lille, France,
[4] Arts et Métiers de Lille (ENSAM), Laboratory of information sciences and systems (LSIS), 8 Boulevard Louis XIV, 59046 Lille, France,
E-mail: elhayek@lne.fr

Abstract. The application of freeform surfaces has increased since their complex shapes closely express a product's functional specifications and their machining is obtained with higher accuracy. In particular, optical surfaces exhibit enhanced performance especially when they take aspheric forms or more complex forms with multi-undulations. This study is mainly focused on the reconstruction of complex shapes such as freeform optical surfaces, and on the characterization of their form. The computer graphics community has proposed various algorithms for constructing a mesh based on the cloud of sample points. The mesh is a piecewise linear approximation of the surface and an interpolation of the point set. The mesh can further be processed for fitting parametric surfaces (Polyworks® or Geomagic®). The metrology community investigates direct fitting approaches. If the surface mathematical model is given, fitting is a straight forward task. Nonetheless, if the surface model is unknown, fitting is only possible through the association of polynomial Spline parametric surfaces. In this paper, a comparative study carried out on methods proposed by the computer graphics community will be presented to elucidate the advantages of these approaches. We stress the importance of the pre-processing phase as well as the significance of initial conditions. We further emphasize the importance of the meshing phase by stating that a proper mesh has two major advantages. First, it organizes the initially unstructured point set and it provides an insight of orientation, neighbourhood and curvature, and infers information on both its geometry and topology. Second, it conveys a better segmentation of the space, leading to a correct patching and association of parametric surfaces.

1. Introduction

Surface reconstruction is an inverse problem. It is the process of creating an approximation of a surface based on a cloud of points measured by optical and tactile probes. Many researchers worked on this subject and proposed a large number of solutions and scientific publications during the last 30 years. Delaunay and Voronoi-based meshing techniques [1] as well as implicit techniques [2] have been developed within the computer graphics community. These techniques relate to the problem of surface reconstruction where only the cloud of points is known and concern applications such as video games, arts and reverse engineering. Nonetheless, when studied closely and tested, these methods are ineffective unless two important initial conditions are fulfilled. The cloud of points must verify a "good" sampling criterion and has to be free of noise [3]. The first Delaunay and Voronoi-based algorithms such as the sculpting algorithm of Boissonnat and the alpha shapes of Edelsbrunner and Mucke did not impose any of these conditions [4,5]. Hence, the resulting meshes were not strictly manifold. Amenta *et al* [6] presents proofs and guarantees for a "good" quality mesh reconstruction for the first time. Implicit techniques are based on a different philosophy and create an implicit representation of a surface. With the absence of model parameters, these approaches are not valid for metrology since no uncertainty budget can be established later on. Moreover, implicit surfaces techniques have no guarantees for surface reconstruction, meaning that no quality assessment can be done.

Freeform surfaces have seen enlarged applications since their complex shapes closely express a product's functional specifications and their machining is obtained with higher accuracy [7]. A freeform is a surface hardly defined by an equation or any generic form and is characterized by its asymmetric profile. According to ISO 17450-1 [8] freeform surfaces are complex geometrical features which have no invariance degree. However, these surfaces have proven their complex design functionalities [9]. This paper deals with datasets of an aspheric optical lens. The surface reconstruction algorithms tested on these datasets can have an extended usage on freeform surfaces. The datasets are measured at two metrology institutes by two different measurement strategies. LNE used a point-to-point measurement strategy on an XY grid. TNO (Netherlands Organisation for Applied Scientific Research) has measured the same optical lens on its high precision measuring machine Nanomefos with a spiral-like measurement strategy.

The 3D metrology of freeform surfaces is not yet normalized as no unified approach exists [10]. It differs from one application to another. Aspheric lenses inspection requires form characterization in some cases [11] [12] and simple profile characterization in some other cases. In form characterization the global surface is of interest. Whereas in profile characterization, only points along a cross-sectional plane at a specific location are considered. If not enough points can be found within the intersection of the cutting plane and the point set or if the level of the measurement noise is large, which is typically the case, surface reconstruction becomes mandatory. Either ways, form and/or dimensional characteristics are checked for conformity with respect to the design specifications.

This paper is focused on the study of meshes based on the Delaunay and Voronoi combinatorial structures. A mesh is a linear interpolation of the points that builds a structure on them. The initially unstructured point set becomes organized and structured. The mesh is a linear approximation of the underlying surface, which gives an insight of its topology and geometry. Discrete differential metrics can be calculated and used for further processing such as filtration, partition and association. Association is of various nature depending on the type of input provided with the cloud of points. If an analytical implicit model of the surface, such as the case of aspheric surfaces, is given, the association of that model to the structured points can be performed [11]. Whereas if no model is given, parametric models are fitted.

2. Delaunay and Voronoi structures

The Delaunay triangulation and Voronoi diagram are dual representations that infer complete knowledge about spatial neighbourhood. They build a structure that allows determining a point's exclusive neighbours. The Voronoi diagram is built based on this notion of proximity and cuts space into neighbourhood cells (3). Each point x in the space is assigned to one single point p in the dataset P as being in its exclusive neighbourhood. The Delaunay triangulation is a simplicial complex that connects the points that are exclusive neighbours among themselves. It is also known as a triangulation of the convex hull of the dataset. Triangulation is a broad word referring to tetrahedralization in 3D and triangulation in 2D, but we use it for simplicity. The simplices forming the 3D Delaunay complex are tetrahedra as opposed to triangles in two dimensional Delaunay complexes. Figure 1 shows how a Delaunay simplex is circumscribed in a circle empty of other points in the dataset. The general course of the surface reconstruction algorithms hereafter starts with the 3D Delaunay triangulation of the point set and then extracts a triangular surface mesh from it. Only triangles that approximate the underlying surface are selected in the output mesh. These triangles constitute the set of restricted Delaunay simplices. The Delaunay triangulation grants the most regular triangulation, as it builds the most equiangular triangles possible [13]. The duality between Voronoi and Delaunay is done according to their simplices dimensions. By taking n=3 as the dimension of the ambient space, a 0-simplex (vertex) of the Delaunay triangulation corresponds to a n-simplex (Voronoi cell) of the Voronoi diagram. Similarly, a 1-simplex (edge), 2-simplex (triangle) and n-simplex (tetrahedron) in the Delaunay triangulation, correspond to a (n-1)-simplex (Voronoi facet), a (n-2)-simplex (Voronoi edge) and a 0-simplex (Voronoi vertex) in the Voronoi diagram, respectively.

$$Vor(p)=\left\{x\in\Re^{d}\,\middle|\,d(x,p)\leq d(x,q),\forall q\in P\right\} \quad (1)$$

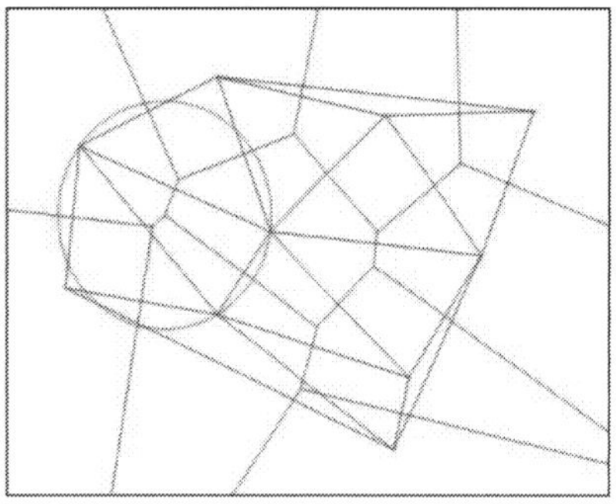

Figure 1. 2D point set (red dots), its Voronoi diagram (blue) and the Delaunay triangulation (black)

3. Expected mesh quality and algorithm requirements

The mesh of the dataset is a linear interpolation of the points, and is based on the Delaunay structure. The common approach for 3D datasets is to build a 3D Delaunay triangulation and extract triangular facets that are a linear approximation of the underlying surface. A good mesh quality is a mesh that approximates well the underlying surface and that connects the points correctly in accordance with the topology of the surface. The quality of the mesh sought is based on well-defined criteria. First, the reconstructed surface should be topologically equivalent to the underlying surface of the points set. This means that the reconstructed surface should be homeomorphic to a 2-manifold and to $\Re^2$. In topology we usually refer to basic topological elements such as disks. A mesh is a discrete 2-manifold only if each vertex umbrella is homeomorphic to a 2-disk [13]. The umbrella of a point on a mesh is the set of triangles (and their vertices) that are incident to that point. Second, the reconstructed surface should also be geometrically equivalent to the underlying surface (figure 2). Both facet normals and surface normals should be similar in direction. Both representations should be geometrically close in terms of Hausdorff distance. The chord error between the facets and the surface should not exceed a given tolerance and be proportional to the sampling density. Surface reconstruction is the process of extracting the restricted Delaunay facets that approximate the first order of the surface out of the 3D Delaunay complex.

Algorithmic complexity is also a major issue to our application. The reconstruction algorithm is to be optimized with regard to space and time complexities. Space, because very large data need to be processed, and time because the characterization of the parts is an on-line process that should not exceed -at least- the measurement time. In fact, the Delaunay triangulation algorithm is incremental so it can be applied incrementally with a sequential insertion scheme as the points are being measured. Furthermore, in the case of an on-line metrology application, an automatic reconstruction process is required. No human intervention shall be necessary.

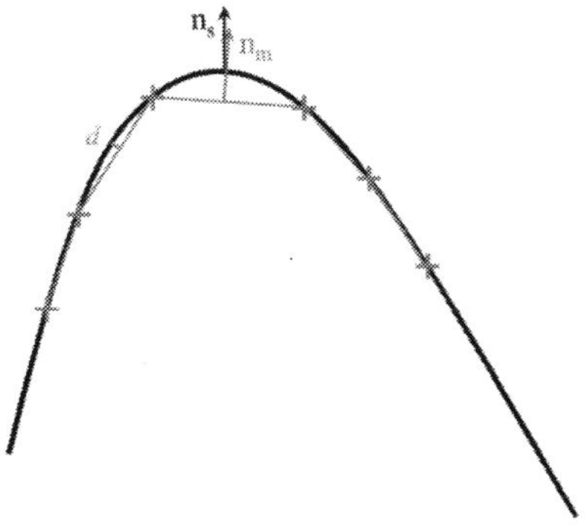

Figure 2. Geometrical equivalence between mesh and theoretical surface

4. Surface reconstruction

4.1 Alpha shapes

The Alpha shapes is a kick-off algorithm for surface reconstruction and is used as a data filter in applications such as in form fitting of freeform surfaces [5,11]. From the 3D Delaunay triangulation of the dataset, the α-shape is an extracted sub-complex formed by all simplices that have the property of being α-exposed. Figure 3 shows the difference between an α-exposed simplex and a non α-exposed simplex. The simplices considered are triangular facets. The algorithm passes a sphere of constant diameter α across the facets and checks for its content. Either the sphere is empty of points and the facet is α-exposed or the sphere contains points from the dataset and the facet is non α-exposed. All the facets that are α-exposed are selected and constitute the simplices of the output mesh. Alpha shapes are not designed to build manifold meshes but only an approximation of the shape inherent to the points. It can be shown that a non-uniform point distribution causes the result to be non manifold.

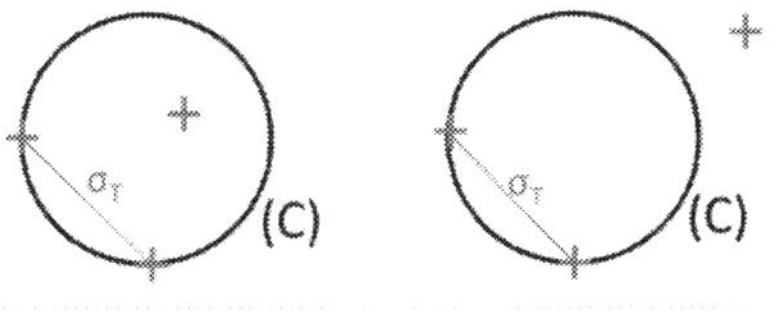

Figure 3. 2D example, a constant diameter circle. Non α-exposed simplex (left), α-exposed simplex (right)

4.2 Cocone

The Cocone algorithm was developed by Amenta *et al* for 2D curves and expanded to surfaces in 3D [3,6]. Cocone is the first algorithm to provide geometrical and topological guarantees which ensure the output triangular mesh is a good approximation of the surface to be reconstructed. These guarantees are conditioned by an ε-sampling condition on the dataset which is defined with respect to the local feature size at a given location on the surface. The condition states that any point x taken on the surface should have at least one sample point within the range of ε times the perpendicular distance from x to the medial axis of the surface. This perpendicular distance is called the local feature size of x and denoted LFS(x). Therefore the ε condition requires that the medial axis of the surface exists and the surface is closed (2). Ultimately, it also implies that the sampling is denser in regions where the curvature is high, and inversely.

$$\|x - p\| \leq \varepsilon.LFS(x), \qquad \text{with } \varepsilon \leq 0{,}1 \tag{2}$$

The robustness of the Cocone algorithm highly depends on the ε-sampling condition. In the absence of it, the guarantees are dropped. The algorithm is based on the Delaunay triangulation and its dual representation, the Voronoi diagram of the point set. Once the Delaunay triangulation has been constructed, the algorithm works on the triangular facets constituting the *(n-1)* simplices of the triangulation. The idea derives from the fact that the triangular facets of the output mesh are a first order approximation of the underlying surface. Each facet is thus an approximation of the tangent plane to the surface at the facet's location up to a chord error. It follows that the normal to the facet is an approximation of the surface normal at the same location. With this property, Cocone searches for the restricted Delaunay facets by stating that these are within a given tolerance zone delimited by a double cone. So if, within the conical zone, a facet should exist, it means that this facet belongs to the restricted Delaunay complex and that it is part of the output mesh.

The Cocone is defined by two geometrical elements and an angle: the axis, the apex and the opening angle. The cone axis is defined as being orthogonal to the approximate tangent plane. Let's assume that a facet *f* is a good approximation of the tangent plane to the surface at a certain location. The dual Voronoi element to a Delaunay facet is an edge that is orthogonal to the facet. Therefore, the dual Voronoi edge to *f* is in theory both, the approximation of the normal to the surface, and the axis of the Cocone of *f*. Since only some Voronoi edges are in the correct direction, the algorithm seeks those ones which are directionally close to the vector that connects a query point *p* to its pole. The poles of a Voronoi cell are the farthest two Voronoi vertices the cell. Amenta *et al* [6] have proven that these poles approximate the normal directions inwards and outwards at *p*. Consequently, for a given facet and its dual Voronoi edge, the Cocone test is applied. It consists of simulating the plot of a double cone at each vertex of the facet (figure 4). The normal approximation of a facet is approximated by its three vertices normals. If the Cocone condition is verified for all three vertices of a facet, the facet is selected in the output mesh.

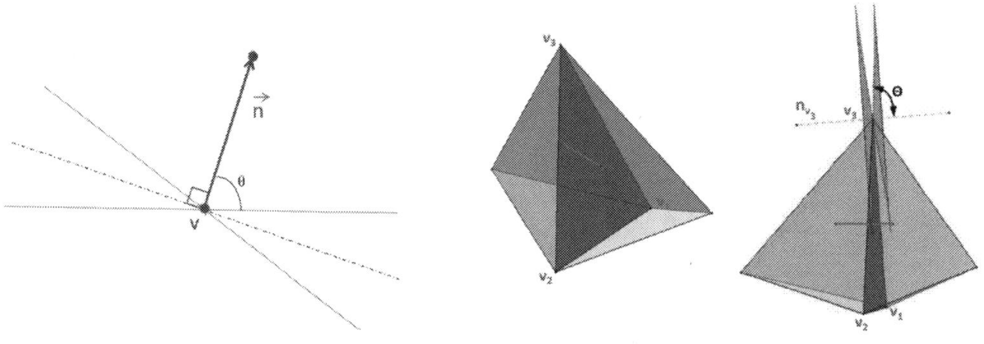

2D scheme 3D scheme

Figure 4. The Cocone condition at one vertex of the considered facet

4.3 Other Methods

Other methods that were developed are also based on the ε-sampling condition. The Natural Neighbours Interpolation of Boissonnat and Cazals [14] follows, by some means, the same principle. It also seeks the restricted Delaunay facets and the dual Voronoi edges that approximate the normal direction. However, the detection of Voronoi edges is different. The algorithm calculates an implicit distance function on the points. This function would be the association of a scalar to a three dimensional point. The function returns a zero at the points of the dataset, a negative value inside the surface and a positive value outside. The Voronoi edges that have mixed vertex signs are edges that cross the implicit surface and are considered as a good approximation of surface normals. Figure 5 illustrates the logic behind this algorithm and shows Voronoi edges that are in the configuration sought and their dual restricted Delaunay elements.

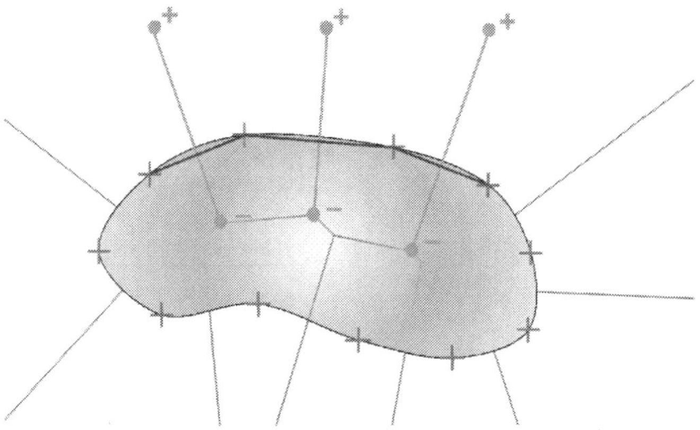

Figure 5. 2D example of the Natural Neighbours Interpolation algorithm

The Wrap algorithm of Edelsbrunner [15] follows a different approach to what preceded. It calculates a distance function on the points, identifies critical points through the gradient of that function and then builds a flow relation following the steepest ascent of the gradient. Grove [16] has previously shown that there exists a duality between critical points and the Delaunay and Voronoi simplices and their intersections. Based on this fact, flow relations congregate Delaunay simplices in clusters. Some clusters are stable manifolds and some others are unstable manifolds that need to be eliminated. The Wrap algorithm does not provide global guarantees, only locally at the stable manifold clusters level. The works of [17] and [18] have linked Wrap to the ε-sampling theory and by that have demonstrated that Wrap can supply global topological and geometrical guarantees, again only if the ε-sampling condition is satisfied.

Dey *et al* [19] continued working on the Cocone algorithm and developed what is known by enhanced Cocone algorithms. They discussed a version of the Cocone algorithm in which large datasets are processed via an Octree subdivision of the points space. Dey *et al* [20] proposes an algorithm with provable guarantees for noisy data. In theory, the noise is limited to a given limit proportional to that of the sampling. Practically, it performs poorer than the basic Cocone algorithm on the same non noisy dataset. The reason is that RobustCocone filters out tetrahedra associated to points that do not verify the noise level condition which is in turn very tight. The last publication on Cocone [21] puts forward a local approach to algorithmically reconstruct a surface from the 3D Delaunay triangulation. In fact, even the Delaunay triangulation is localized. Points are clustered into cells. The Delaunay triangulation is built piecewise, cell by cell, by considering the points of a cell and the points of a portion of the neighbouring cells.

Boissonnat and Ghosh proposed an algorithm based on tangential Delaunay complexes [22]. Instead of natural neighbours, tangential neighbours are computed and the approximation of Delaunay restricted simplices follows this kind of localized neighbourhood. The algorithm assumes the knowledge of normal directions and that the local feature size at any point is positively defined. So these additional constraining conditions make of this algorithm a specific one.

5. Comparison on a simple 2D point set

The sculpture algorithm as well as the alpha shapes and Cocone are compared in 2D. Matlab codes are generated on the same simple dataset showing the difference in algorithmic behaviour. Figure 6a shows a random dataset representing a curve in 2D and figure 6b shows the Delaunay triangulation of the set. The algorithms are all applied on the same dataset.

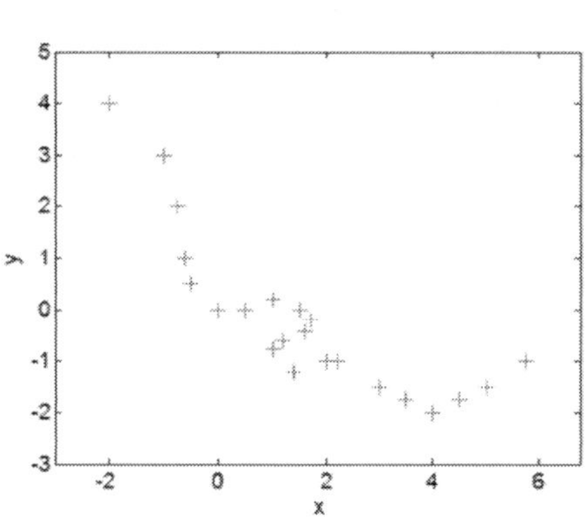

Figure 6a. 2D dataset of a curve

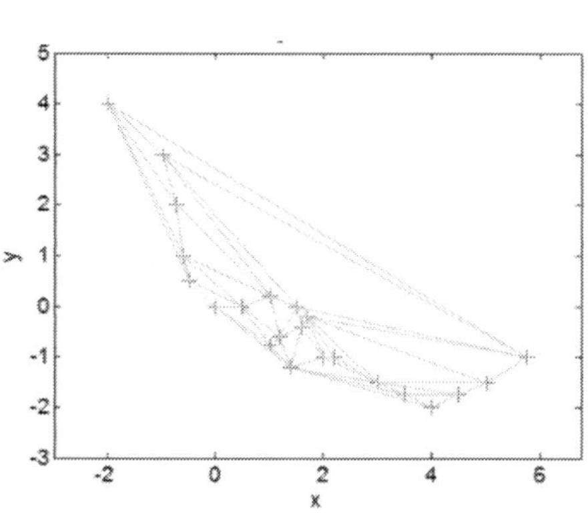

Figure 6b. Delaunay triangulation of the point set

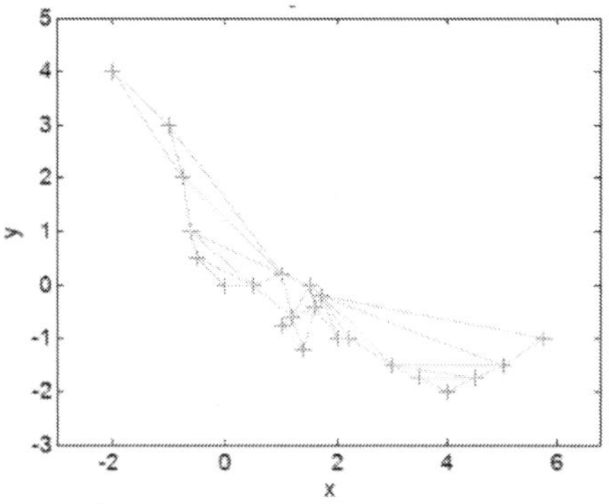

Figure 6c. Sculpture algorithm

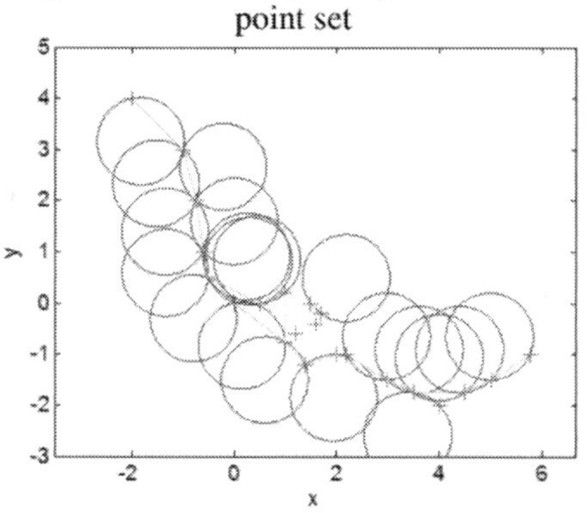

Figure 6d. Circles of constant diameter α traversing Delaunay edges

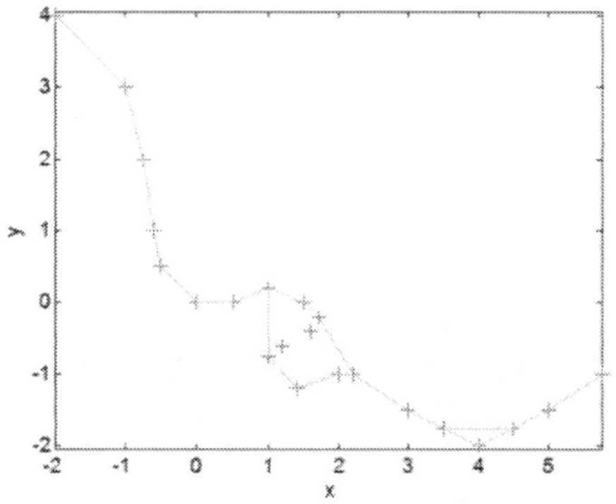

Figure 6e. Alpha shape

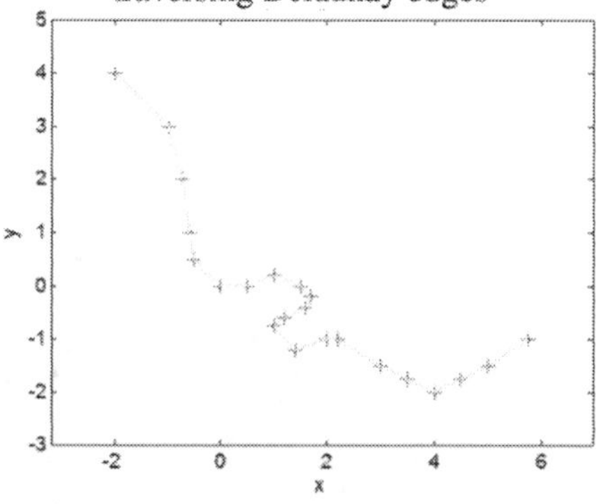

Figure 6f. Cocone

Figure 6c is the result of the sculpture algorithm of Boissonnat applied on the Delaunay triangles. By detecting interior points one after the other, associated triangles are sculpted out. Figure 6d shows the plot of circles of diameter $\alpha=0.85$ tested at each segment. Delaunay edges are traversed because it is a 2D application. The circles that are empty of other points from the dataset and that cross a Delaunay edge are called α-exposed edges. These edges constitute the α-shape of the points shown in figure 6e. It is to notice how the α-shape is not necessarily manifold. The Cocone result shown in figure 6f is manifold. Cocone reconstructs the curve only because it respects the ε-sampling condition. In the central region where the curvature is high a high point density is required, unlike low curvature regions.

Based on the above comparison, the comparison in table 1 and the abundance of the Cocone citations in literature, the Cocone algorithm best fits our application. Guarantees and algorithmic complexity are the two major criteria on which the comparison is made. Another decisive criterion is the possibility to process large data. The Wrap algorithm requires that a flow relation is calculated over the Delaunay simplices and this is a relationship-based computation. Simplices cannot be treated independently. So provided that the number of Delaunay simplices is very large, the algorithm comes to a crash for excessive memory load. Unlike Wrap, the Natural Neighbours Interpolation algorithm calculates a distance function over the points and then treats the Voronoi edges separately. However, this algorithm works only on closed surfaces since it requires the knowledge of inside and outside spaces with respect to the surface. With Cocone algorithm, each facet and its dual Voronoi edge undergo the Cocone test one by one which gives it the advantage of hashing and parallelization. In addition, it does not need any information regarding inside and outside spaces. The only input, which can be generalized to an acceptable value, is the angle Θ of the Cocone.

Table 1: Comparison of surface reconstruction algorithms

	Complexity	**Guarantees**	**Large datasets**	**Sampling**	**Input information**
Sculpture	$N^2.\log N$	x	x	x	x
α-shapes	N^2	x	x	x	x
Cocone	N^2	✓	x	$\varepsilon \leq 0.05$	Θ
N.N.Interp	N^2	✓	✓	$\varepsilon \leq 0.25$	Inside & outside spaces
Wrap	N^2	✓	x	$\varepsilon \leq 0.01$	x
Convection	N^2 (local)	x	✓	x	Normal direction
Tangential complexes	N^2 (local)	✓	✓	$\varepsilon \leq 0.09$	Normals
LocCocone	N^2 (local)	✓	✓	$\varepsilon \leq 0.05$	Θ

6. Example of application of the Cocone method to measured optical lens surfaces

The idea here is to investigate surface reconstruction methods and to apply an algorithm that would build a data structure on the initially unstructured 3D point sets and that would transform the discrete point clouds into continuous piecewise linear representations. For applications such as profile metrology, it is crucial to have a continuous representation. Profile metrology is an application where dimensional characteristics are estimated on an extracted cross section of the surface. Yet, another application to surface reconstruction relates to ray-tracing simulation of optical parts. An this application requires to have a mesh representation [23].

To be representative of most common applications in 3D metrology of complex optical surfaces, an optical lens measured in two different strategies is considered. The optical lens is an asphere defined by the analytical equation (3).

The first set of measured points on an XY grid basis contains 2,782,224 points over a squared area of 5x5mm. The measurement is done using the reference profilometer developed at LNE. It ensures three degrees of freedom, x, y and z linear motions and perfectly respects the Abbe principle. Its metrology loop is optimized to be as short as possible. The part only moves in the XY-plane and the probe moves only along the z axis. The second set of points contains more than 1,000,000 points covering the entire surface and is measured using the Nanomefos machine. The part is mounted on a rotating table. An optical sensor is focused at the centre of the part and while the latter undergoes rotation, the measuring head moves in the radial direction from the centre outwards.

$$Z(r) = \frac{\frac{r^2}{R}}{1+\sqrt{1-\frac{r^2}{R^2(1+\kappa)}}} + \sum_{n=1}^{m} \alpha_{2n}.r^{2n} \tag{3}$$

The Cocone algorithm was implemented and tested on both datasets. These datasets are known to be excessively dense. Figures 7 and 8 (left) show the failure of this algorithm for the fact that the ε-sampling condition is not fulfilled on the datasets. Figures 7 and 8 (right) show that when the sampling respects the ε condition, the reconstruction is proper. The reconstruction algorithms cited in the previous section are very sensitive to the ε-sampling condition. The implementation of the Cocone algorithm has put into evidence the limitation related to having overly dense sampled data.

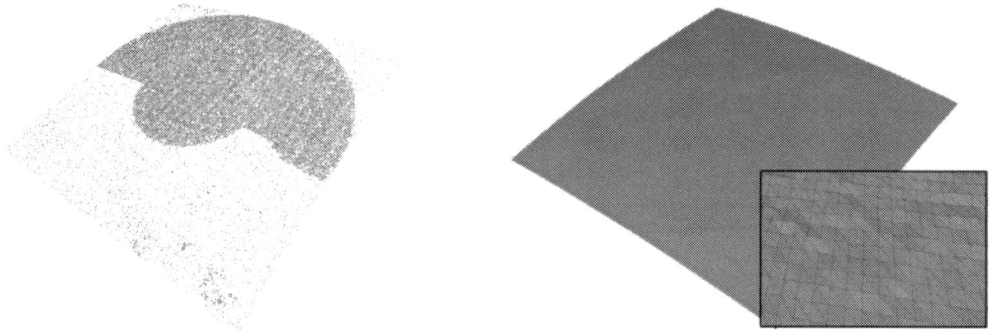

Figure 7. Optical lens reconstruction from measurement made by LNE (XY grid): Original dataset not respecting ε-sampling (left, 2,782,224 points), filtered dataset respecting ε-sampling (right, 1,023,300 points)

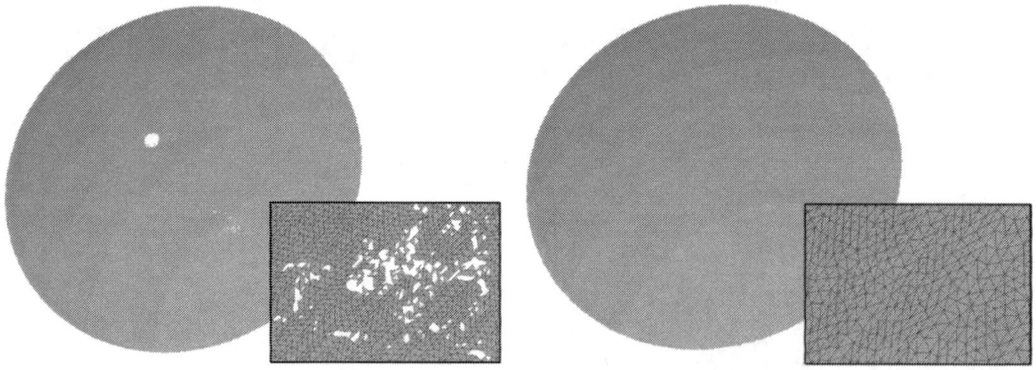

Figure 8. Optical lens reconstruction from measurement made by TNO (spiral scan): Original dataset not respecting ε-sampling (left, 1,044,336 points), filtered dataset respecting ε-sampling (right, 301,698 points)

Optical surfaces, whether they have aspheric or freeform shapes, are usually almost flat surfaces. Irrespective of the measurement strategy, the resulting cloud of points is a set of points that can be projected onto a plane following a bijective mapping without any superposition of points from different sides of the surface. Non bijective projections result from freeform surfaces that represent superposition of points once projected onto a plane. An example of such surfaces is turbine blades. The bijection property offers the advantage of tracing back the points to their original position without modifying either the geometry or the topology of the underlying surface. Since topology is preserved on the map, neighbourhood is conserved and points can thus be meshed in a 2D-like fashion. The application of a 2D Delaunay triangulation is the best solution as it ensures the most regular and complete triangulation possible. The data structure is built in 2D and then mapped back to 3D as simplified in figure 9. The links created among the points are preserved. Figures 10 and 11 show the resulting triangulation based on this technique on both datasets without any filtering or modification of the points. The reconstruction is proper.

With the data structure created on the points, the association of the dataset to the theoretical model of aspherical surfaces is improved. Normal and curvature estimations are more precise since they are calculated based on the true neighbourhood of a point which is the neighbourhood found on the manifold and not in the 3D neighbourhood of the point [24].

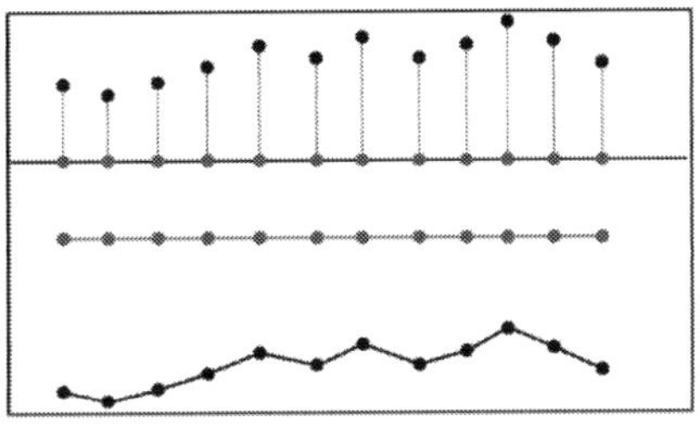

Figure 9. Top to bottom: projectable 3D point set, 2D Delaunay triangulation and inverse mapping

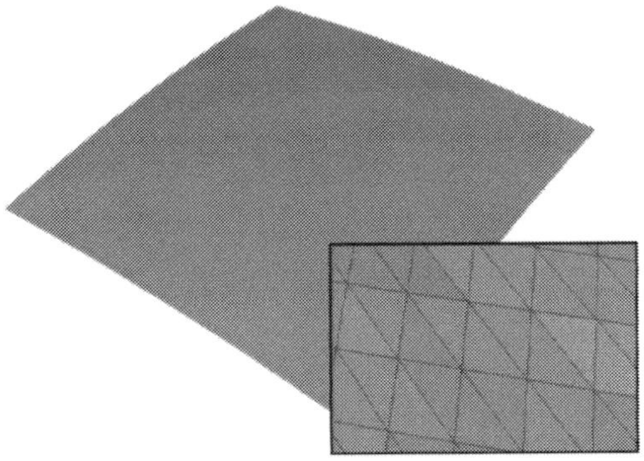

Figure 10. Optical lens reconstruction from original dataset measured by LNE (XY grid)

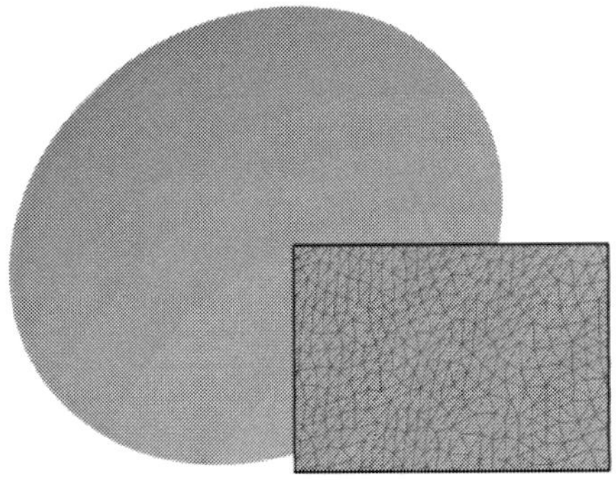

Figure 11. Optical lens reconstruction from original dataset measured by TNO (spiral scan)

7. Conclusion

In this paper, a comparison of surface reconstruction algorithms is presented and lead to the choice of Cocone which fits at best our application. The chosen algorithm possesses inherent limitations related to sampling density. The tests run on real measured optical lenses highlighted the necessity of having an ε-sampled point set. If the cloud of points does not respect the ε condition, a filter is required in order to reduce the overly dense dataset. With surface reconstruction, the underlying surface can be approximated by a first order representation called the mesh. The mesh builds a data structure for initially unstructured points and most importantly creates a continuous representation of the surface. This is required whenever profile extraction and dimensional metrology are to be applied at some specified location on the surface. The mesh also offers the possibility to perform partition of the mesh vertices and thus the possibility to either apply reverse engineering or geometric feature extraction. However, the mesh is used here to structure data and to create a continuous representation of the underlying surface. A mesh is a more substantial structure to approximate normals and curvatures. Curvature estimation is an important tool for surface defects detection and for filtration which can be applied on the mesh vertices in order to reduce their number. Consequently, fitting becomes less complex. A future extension of this work will cover freeform surfaces such as freeform optics or turbine blades.

8. Acknowledgment

1-The authors sincerely thank the EMRP organization. The EMRP is jointly funded by the EMRP participating countries within EURAMET and the European Union (IND10).

2-The authors also thank TNO for the experimental dataset provided, especially Dr. Rens Henselmans.

9. References

[1] Cazals F and Giesen J 2004 Delaunay triangulation based surface reconstruction: ideas and algorithms *In Effective Computational Geometry for Curves and Surfaces, J Boissonnat and M Teillaud (Eds.)* Springer-Verlag (2006) pp 231–276

[2] Osher S and Fedkiw R 2002 *Level Set Methods and Dynamic Implicit Surfaces*. Applied Mathematical Sciences **153** (New York: Springer-Verlag)

[3] Amenta N, Bern M and Eppstein D 1997 The crust and β-skeleton : combinatorial curve reconstruction *Graph. Mod. and Im. Proc.* **60** 125-135

[4] Boissonnat J-D 1984 Geometric structures for three-dimensional shape representation *ACM Trans. on Graph.* **3** No.4 266-286

[5] Edelsbrunner H and Mucke E 1994 Three dimensional alpha shapes *ACM Trans. on Graph.* **13** 43-72

[6] Amenta N, Choi S, Dey T and Leekha N 2000 A Simple Algorithm for homeomorphic surface reconstruction. *Internat. J. Comput. Geom. Appl.* **12** 125-141

[7] Savio E, De Chiffre L and Schmitt R 2007 Metrology of freeform shaped parts *CIRP Annals-Manuf. Tech.* **56** (2) 810-835 Elsevier

[8] ISO/TS 17450-1:2005 Geometrical product specifications (GPS) - General concepts - Part 1: Model for geometrical specification and verification

[9] Klocke F and Dambon O 2003 Precision machining of glass for optical applications *Proc. Of Int. Work. on Extr. Opt. and Sens. Tokyo, Japan* 185-193

[10] Jiang X, Scott P, Whitehouse D and Blunt L 2007 Paradigm shifts in surface metrology part II the current shift *Proc. R. Soc. A* **463** 2071-99

[11] Zhang X, Jiang X and Scott P 2011 A minimax fitting algorithm for ultra-precision aspheric surfaces. *Proc. 13th Intern. Conf. on Met. and Prop. of Eng. Surf.* 285-289, *National Physical Laboratory, Teddington, UK*

[12] ISO 10110-12:2007 Optics and photonics -- preparation of drawings for optical elements and systems -- part 12: aspheric surfaces

[13] Dyer R, Zhang H and Möller T 2009 A survey of Delaunay structures for surface representation *Tech. Rep. 2009-01, GrUVi Lab, School of Computing Science*

[14] Boissonnat J-D and Cazals F 2000 Smooth surface reconstruction via Natural Neighbour interpolation of distance functions *Proc. 16th Ann. Symp. Comput. Geom. 2000,* 223-232

[15] Edelsbrunner H 2003 Surface reconstruction by wrapping finite sets in space *Disc. & Comput. Geom.* **32** (2004) 231–244

[16] Grove K 1993 Critical Point Theory for Distance Functions *Proc. of Symp. In Pure Math.* **54** (3) 357-385

[17] Dey T, Giesen J, Ramos E and Sadri B 2005 Critical points of the distance to an epsilon-sampling of a surface and flow-complex-based surface reconstruction *Proc. of the 21st Ann. Symp. on Comput. Geom.* 218-227

[18] Ramos E and Sadri B 2007 Geometric and topological guarantees for the Wrap reconstruction algorithm *Proc. of the 18th Ann. ACM-SIAM Symp. on Discr. Algo.*

[19] Dey T, Giesen J and Hudson J 2001 Delaunay based shape reconstruction from large data *Proc. IEEE Symp. in Paral. and Large Data Visual. and Graph.* 19-27

[20] Dey T and Goswami S 2004 Provable surface reconstruction from noisy samples *In Comput. Geom.: Theory & Appl.* **62**

[21] Dey T, Dyer R and Wang L 2011 Localized cocone surface reconstruction *In Computers & Graphics* **35** 483-491

[22] Boissonnat J-D and Ghosh A 2009 Manifold reconstruction using tangential Delaunay complexes *Proc. of ACM symp. on Comput. Geom.* 324-333

[23] Morita S *et al* 2010 Ray-tracing simulation method using piecewise quadratic interpolant for aspheric optical systems *Appl.Opt.* **49** 3442-51

[24] Smith T and Farouki R 2001 Gauss map computation for free-from surfaces *Comp. Aid. Geom. Des.* **18** 831-850

Freeform surface measurement and characterisation using a toolmakers microscope

Francis Seung-yin Wong, Kong-Bieng Chauh[1] and Patri K. Venuvinod

City University of Hong Kong
[1]E-mail: mebchuah@cityu.edu.hk

Key Words: Freeform surface design, Freeform surface production, Surface inspection, Dimension feature recognition, Toolmakers microscope, Artificial intelligence

Abstract. Current freeform surface (FFS) characterization systems mainly cover aspects related to computer-aided design/manufacture (CAD/CAM). This paper describes a new approach that extends into computer-aided inspection (CAI).The following novel features are addressed:

- Feature recognition and extraction from surface data
- Characterisation of properties of the surface's M and N vectors at individual vertex
- Development of a measuring plan using a toolmakers microscope for the inspection of the FFS
- Inspection of the actual FFS produced by CNC milling
- Verification of the measurement results and comparison with the CAD design data

Tests have shown that the deviations between the CAI and CAD data were within the estimated uncertainty limits.

1. Introduction

There is a need to compare the actual produced with those specified by the CAD (Computer Aided Design) part model freeform surfaces. Several investigators have reported on the question of the measurement of free form surfaces on a CMM ([1], [2], [3], and [4]), specialized measuring machine Talysurf PGI1240 [5] and special-purpose-built NANOMEFOS measurement machine [6] using the CAD part model as the reference. Another approach of three dimensional measurement is through the X-Y stage of a TMM (toolmakers microscope) with a Z attachment. A laboratory grade [7] provides ZKM 01-250C yielded an accuracy of $\pm(1.8+L/120+HL/17000)\mu m$ along both x and y axes, where L is the measured length in mm and H is the height in mm of the measuring plane above the measuring table. Likewise, it has been claimed that one can achieve an accuracy of $\pm(3+L/200)\mu m$ along X, Y axes just by using a typical workshop grade measuring microscope. A 60 mm long range probe with errors under 0.1µm without compensation and 0.05µm with linear length-error compensation [9] which could work for the measuring system of Z attachment of a TMM. According to a review conducted by Carbone [1] on CMM accuracy, the typical 1D length measuring uncertainty is $U_1 = (2.2+L/300)\mu m$, where L is the measured length in mm. A workshop grade TMM with an appropriate Z attachment would yield comparable accuracy levels of a CMM. TMM uses a non-contact approach that sidesteps problems associated with the probing geometry in CMM. The goal of the work reported in this paper concerns mainly the task of measuring a freeform surface against its CAD model by means of a non-CMM measurement approach (using a TMM) which has long been ignored.

2 Feature recognition and extraction from surface data

In our tests, the freeform surfaces were modelled using an AutoCAD system. Each freeform surface was subdivided into m and n patches in u and v directions respectively. Figure 1 shows the flow diagram for feature recognition and extraction from surface data.

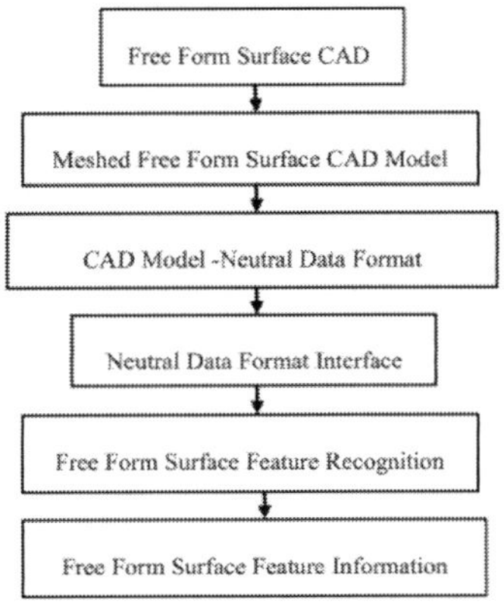

Figure 1 Flow diagram for feature recognition and extraction from surface *data*

2.1 Neutral Data Format File Interface

The objective of this module is to translate the CAD data of the target part to a computer readable neutral data file. The input to the module consists of a CAD file of part to be inspected. IGES (Initial Graphics Exchange Specification) is the first and still the most popular amongst neutral formats. Subsequently, a series of data exchange standards have been (or being) developed, e.g., VDAFS, SET, PDDI, XBF, PDES, ESPRIT Project 322: CAD*I, STEP and DXF.

2.1.1 DXF neutral format

The DXF (Drawing Interchange File) interface is an important part of an Autodesk system. DXF translators have been built into all versions of AutoCAD. DXF is an open sourced CAD data file format supporting both ASCII and binary forms. DXF output created with the earlier versions can also be opened with later releases of DXF. DXF provide flexibility in managing data and translating AutoCAD drawings into a file format that could be read and used by other CAD systems. In the present work, AutoCAD and DXF file format were selected as the CAD system and neutral data format in view of their popularity.

2.1.2 Data structure of DXF

A typical DXF output is composed of a multiplicity of groups, each of which occupies two lines [10]. The first line of the group consists of the group code whereas the second line encodes the group data. The format used depends on the type of group as specified by the group code. In turn, the specific assignment of the group code depends upon the item being described in the file. The type of the value of the group may be derived from the group code.

2.1.3 Development of DXF interface program

A crucial phase in CAD and inspection process planning (CAIPP) integration is the ability to automatically dump, read, and interpret the CAD database. The DXF file structure consists of two-line groups. The first line of each group is an integer representing the group code. The second line captures the data. Predicates and clauses were developed to read the DXF output. Some examples of these predicates and clauses are shown below.

Journal of Physics: Conference Series **483** (2014) 012004 doi:10.1088/1742-6596/483/1/012004

Read integer group	Read real group	Read string group
check(A,i(B)):- 60<=A, A<=79, readint(B).	check(A,r(B)):- 10<=A, A<=59, readreal(B).	check(A,s(B)):- 0<=A, A<=9, readln(B).

2.2 Free Form Surface Feature Recognition

This process concerns the extraction of the freeform surface information from the DXF file input: (i) M: the number of vertices in the M direction, (ii) N: the number of vertices in the N direction, and (iii) the coordinates of all vertices on the freeform surfaces. Rules were developed to match the following conditions to enable the extraction of freeform surface information from DXF.

Freeform face *if* the following conditions are met: **data** are within "**Entity Section**" *and* the group code is **100** and the data type next line is **string** and the string is "**AcDPolygonMesh**" *and* the group code is **10** and the data type next line is **real** number and the real number is the **X** coordinate value of its first vertex in M direction and N direction *and* the group code is **20** *and* the data type next line is **real** number *and* the real number is the **Y** coordinate value of its first vertex in M direction *and* N direction and the group *code is* ***30*** *and* the data type next line is real number *and* the real number is the **Z** coordinate value of its first vertex in M direction and N direction *and* the group code is **71** *and* the data type next line is **integer** *and* the integer is the number of vertexes in M direction *and* the group code is 72 *and* the data type next line is **integer** *and* the integer is the number of vertexes in N direction.

Vertex of a freeform surface *if the following conditions are met*: **data** are within "**Entity Section**" *and* is a member of the freeform surface *and* the group code is **100** *and* the data type next line is **string** *and* the string is "**AcDPolygonMeshVertex**" *and* the group code is **10** *and* the data type next line is **real** number *and* the real number is the **X** coordinate value of this vertex *and* the group code is **20** *and* the data type next line is **real** number *and* the real number is the **Y** coordinate value of this *and* the group code is **30** *and* the data type next line is real number *and* the real number is the **Z** coordinate value of this vertex.

2.3 Free form surface feature information report

The content of the internal database "f" storing the data obtained from the feature recognition process consists of "no. of faces", "no. of vertices", and "indexed vertex with coordinates". The database is stored and saved as an external text file

3. Post-processing the data obtained from feature recognition

Next the data obtained from feature recognition are further processed for (i) inspection process planning, and (ii) the calculation of the M and N vectors at each vertex.

3.1 Generating the inspection process plan

The X, Y, Z coordinates of each vertex point are generated from the database as a text file or in the form of a spreadsheet for inspection process planning.

3.2 Calculation of the M and N vectors

$M_{(i,j),(i+1,j)}$ is the vector within the boundary of the surface from vertex(I, J) to vertex(I+1,J) in direction M, where $M_{(i,j),(i+1,j)} = A_i+Bj + C_k[X_{(I,J)},Y_{(I,J)},Z_{(I,J)}]$ are the X,Y and Z coordinates of vertex(I,J).

[X((I+1),J),Y((I+1),J),Z((I+1),J)] are the X,Y and Z coordinates of vertex((I+1),J), where $A= X_{((I+1),J)}-X_{(I,J)}$, $B= Y_{((I+1),J)}-Y_{(I,J)}$, and $C= Z_{((I+1),J)}- Z_{(I,J)}$.

$N_{(i,j),(i,j+1)}$ is the vector within the boundary of the surface from vertex(I,J) to vertex(I,J+1) in direction N, where $N_{(i,j),(i,j+1)} = E_i + F_j + G_k\ [X_{(I,J)},Y_{(I,J)},Z_{(I,J)}]$ are the X,Y and Z coordinates of vertex(I, J).

[X(I,(J+1)),Y((I,(J+1)),Z((I,(J+1))] are the X,Y and Z coordinates of vertex((I,(J+1)), where $E=X_{((I,(J+1))}-X_{(I,J)}$, $F=Y_{(I,(J+1))}-Y_{(I,J)}$ and $G=Z_{(I,(J+1))}-Z_{(I,J)}$.

4. Implementation and testing

The algorithms were implemented, tested, and verified using the following steps:

1. Design of a test part with a freeform surface
2. Manufacture of a real part by the CAM process.
3. Development software for feature recognition
4. Generation of the inspection process plan
5. Measurement of the test parts on a toolmakers microscope
6. Measurement of the test parts on a CMM

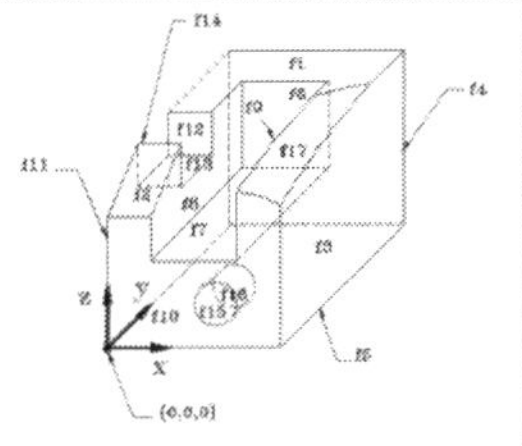

Figure 2

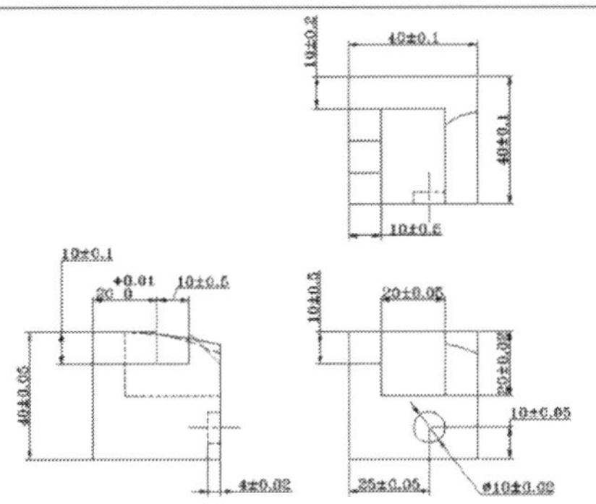

Figure 3

Actual part produced by CNC machining

Figure 4

4.1 Design of a test part with a freeform surface

The test part (setting gauge) shown in Figures 2 and 3 was designed and modelled on an AutoCAD Mechanical workstation. The test part consisted of 17 faces. Note that the part includes a freeform surface (f17).

4.2 Manufacture of a real part using CAM

The CAD model was exported as an IGES file. The CNC part programme was generated by Mastercam software based on the IGES file of the part model. The program was implemented on a CNC milling machine: Mikron WF21 which is easily readable for further processing by a computer program. AutoCAD databases provide curved surface data in a polygon mesh form arranged in M rows and N columns in u and v directions respectively. Thus, the database of each polygon mesh has M×N data. In our work, the polygon mesh was automatically extracted by means of the Prolog algorithm presented in section 2.2. The machining conditions were: cutter diameter 10mm, tool radius 5mm, spindle speed 1000 RPM, and feed rate 300 mm/min.

4.3 Development of software for feature recognition

The CAD model of the test part was exported in DXF format. AutoCAD databases provide curved surface data in a polygon mesh form arranged in M rows and N columns in u and v directions respectively. Thus, the database of each polygon mesh has M×N data. In our work, the polygon mesh was automatically extracted by means of the Prolog algorithm presented in section 2.2.

The free form surface (f17) was modelled with 6 meshes in the u direction and 6 meshes in the v direction. Thus, there were 36 meshes in total with 49 vertices (see Figure 5). The coordinates of the individual mesh vertices of the free form surface as well as the corresponding individual relative positions within the free form surface were captured as described below. Mesh vertex (0,0) was used as the reference position of the surface. The location of this vertex was directly extractable (in 14 decimal places) from the CAD model of the part in 14 decimal places as: mesh vertex (0, 0) = 30.00000000000000, 25.00000000000000, 39.99999999999999.

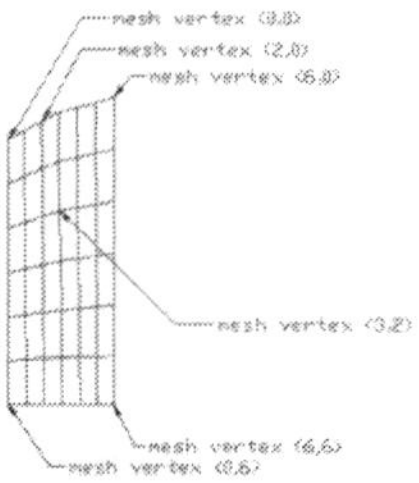

Figure 5 Meshes of the freeform face f17

4.4 Generation of inspection process plan

The inspection process plans for the measurement of the free form surface on a toolmakers microscope for measurement of the Z coordinate at (X, Y) location of each vertex was generated in table form. The inspection process plans for the measurement of the free form surface on a CMM for the measurement of the Z coordinate at the (X,Y) location of each vertex was generated using the coordinates of the vertices extracted from the CAD model.

4.5 Measurement of the test part by TMM

All the 49 vertices of the free form surface f17 were measured on a TMM: Make/Model: Mitutoyo/TM321.

Specifications of the measuring microscope: Measuring range: 100mm (X) × 50mm (Y) × 150mm (Z); Resolution: 1μm; Linear scale accuracy for X and Y axis: $3+3L/1000$μm, L in mm.
Accuracy of the Z axis: 0.00015" by a digital comparator (= 3.81μm).
Estimated linear scale accuracy in X, Y and Z axis over the range of the test piece: 4μm
Estimated volumetric error of the part: $(4^2+4^2+4^2)^{1/2}=7$μm

4.6 Measurement of the test part on a CMM

All the 49 vertices of the free form surface (f17) were measured on a Brown & Sharpe/MicroVal PFx CMM with the following specifications:
Measuring range: 457mm (X) x 508 mm (Y) x 432 mm (Z)
Probe: 1μm resolution; $3+3L/1000$μm linear accuracy (L in mm)
Estimated linear scale accuracy for the part: 4μm
Volumetric performance: 10μm over 330mm

5. Results

The machine specifications indicated a linear accuracy of 20μm along X, Y, and Z axes. Hence, the error in point-t-o point in space within the work envelope was estimated as $(20^2+20^2+20^2)^{1/2}=35$μm. The profile error due to the translation error of the machine and other factors was therefore estimated to be of the order of 50μm. All the 49 vertices of the free form surface f17 were measured on a Mitutoyo/TM321 TMM. It was found that the volumetric error of measuring equipment was 7μm. Thus, the overall profile error limit of the vertex obtained from toolmakers' microscope could be taken to be is $(50^2+7^2)^{1/2}=50.5$ μm≈51μm. An almost identical result of 50.99 μm≈51μm was obtained when the procedure was repeated on a Brown & Sharpe MicroVal PFx CMM.

6 Conclusion and discussion

Traditional CAI (Computer Aided Inspection) practices directed at free form surface production have required CNC machining followed by the measurement of the resulting surface(s) on a CMM [11]. However, owing to the high price and skill requirement associated, CMM use is rare even in modern CAM shops. This paper has outlined an effective alternative that sidesteps the need for a CMM by utilizing a TMM—a much less expensive and more easily managed alternative. Among the issues addressed in this paper are the development of software for geometric feature recognition and the generation of a computer aided inspection process plan (CAIPP). Measurements conducted on identical test parts containing free form surfaces have shown that, notwithstanding the fact that it is less expensive and managed more easily, the new methodology based on tool-room microscopy is as effective technically as the traditional approach based on the use of a CMM.

References

[1] Carbone, V., Carocci, M., Savio, E., Sansoni, G. and Chiffre, L.D., Combination of a Vision System and a Coordinate Measuring Machine for the Reverse Engineering of Freeform Surfaces. *International Journal of Advanced Manufacturing Technology* (2001) 17: 263-271.

[2] Ainsworth, I., Ristic, M. and Brujic, D., CAD-Based Measurement Path Planning for Free-Form Shapes Using Contact Probes. *International Journal of Advanced Manufacturing Technology* (2000) 16: 23-31.

[3] Kase, K., Makinouchi, M., Nakagawa, T, Suzuki, H. and Kimura, F., Shape Error Evaluation Method of Free-form Surfaces. *Computer-Aided Design* (1999) 31: 495-505.

[4] Várady, T., Martin, R.R. and Cox, J., Reverse Engineering of Geometric Models - An Introduction. *Computer-Aided Design* (1997), 29, 4: 255-268.

[5] Cheung C.F., Kong L.B. and Lee W.B., Measuring Ultra-precision Freeform Surfaces Using a Hybrid Fitting and Matching Method. *Measurement Science Technology* (2009), 20, 105103 (12p).

[6] Henselmans R. et al, The NANOMEFOS Non-contact Measurement Machine for Freeform Optics. *Precision Engineering* (2011), 35, 607-624.

[7] ZKM 01-250C, Operation Instructions of Measuring Microscope ZKM 01-250C from CARL ZEISS JENA.

[8] Heidenhain CT6001, Specifications of CT6001, New Length Gauges, Heidenhain catalog 208945-22.70.12/198.E.

[10] DXF File Reference, Autodesk. http://usa.autodesk.com/adsk/servlet/item?linkID=10809853&id=12272454&siteID=123112

[11] Wong, F.S.Y., Case-based reasoning for computer-aided inspection process planning, PhD thesis, City University of Hong Kong, 2005.

14th International Conference on Metrology and Properties of Engineering Surfaces IOP Publishing
Journal of Physics: Conference Series **483** (2014) 012005 doi:10.1088/1742-6596/483/1/012005

Freeform surface characterisation: theory and practice

P J Scott[1] **and X Jiang**
EPSRC Centre for Advanced Metrology
University of Huddersfield, Queensgate, Huddersfield, HD1 3HD

E-mail: p.j.scott@hud.ac.uk

Abstract. The specification and characterisation of freeform surfaces is immature: many CAD packages only allow the nominal freeform geometry to be specified. To control manufacturing and function, the allowable geometric variability also needs to be specified. Currently International standards specify allowable geometry through geometrical tolerancing. Geometrical tolerancing has proved to be a very blunt instrument with many cases, particularly in the aerospace and biomedical industries, of failure of the function of freeform surfaces due to inadequate specification. The paper begins by discussing the importance of the decomposition of the surface geometry into different scales for both specification and characterisation. The three main types of geometrical decomposition: linear, morphological and segmentation are briefly discussed. The Laplace-Beltrami operator (LBO) is the generalization of the Laplace operator to manifolds (i.e. freeform surfaces). By taking Eigenfunctions/Eigenvalues of the LBO a spectral decomposition of the freeform geometry can be realized. The Eigenfunctions are called manifold harmonics and are a direct generalization of Fourier harmonics for freeform surfaces. Tolerancing the Eigenfunctions provides a very powerful and flexible system to control geometrical variability of freeform surfaces; extending the geometrical tolerancing system by allowing decomposition of the geometrical shapes within tolerance zones. Further a generalized Gaussian filter can also be realized by Gaussian weighting the LBO spectra and reconstructing the weighted Eigenfunctions.

1. Introduction

Geometrical tolerancing consists of a system of specifying tolerance zones in engineering drawings in order to control workpiece variability. It was developed to improve the weakness of previous tolerance systems to handle imperfect form and ambiguous references and was primarily developed for assembly of components. Geometrical tolerancing is still at the core of today's tolerancing systems: *Geometrical Product Specification and Verification* (GPS) defined by the International Organisation for Standardization (ISO). The use of tolerance zones is causing real industrial problems in the specification of high valued precision products; two illustrative examples are:

1. In the aerospace industry, the geometry of the tips of turbine blades cannot be controlled by specifying a tolerance zone alone. To distinguish between good and failing functional geometries,

[1] To whom any correspondence should be addressed.

that both satisfy the tolerance zone, requires further mathematical specification to identify allowable geometries from those that are not allowed [1].

2. In healthcare, the specification of the geometrical shape of the cup in total hip-joint replacements by a simple tolerance zone is allowing some cups to fail, by dislocating out of position, prematurely. A new design of non-spherical head is beginning to appear and the market requires improved specification. Again further mathematical specification is required to distinguish between good and failing functional geometries [2,3].

Traditional metrology is essentially analogue, using hard and hand gauges to verify an artifact is to specification. The mathematics is encapsulated within the mechanical set-up (jigs and fixtures) and the relationships of the gauges (see figure 1). Post-measurement calculation is relatively simple and, traditionally, carried out by hand. Traditional metrology is restricted mainly to simple geometrical shapes (planes, spheres, cylinders, cones, etc.) that described the idealised geometry of artifacts of the day [4].

Digital metrology, through sampling in a co-ordinate system, provides a digital representation of an artifact, enabling computational techniques to be applied (see figure 2). The mathematics is encapsulated in the post-measurement calculations through software implementing algorithms that are becoming increasingly complex. Digital techniques thus enable GPS specification (and artifact manufacture) to move beyond the traditional metrology domain of simple geometrical shapes: to manufacture, measure and analyse almost any desired geometry (including freeform geometries).

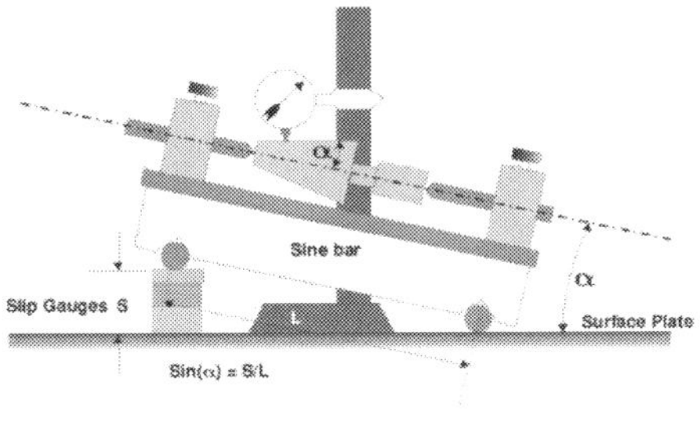

Figure 1 Measuring taper angle with traditional metrology

Figure 2 Measuring taper angle with the digital paradigm

Although there have been recent modifications to increase GPS's utility, specification is still basically a go/no-go system with components being in or out of tolerance. It still reflects the language, methodologies, and concepts of traditional metrology and has not truly entered the modern digital era.

The Grand Challenge is to make a step change to the geometrical tools in the specification toolbox, beyond tolerance zones, to break the formidable barriers facing a designer in specifying geometries (particularly freeform geometries) and their allowable variability optimised for final functionality and manufacturing processes. In this paper a possible solution to the challenge is discussed and mathematical tools developed to realize a geometrical specification toolbox that has potential utility for current and future manufacturing requirements.

2. The structure of GPS specification

The GPS specification process starts by identifying the geometrical features of interest before putting limits on both intrinsic characteristics of the identified features and/or relationships between these features. Geometrical features are defined by a series of four basic operations carried out on a geometrical surface: **sampling**, moving from a continuous object to a discrete object, its dual **reconstruction**, moving from a discrete object to a continuous object, **decomposition** of a surface into different features, and its dual **composition**, building a surface up from different features.

Decomposition is fundamental for specification: it is the operator that creates the different geometrical features of interest. There are currently many different sub-types of decomposition within ISO standards [5] including: **partitioning** an artefact into different functional surfaces; **filtering** a surface into features of different sizes; **association,** i.e., determining the best-fit specified nominal geometrical surface to the sampled points, etc. It is the structure of decomposition that is focused on in this paper. The basic idea is to decompose a tolerance zone into features of different scales.

2.1. Geometrical scales

It is the concept of geometrical scales and their association with decomposition that is introduced in this paper. So what is a geometrical scale and why are they so useful for decomposition? An analogy between GPS and music is used here to illustrate the basic idea.

Beethoven specifies some music (composes). This is manufactured (by playing), see figure 3. How do we know it is manufactured to specification?

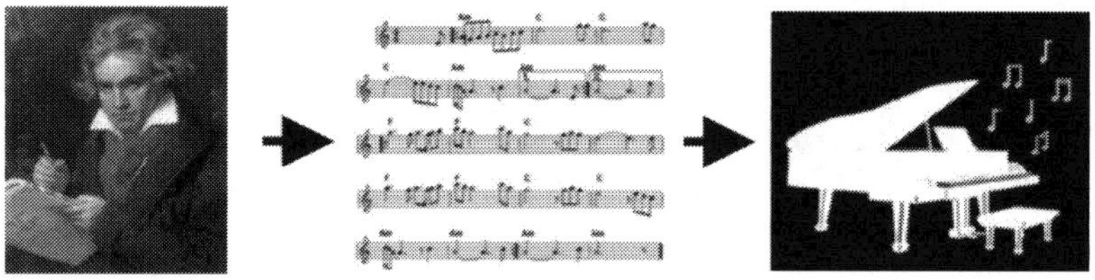

Figure 3 Beethoven composes some music which is played

The played music is measured (by listening) and the notes written down on a MUSICAL SCALE (decomposition) to produce a musical score, see figure 4. This score can be then compared to the original specification to check for conformity. The important concept here is that the decomposition requires a scale in order to carry out the decomposition.

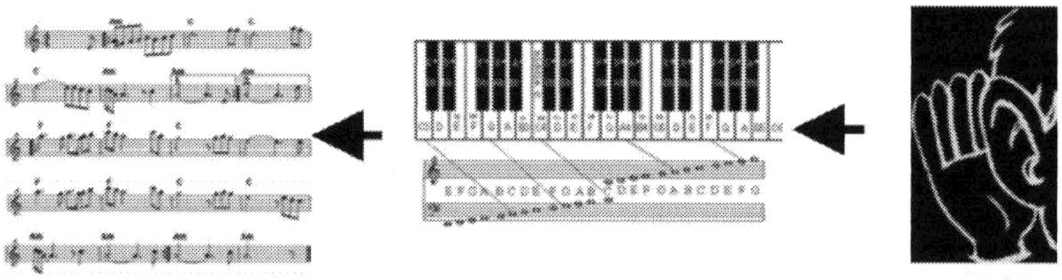

Figure 4 Identifying the individual notes on a musical scale to produce a musical score

One very well-known scale is the Fourier scale based on the harmonics of sine and cosine functions. Unfortunately the Fourier transform can only be applied, without distortion, to surfaces with zero Gaussian curvature (i.e. planes, cylinders etc.). Fortunately there is an approach to extend Fourier like scales to surfaces with non-zero Gaussian curvature (i.e. freeform surfaces) without distortion. The missing connection is that sine and cosine functions are Eigenfunctions of the Laplace operator. This extension will be used as an illustrative example of the use of geometrical scales for decomposition in the next section.

2.2. The Laplace-Beltrami operator

The Laplace-Beltrami operator (LBO) is the generalization of the Laplace operator (heat equation) to manifolds (i.e. freeform surfaces) [6]. Since the discrete version of the LBO, for triangular meshes, is very well known it is natural to choose the Eigenfunctions of the LBO as a functional basis for decomposition [7,8]. The Eigenfunctions are called manifold harmonics and are a direct generalization of Fourier harmonics for freeform surfaces. The Eigenvalues correspond to the square of the angular frequency of the surface and form a natural geometrical scale. Figure 5 shows a rendered version of triangular mesh data of a measured artificial knee joint (47,638 vertices and

94,243 faces). Figure 6 shows the corresponding first 16 Eigenfunctions from the LBO of this triangular mesh data (2000 Eigenfunctions and Eigenvalues were calculated in total). As can be clearly seen, starting from the top left and preceding row by row, the Eigenfunctions become progressively more detailed and form a basis for describing the form errors, shape and texture of this particular freeform surface. The one disadvantage of using this approach with the LBO is that the form needs to be removed first before calculating the LBO Eigenfunctions and Eigenvalues otherwise the size changes.

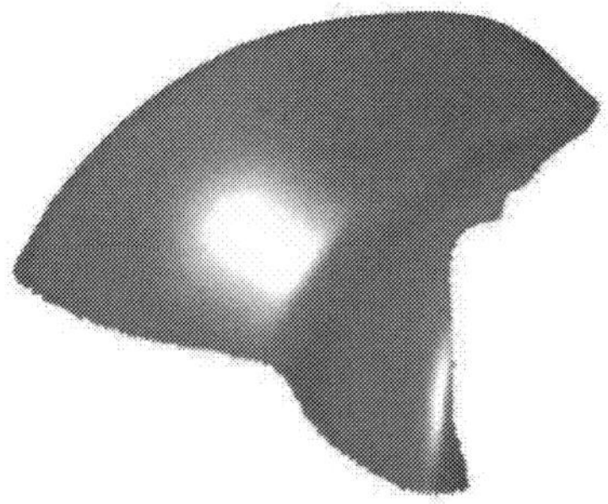

Figure 5 Measured knee joint

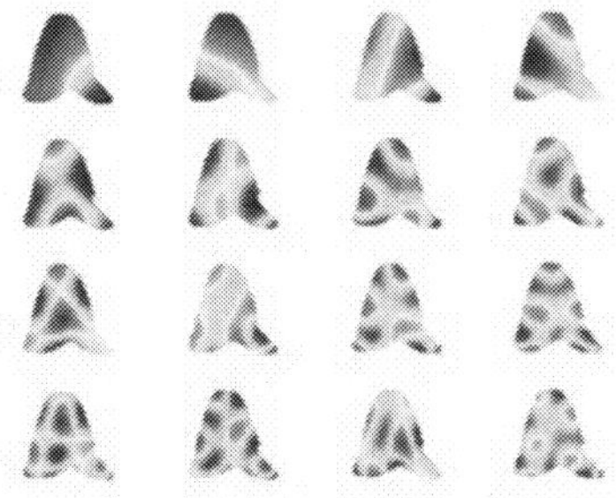

Figure 6 First 16 Eigenfunctions from the LBO on the knee joint data

Tolerancing the Eigenfunctions provides a very powerful and flexible system to control geometrical variability of freeform surfaces; extending the geometrical tolerancing system by allowing decomposition of the geometrical shapes within tolerance zones. Further The Eigenfunctions can be used to filter noisy data through a weighted reconstruction (such as Gaussian weights) of the Eigenfunctions using the Eigenvalues as a nesting index [7,8]. When Gaussian weights are used the resulting filter generalizes the Gaussian filter to freeform surfaces.

2.3. Other Geometrical Scales

There are many different types of geometrical scales that have potential utility in GPS from scales based on morphological operations (scale space), segmentation (Wolf pruning), to linear operators (Partial Differential Equations (PDE), Wavelets, Spectral polynomials etc.), see figure 7. Different scales will correlate with different surface function for example it is anticipated that morphological scales will correspond to contact type phenomenon. With all the above scales once the required surface features have been identified it is fairly straightforward to define both intrinsic characteristics of the identified features and/or relationships between these features and to calculate statistics of these characteristics to define measurands for specification and verification.

3. Summary and further work

This paper has elucidated that whereas metrology and manufacturing have both moved beyond the traditional metrology domain of simple geometrical shapes, specification still reflects the language, methodologies, and concepts of traditional metrology and has not truly entered the modern digital era. This is a severe limitation on the specification of complex geometries such as freeform surfaces. The structure of specification is then discussed showing the importance of decomposition to identify features of interest. The connection between decomposition and geometrical scales is introduced before an example of a geometrical scale, based on the Eigenfunctions and Eigenvalues of the Laplace-Beltrami operator, is illustrated. Finally the characterisation of identified features together with suitable statistics is outlined to define measurands for specification and verification is briefly mentioned.

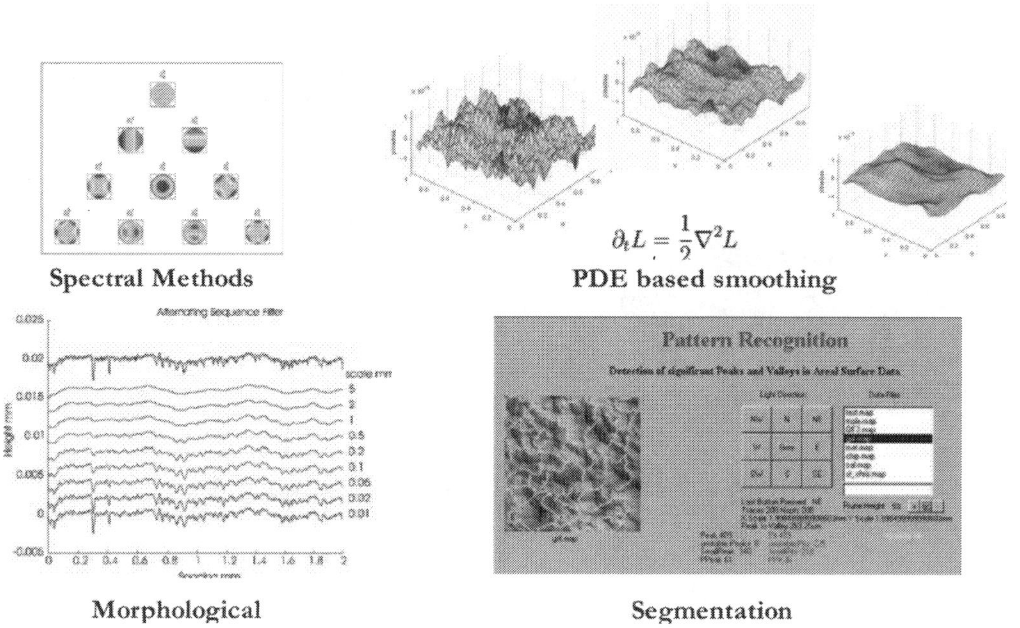

Figure 7 Some examples of geometrical scales

This paper is a manifesto for an innovative approach to allow specification of complex geometries, particularly specification of the allowable geometrical variation in shape and texture beyond tolerance zones. There are still details to be explored, particularly the correlation between the different scales and function. Another unanswered question is: do PDE scales exist that do not require the form to be removed first before being applied? The future is going to be very exciting developing the concepts and addressing the challenges presented by this manifesto for specification and verification of complex geometries.

4. References

[1]. http://www.npl.co.uk/1-oct-09-national-freeform-centre-launch-orchard

[2]. Underwood R J, Cann P M, Ilo K, Zografos A, Fowell M, Skinner J, and Hart A 2010 *A study of explanted durum metal On metal hips*, ORS Annual Meeting, March 6-9 2010, New Orleans, USA

[3]. National Joint Registry for England and Wales 2011 *8th Annual report,* ISSN 1745-1450, www.njrcentre.org.uk

[4]. Rolt F H 1928 *Gauges and Fine Measurements*, MacMillan and Co Ltd, London

[5]. ISO 17450-1:2011 *Geometrical Product Specification (GPS) – General Concepts Part 1: Model for geometrical specification and verification* International Organization for Standardization

[6]. Jiang X, Cooper P, and Scott P J 2011 *Freeform surface filtering using the diffusion equation* Proceedings of the Royal Society *A* **467** (2127), 841-859

[7]. Seo S, Chung M K, and Vorperian H K, 2010 *Heat Kernel Smoothing Using Laplace-Beltrami Eigenfunctions*, Med Image Comput Assist Interv. **13** 505–512

[8]. Botsch M, Kobbelt L, Pauly M, Alliez P and Levy B 2010 *Polygon Mesh Processing* A K Peters Ltd, Natick, Massachusetts

Acknowledgement

The authors gratefully acknowledge the UK's Engineering and Physical Sciences Research Council (EPSRC) funding of the EPSRC centre for Innovative Manufacturing in Advanced Metrology.

Study of the effect of nano surface morphology on the stain-resistant property of ceramic tiles

S P Pan[1], J K Hung, Y T Liu
Center for Measurement Standards, Industrial Technology Research Institute, Hsinchu 30011, Taiwan

E-mail: panshanpeng@itri.org.tw

Abstract. In this study, six types of commercially available ceramic tiles, including nano-structured ceramic tiles and regular ceramic tiles, were selected to investigate the effect of surface morphology on their stain-resistant property. The stain-resistant efficiencies of various ceramic tiles with nano-size surface were measured in order to determine the appropriate method for testing ceramic tiles with nano-structure surface.

1. Introduction

Taiwan is warm and humid due to its location in the subtropical area. Owing to the rapid industrial development as well as the increasing number of motor vehicles in recent years, Taiwan's air pollution is becoming worse, leading to many problems including the fouling of ceramic tiles used by metropolitan buildings. As a result, the Professional Executing Agencies of Nano Mark has received several public suggestions regarding this matter, such as applying nano technology to the treatment of ceramic tile surface in order to solve the fouling problem of ceramic tiles. For this reason, the Professional Executing Agencies of Nano Mark began to study methods for testing the stain-resistant efficiency of ceramic tiles with nano-structure surface. It is hoped that the appropriate method found can be used as the standard testing method for the verification of ceramic tiles with nano-structure surface.

To transform regular ceramic tile into nano-structure ceramic tile, the key technology is to deposit a layer of hydrophilic or hydrophobic nano material on the ceramic tile surface via coating or spraying technique. The stain-resistant effect of ceramic tile with hydrophilic nano surface is resulted from the fact that the hydrophilic nano layer can react with water molecules in air to form a thin layer of water film. Such water film not only can prevent the adhesion of stains (pollutants) on the ceramic tile, but also make the removal of stains on the ceramic tile easier by rain. On the other hand, the stain-resistant effect of ceramic tile with hydrophobic nano surface is due to its self-cleaning nature, which also prevents stains from attaching onto the ceramic tile. Such phenomenon is analogous to the well known Lotus effect. Both nanotechnologies can be applied to improve the stain-resistant effect of ceramic

[1] E-mail address: panshanpeng@itri.org.tw

tiles, which enhances not only its quality but also its competitiveness when the product is promoted to the global market.

2. Experimental design

2.1. Experimental parameters

The test conducted in the beginning of this study to investigate the stain-resistant property of ceramic tiles was based on Chinese National Standard (CNS) 3299-9: Methods of Test for Ceramic Tiles-Part 9[1]. The following stains are requested by CNS 3299-9: (1) stains (Cr_2O_3 or Fe_2O_3) in the form of light oil or test cream that can generate clear marks, (2) stains that can cause chemical reaction (alcohol solution containing 13 g/L of iodine), and (3) stains that can form thin film (olive oil).

Six types of commercially available ceramic tiles were collected in this study. Among them, samples labeled A, B and C claimed that they have been treated by nano process, while samples labeled A1, B1 and C1 were regular ceramic tiles. The methods and stains used in the stain-resistant tests for all the samples were as requested by CNS 3299-9. The results indicate that all ceramic tiles exhibited level-5 stain-resistant efficiency as shown in Table 1, which is the highest level in CNS 3299-9.

Table 1. Testing results of various ceramic tiles according to the method of CNS 3299-9

Testing method	Stain	Ceramic tile sample					
		A	A1	B	B1	C	C1
CNS 3299-9	Iodine	Level 5	Level 5	Level 5	Level 5	Level5	Level 5
	Cr_2O_3	Level 5	Level 5	Level 5	Level 5	Level 5	Level 5
	Olive Oil	Level 5	Level 5	Level 5	Level 5	Level 5	Level 5

According to the results in Table 1, all the ceramic tiles have excellent stain-resistant property. Therefore, it is difficult to determine whether the good stain-resistant property is resulted from nano-structure of the ceramic tile surface or not. To solve the problem, we have studied many literatures (domestic and international) and visited several ceramic tile manufacturers in Taiwan to collect their testing methods and standards for ceramic tiles. Finally, we found that in order for the ceramic tile manufacturers to sell their products in Taiwan, the products must meet the highest stain-resistant level of CNS 3299-9. For this reason, we decided to abandon the testing methods of CNS 3299-9 as an option for the verification of nano products.

2.2. Experimental parameter update

To further understand the composition of stains on the contaminated ceramic tiles, stains from ceramic walls of the buildings near Kuang Fu Road of Hsinchu were collected in this study (as shown in Figure 1). The samples collected were then analyzed by SEM-EDS (Scanning Electron Microscope-Energy Dispersive Spectrometer) to determine the chemical compositions of the stains. From EDS results, elements such as C, K, Si, Al, Mg, Na, Fe and Cu were detected as shown in Table 2 (mostly carbon black compound). In addition, according to the results of the supersite monitoring project reported by the Environmental Protection Administration (EPA), Executive Yuan [2], and the stains of metropolitan ceramic tiles are mainly substances resulted from exhaust emission, dust, coal smoke, and carbon black. Moreover, from the method of JSTM J 7602[3] reported by the Japan Testing Center for Construction Material, carbon black compound is used as a stain for the standard testing procedure. Therefore, we decided to change the experimental parameters of this study by replacing the stains requested by CNS 3299-9 with carbon black, a common stain that is easily observed in the environment. Two types of carbon black compound were selected in this study:

(1) Aqueous type stain: 25 wt % carbon black in pure water (volatile: 18% ± 1%, particle size: 15 nm ± 2 nm, specific surface area: 350 m^2/g ± 40 m^2/g).
(2) Oil (organic) type stain: 25 wt % carbon black in neutral silicon oil (volatile: 3% ± 1%, particle size: 15 nm ± 2 nm, specific surface area: 350 m^2/g ± 40 m^2/g).

Figure 1. Stain on ceramic tiles of building wall.

Table 2. EDS results of stains from ceramic walls of the buildings in Hsinchu

Element	Weight%	Atomic%
C K	39.63	50.47
O K	42.56	40.68
Na K	0.33	0.22
Mg K	0.39	0.25
Al K	2.62	1.49
Si K	10.73	5.85
K K	0.66	0.26
Fe K	1.77	0.48
Cu K	1.31	0.32
Totals	100.00	

3. Method for testing the stain-resistant property of ceramic tile

According to the recommendation (European Committee No. 2011/696/EU) officially passed by the European Union on October 18, 2011[4], "Nanomaterial" refers to material containing particles with any given dimension of less than 100 nm. In order to prove that the stain-resistant property of nanotechnology treated ceramic tiles is resulted from its nano-structure surface, the correlation between the size of the nano structure and the stain-resistant property of ceramic tiles was investigated in this study. The testing methods are described as follows:

3.1 Nano-size measurement

The equipment recommended in this study to carry out nano-size measurement includes Scanning Electron Microscope (SEM), Transmission Electron Microscope (TEM) and Atomic Force

Microscope (AFM). To ensure the accuracy of the electron microscope, magnification test must be performed in accordance with ISO 16700:2004(E)[5] prior to SEM/TEM analysis. Moreover, the analysis of samples can be conducted following the TNS M017001-2012[6] procedures reported by the Taiwan Nano Standard Council (TNSC). For those require using Atomic Force Microscope, the analysis of samples can be conducted following the ASTM E2859-11[7] procedures.

3.2 Stain-resistant efficiency measurement

To ensure the long-term workability of the product since ultraviolet (UV) from sunlight may speed up the aging of ceramic tile surface structure, it is recommended that accelerated weathering test of 500-hour UV irradiation must be implemented for ceramic tiles. Furthermore, to ensure that the product is anti-corrosive under acidic and basic conditions, it is also recommended that ceramic tiles must undergo chemical resistance (anti acid/base corrosion) test. After finishing the above mentioned tests, stain-resistant efficiency measurement can then be conducted for ceramic tiles. Detailed test and measurement procedures are described as follows:

3.2.1 Accelerated weathering test

In this study, the condition of sunlight irradiated on ceramic tile was simulated by carrying out the accelerated UV irradiation test. The UV irradiation tester selected in this study is able to simulate the UV intensity in sunlight and accelerate the weathering effect by employing thermal cycles on the testing material. Testing was conducted in accordance with conditions stated in the ASTM G154:06[8] procedure 1 (cycle 1), in which the wavelength of UV light source selected is 340 nm. The UV exposing conditions are shown in Table 3.

Table 3. ASTM G154:06 procedure 1 (cycle 1) conditions

Cycle	Light source	Intensity	Wavelength	Exposing cycle
1	UVA-340	0.89 $W/m^2/nm$	340 nm	8 h UV at 60 (±3)°C Black Panel Temperature; 4 h Condensation at 50 (±3)°C Black Panel Temperature

3.2.2 Anti corrosion (chemical resistance) test

The anti corrosion (chemical resistance) tests for ceramic tiles were conducted in accordance with the rules specified in Section 7 of CNS 3299-10: the Methods of test for ceramic tiles - Determination of chemical resistance [9]. During the chemical resistance test, the testing samples and the control samples of ceramic tiles were first immersed separately into hydrochloric acid (HCl) solution, followed by potassium hydroxide (KOH) solution. The immersing period for each solution was 24 hours. After removing the samples from KOH solution, they were rinsed with water (flow) and dried by towel to complete the chemical resistance test. The next test to perform is stain-resistant test.

3.2.3 Stain-resistant test

To ensure that ceramic tiles can fulfil their application requirement, stain-resistant test was conducted in this study. During the stain-resistant test, stain (see section 2.2) was coated on the ceramic tile sample to cover an area of approximately 1 cm^2. The sample was then placed into the oven and baked at 50 °C for a period of 24 hours to simulate the scenario of sunlight drying up (hardening) the stain on ceramic tiles. After baking the sample, the washing procedure stated in CNS 3299-9: Methods of test for ceramic tiles - Determination of resistance to stains was conducted.

3.2.3.1 Washing procedure

(1) Procedure A: Rinse the ceramic tile by flowing water (at room temperature) through its surface for 5 minutes, and then clean its surface with a wet towel before making any observations.
(2) Procedure B: Use mild detergent to clean the ceramic tile surface manually with a sponge or soft towel. Rinse the ceramic tile by flowing water through its surface, and then clean its surface with a wet soft towel before making any observations.
(3) Procedure C: Use strong detergent to clean the ceramic tile surface through the help of mechanical equipment. The equipment implemented should be operated in accordance with the rules stated in CNS 3299-9 (Section 5.2.3). The cleaning period is 2 minutes. Rinse the ceramic tile by flowing water through its surface, and then clean its surface with a wet soft towel before making any observations.
(4) Procedure D: Immerse the sample into a prepared chemical solution for a period of 24 hours. The solution should be prepared in accordance with the rules stated in CNS 3299-9 (Section 5.1.4). Rinse the ceramic tile by flowing water through its surface, and then clean its surface with a wet soft towel before making any observations.

3.2.3.2 Stain-resistant level determination
After completing the washing procedure, the stain-resistant level of the sample is determined according to the rules stated in CNS 3299-9. There are 5 stain-resistant levels in CNS 3299-9. If the stain cannot be removed, proceed to the next washing procedure.
(1) If all stains can be completely removed by Procedure A, the sample is rated level 5.
(2) If all stains can be completely removed by Procedure B, the sample is rated level 4.
(3) If all stains can be completely removed by Procedure C, the sample is rated level 3.
(4) If all stains can be completely removed by Procedure D, the sample is rated level 2.
(5) If all stains cannot be removed by Procedure D, the sample is rated level 1.
If no stain is observed on the sample surface, it means that the stain has been completely removed. Record the corresponding washing procedure.

3.3 The basis of verification standards
The testing criteria for nano-size measurement and nano material performance described in Section 3.1 and 3.2 are summarized in Table 4.

Table 4. Criteria of nano-size measurement and nano material performance for stain-resistant ceramic

Item	Characteristic	Criteria	Note
Nano size	Size of the nano structure on the surface of stain-resistant ceramic tile	Any given dimension of nano surface structure with size of less than 100 nm	
Nano performance	Stain-resistant performance	Stain-resistant performance is above level 4	
Other requirement	Weathering resistance, chemical resistance	After 500 hours of UV irradiation test and chemical resistance test, the stain-resistant performance is above level 4	

4. Measurement results
In this study, six types of commercially available ceramic tiles were collected. Among them, samples labeled A, B and C claimed that they have been treated by nano process, while samples labeled A1, B1 and C1 were regular ceramic tiles. Tests were conducted according to the methods planed out in this study. The results of the stain-resistant tests are shown in Table 5. From the results, it is noted that

sample A is a commercially available ceramic tile with stain-resistant performance of level 4. By analyzing its surface structure with AFM, it was found that particles on the surface of sample A are closely arranged, with an average size of 20 nm as shown in Figure 2. This is believed to be the reason for its high stain-resistant performance.

Table 5. Results of the TN-041 stain-resistant test for various ceramic tiles with nano surface treatment

Testing Methods	Stain	Ceramic tile sample					
		A	A1	B	B1	C	C1
TN-041	Hydrophilic carbon black	Level 4	Level 1	Level 1	Level 1	Level 1	Level 1
	Hydrophobic carbon black	Level 4	Level 1	Level 1	Level 1	Level 1	Level 1

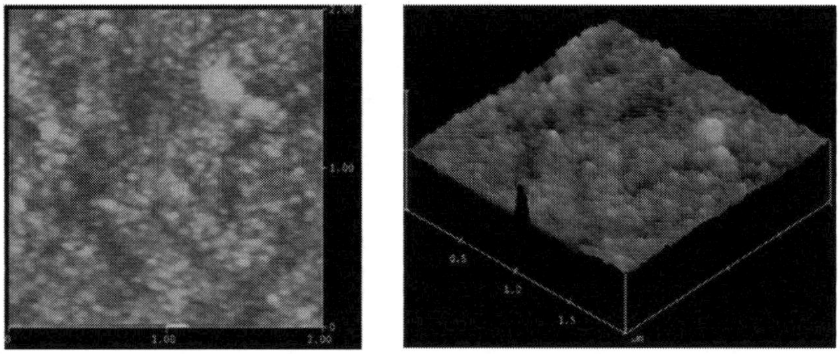

Figure 2. Surface structure of sample A (AFM analysis).

The stain-resistant performance of sample B and C is level 1 only, suggesting that they are more vulnerable to the attack of stain with smaller particle sizes. From the AFM analysis results shown in Figure 3 and 4, it is apparent that the size of particles on the surface of sample B and C is not consistent and the particles are not closely distributed. Such poor surface morphology is believed to be the reason lowering the stain-resistant performance of sample B and C.

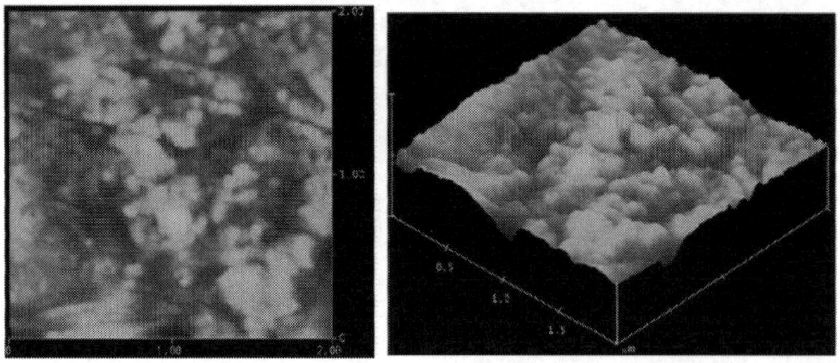

Figure 3. Surface structure of sample B (AFM analysis).

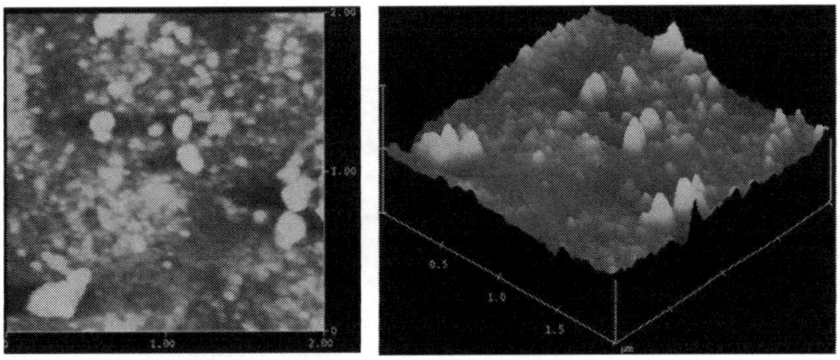

Figure 4. Surface structure of sample C (AFM analysis).

5. Discussion

To investigate the effect of surface morphology on the stain-resistant performance of ceramic tiles, six types of commercially available ceramic tiles, including nano-structured ceramic tiles and regular ceramic tiles, were selected in this study. The testing results show that ceramic tiles with stain-resistant performance of level 4 exhibits closely distributed surface particles and an average particle size of 20 nm, which is believed to benefit the stain-resistant performance of ceramic tiles. On the other hand, the particles on the surface of other ceramic tile samples show uneven distribution and sizes, making these ceramic tiles to have lower stain-resistant performance.

6. Reference

[1] CNS 3299-9: 2011 Methods of test for ceramic tiles. Part 9: Determination of resistance to stains.
[2] Yuan, 2008 The stains of metropolitan ceramic tiles are mainly substances resulted from exhaust emission.
[3] JSTM J 7602T: 1992 Pollution inspection methods of Construction with ceramic tiles.
[4] Recommendation, commission recommendation of 18 October 2011, Official Journal of the European Union.
[5] ISO 16700: 2004(E) Microbeam analysis – Scanning electron microscopy – Guidelines for calibrating image magnification.
[6] TNS M017001: 2012NanoProduct: Method of Dimensional Measurement of Nanoparticles on Surface - Scanning Electron Microscope
[7] ASTM E2859: 2011 Standard guide for size measurement of nanoparticles using atomic force microscopy。
[8] ASTM G154:2006 Standard practice for operating fluorescence light apparatus for UV exposure of nonmetallic materials.
[9] CNS 3299-10: 2011 Methods of test for ceramic tiles-Part 10: Determination of chemical resistance.

Problems During Milling and Roughness Registration of Free-form Surfaces

M Rybicki[1]
Poznan University of Technology, Faculty of Mechanical Engineering,
Piotrowo 3, 60-965 Poznan, Poland

E-mail: marek.rybicki@put.poznan.pl

Abstract. Finishing machining of free-form surfaces is usually performed using ball mills. Constant rotational speed n and feed rate per wedge in axis of the mill f_z are applied most often for the whole surface. It causes long machining time and theoretical machined surface roughness Rz_t heterogeneity. In the paper we propose a milling method with variable parameters n and f_z which gives greater productivity and more homogenous roughness of free-form surfaces. Based on roughness parameter measurement of the samples, many problems have been enumerated: necessity of the surface curvature pre-filtering from primary profile, necessity of 2D roughness assessment in two perpendicular directions, requirement for a high vertical measurement range and special long tips in contact profilometers most often, roughness variability and necessity of milling technology knowledge, to find a region with maximal roughness.

1. Introduction

Free-form machined surfaces have more and more participation in machining. Applying of the HSM strategy in finish machining has allowed e.g. to decrease machining time and to improve of surface quality. For the sake of small passes interval and small federate, time of finish machining is still amounted to almost or over 50% of total machining time of the parts, and surfaces can required lapping in some places.

Constant allowance q_0 for finish machining are applied. Nowadays the whole machined surface are milled with constant feedrate vf in axis of the mill and constant rotational speed n most often (in whole machining cycle). It causes increasing of such quantities like v_{f0}, ψ, $h_m(h_{zsr})$ on concave surface and decreasing of the quantities on convex surface (figure. 1). When milling free-form surface with constant rotational speed n by ball mill cutting speed v_c is changed additionally due to change of effective diameter D_{ef}. I causes local changes of tool's load (form errors) and kinematic-geometrical mapping (roughness) on milling length. Besides, constant values of the v_f and n are selected for the most difficult to cutting region of the surface, what causes ineffective machining most time (in easier regions) and long machining time.

[1] Corresponding author.

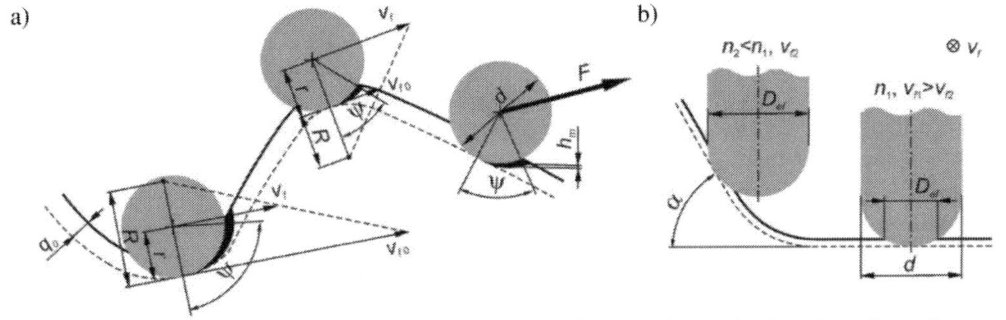

Figure 1. Kinematics and dimensions of cut a) as well as impact of machined surface tilt angle α on effective ball mill diameter D_{ef} with recommended speeds n and v_f b) during milling of curvilinear profile

During milling by ball mill roughness can be considered in feed and pick direction. In finish machining tool path assuring constant theoretical roughness in pick direction are applied mainly [5,19]. Many sources suggest that in order to keep constant theoretical roughness feedrate per revolution f or per wedge f_z should be equal to distance between paths (pick f_p) in finish machining. It explained by possibility to keep constant 3D roughness and high material removal rate, assuming that ball maps in machined material. In reality wedge path is epicyclical and when $p=f$ component of roughness coming from feedrate is 3 to 4 times higher (depending on the mill diameter d and angle α of its axis til relative to machined surface) that component coming from pick [2].

In the paper is proposed original method of simultaneous controlling of feedrate v_f and rotational speed n on milling length giving more homogenous roughness and shorter machining time.

Strategy of changeable rotational speed, in order to assure constant cutting speed v_c=const., is known in turning. It causes shorter machining time and improvement of machined surface roughness. But the strategy is not applied during milling by ball mill. Producers of cutting tools recommend cutting speed v_c in narrow range (or rotational speed n when diameter D_{ef} of the mill is known). For various angles of machined surface tilt α various rotational speeds n are recommended usually. During milling with constant rotational speed n=const its smallest recommended value are selected. It causes underrating of the n value in regions it should be higher. And then feedrate $v_f \sim n$ is also small and machining time $t_s \sim 1/v_f$ is long in a given toll path. Small cutting speed (for $\alpha=0°$) can be avoided by numerical control of machine tool's rotational axes but during milling of deep cavities it is possible in very narrow range.

Also during milling with variable feedrate v_f is connected with considerable time saving, what results from increasing of its value on small curvature segments of milling path. If concave surface has the highest roughness Rz_{max} and force F_{max}, feedrate v_f in axis of the mill can be increased in small curvature regions to obtain the same roughness and force as in the concave one. Then maximal roughness and force will not be higher but machining time will be shorter. Feedrate can be changed in way assuring constant deflection of the mill [11,18] or roughness [3], what can minimize or eliminate manual lapping of cavities in dies and moulds.

The most often applied in literature method of the feedrate changes gives constant feedrate v_{f0} of wedges, which is a real rate of work material forming. In order to obtain the v_{f0} value feedrate v_f pr should be programmed in axis, often given in in form of relationship (1) in which v_f is feedrate in axis when tool path is straight. Quantity R_0 is path radius of the mill's axis, which can be write as $R_0=R+r$ for convex surface and $R_0=R-r$ for concave one.

$$v_{fpr} = v_f \frac{R_0}{R} \tag{1}$$

2. Range, conditions and technique of research

The experimental research will be carried out onto the Cincinnati VMC-500 Arrow milling machine. The Mastercam X3 and option *Surface Finish Blend* was used for making NC program. In the machining strategy pick is kept constant along curvilinear surface if possible.

Machining allowance q_0=0,06mm, rotational speed n=8000rev/min and constant feedrate in axis of the mill f_z=0,06mm/wedge (variable v_{f0}) as well as pick f_p=0,12mm equal to feedrate per revolution $f=f_z \cdot z$ were applied.

Sample was made from hardened hot work tool steel 55NiCrMoV of hardness 57 HRC and milled down without cooling.

Ball mill made of Fraisa company with designation D15140.300, diameter d=6mm, number of wedges z=2, angles λs=30° and γ = -10° was used.

a) b)

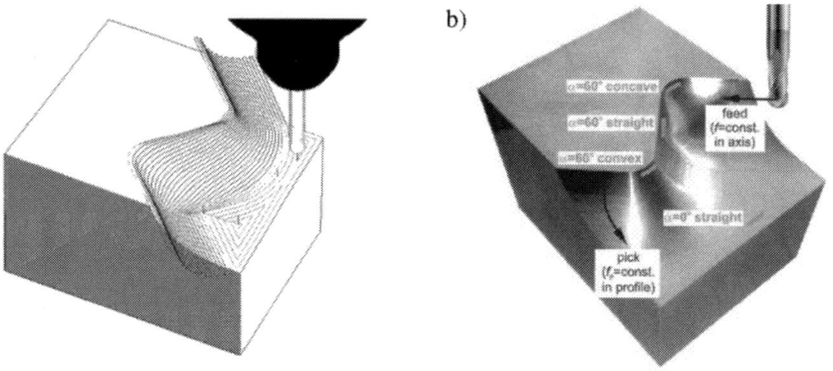

Figure. 2. Machining strategy a) and places of roughness registration b)

Roughness measurements were done on to the HOMMEL roundscan 535 parallel to the v_f vector. In earlier measurements on the T500 profilometer made of Hommewerke measurements were not possible in all places due to too short measuring tip and vertical range of the device. Traverse length l_t=4,8mm and double filtration was used. First radius of machined surface R=9mm was filtered from primary profile. Measuremet tests without the filtration had gave another values of roughness parameters Next filter ISO 11562(M1) with wavelength of cut-off λ_c=0,25 mm was used. Value of the λ_c>f was applied intentionally to assessment kinematic-geometrical mapping of the mills wedges.

3. Results and analysis of research

3.1. Machined surface roughness

It can be concluded from figure 1 that during milling with constant feedrate in mill's axis roughness should be the highest on concave surface due to the highest value of the v_{f0}. Besides the feedrate value, roughness is also resulted from the same machined surface curvature [22], chatter etc. Roughness is the highest when milling surface perpendicular to axis of the mill, when effective diameter D_{ef} and cutting speed v_c are small. Theoretical roughness height shown at the figure, assumes that in work material maps a ball and do not gives consideration to tilt of axis of the mill.

Theoretical space RSm_t between roughness elements was assumed to be equal to feed per revolution of the mill's wedges (figure 4). From the figure follows that for high α angle micro irregularities arises at space equal to the f_0, like during cylindrical milling, and roughness profile has no visible mapping space equal to feed per wedge f_{z0}. Frequency spectrum of roughness profile when milling with angle α=0° is very scattered.

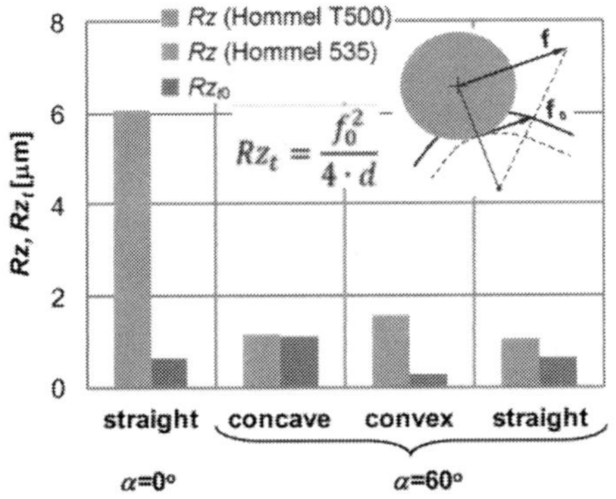

Figure 3. Theoretical and real roughness height (Rz_t and Rz acc. to DIN) in various places of machined surface

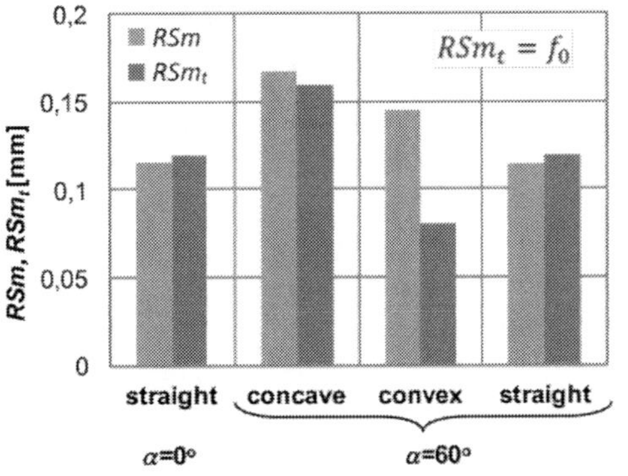

Figure 4. Theoretical and real distance beetwen roughness elements (RSm and RSm_t) in various places of machined surface

3.2. Proposal of variable cutting parameters applying

For ball mill roughness and form errors are affected by angle α of the mill's axis tilt relative to machined surface [3,7,11,18] and local radius R of the surface. So it has been proposed that changes of machining parameters v_f and n took place when the α and R values had change. Changeable (depending on machined surface geometry) machining parameters v_f and n were calculated from equations (3) and (4). Equation (3) consist feed per wedge f_{zpr} calculated by formula (1), but it can also be used formula from [3] which assure constant three dimensional theoretical roughness height R_{max}. Equation (4) consists diameter D_{ef} connected with angle α.

$$f = f_{zpr} \cdot z \cdot n \tag{3}$$

$$n = \frac{v_c}{\pi \cdot D_{ef}} \tag{4}$$

$$D_{ef} = d \cdot \sin\left(\alpha + arccos\left(\frac{d-a_p}{d}\right)\right) \tag{5}$$

At figure 5 is shown theoretical example how many percent of time can be save thanks to changeable feedrate and rotational speed in various places of machined surface applying. Except time saving more homogenous roughness is expected. It was assumed that machining time during milling with constant parameters is a 100%. For analysis was assumed machined surface model with radius R=13mm, length of straight feed intervals in in single path 180mm and maximal angle α=90° as well as diameter d=10mm and wedges number z=2. Recommended machining parameters were set f_z=0,035mm/wedge and v_c=500m/min. It was assumed that maximal rotational speed of machine tool n_{max}=40000rev/min and maximal feedrate v_{fmax}=5000mm/min.

It is proposed modification of NC program for machining with constant parameters v_f and n. Blocs of the program can be divided in places where exact changes of radius R of angle α has occurred, and new parameters calculated from equations (3) and (4) can be loaded. The places can be marked basis on coordinates X, Y and Z of characteristic tool point and vector defining contact place of the tool with workpiece.

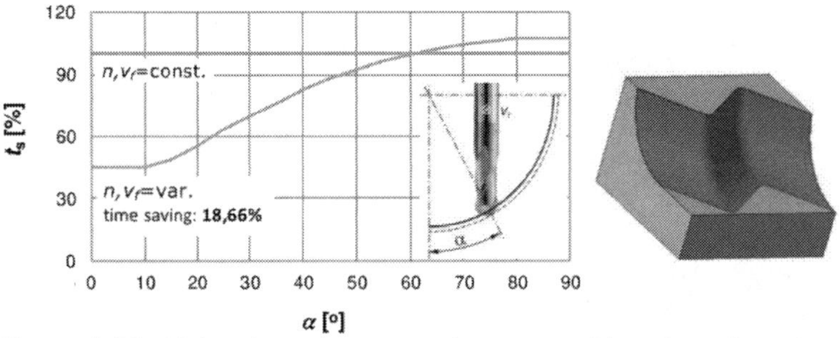

Figure. 5. Machining time t_s in consecutive passes with various tilt angle α during milling with constant and variable parameters v_f and n

4. Conclusions

Basis on roughness parameters measurement of sample, many problems have been enumerated, as: necessity of the surface curvature pre-filtering from primary profile, necessity of 2D roughness assessment in two perpendicular directions, required high vertical measurement range and special long tips in contact profilometers most often, roughness variability and necessity of milling technology knowledge to find region with maximal roughness. The regions are concave surfaces and surfaces cutting by edges of the mill near of its axis.

During ball milling roughness is affected by angle α of the mill's axis tilt relative to machined surface and local radius R of the surface in feed direction. For high angle α micro irregularities arises at space RSm equal to feedrate per revolution f_0, like during cylindrical milling

By applying of variable cutting parameters on curvilinear tool path machining time can be reduced and constant roughness or constant load can be obtained.

Thanks to Division of Metrology and Measurement Systems of Poznan University of Technology for making measurements of machined surface texture parameters.

[1] Bouzakis K D, Aichouh P and Efstathiou K 2003 Determination of the chip geometry, cutting force and roughness in free form surfaces finishing milling, with ball end tools *International Journal of Machine Tools & Manufacture* **43** 499

[2] Chen J S, Huang Y K and Chen M S 2005 A study of the surface scallop generating mechanism in the ball-end milling process *International Journal of Machine Tools and Manufacture* **45** 1077

[3] Choi I H, Yang M Y, Hong W P and Jung T S 2005 Curve interpolation with variable feedrate for surface requirement *International Journal of Advanced Manufacturing Technology* **25** 325

[4] Chu C N, Kim S Y and Lee J M 1997 Feed-rate optimization of ball end milling considering local shape features *Annals of the CIRP* **46** 433

[5] Docker S and Dickin P J 2007 Modern machining techniques for mouldmaking. *4th International Conference and Exhibition on Design and Production of MACHINES and DIES/MOULDS*, Cesme, TURKEY, 21-23/6/2007

[6] Heo E Y, Merdol D and Altintas Y 2010 High speed pocketing strategy CIRP *Journal of Manufacturing Science and Technology* **3** 1

[7] Kecelj B, Kopač J, Kampuš Z and Kuzman K 2004 Speciality of HSC in manufacturing of forging dies *Journal of Materials Processing Technology* **157-158** 536

[8] Kim G M, Chub C N 2004 Mean cutting force prediction in ball-end milling using force map method *Journal of Materials Processing Technology* **146** 303

[9] Kim K K, Kang M C, Kim J S, Jung Y H and Kim N K 2002 A study on the precision machinability of ball end milling by cutting speed optimization *Journal of Materials Processing Technology* **130-131** 357

[10] Kloypayan J and Lee Y S 2002 Material engagement analysis of different end mills for adaptive feedrate control in milling processes *Computers in Industry* **47** 55

[11] Lin J C and Tai C C 1999 Accuracy optimization for mould surface profile milling *The International Journal of Advanced Manufacturing Technology* **15** 15

[12] Lotfi B, Zhong Z W and Khoo L P 2009 Variable feed rates and variable machine forces for a constant material removal rate and constant cutting force along Pythagorean-hodograph curves *International Journal of Advanced Manufacturing Technology* **40** 171

[13] Pateloup V, Duc E and Ray P 2000 Corner optimization for pocket machining. International *Journal of Machine Tools & Manufacture* **44** 1343

[14] Rybicki M 2012 Wybrane problemy dokładnego kształtowania powierzchni swobodnych na obrabiarkach CNC *Mechanik* **1** 31 (in CD)

[15] Schuett T J 2002 Look-ahead for faster machining. Rapid acceleration/deceleration key to high-speed machining *Cutting Tool Engineering* **54**

[16] Sencer B, Altintas Y and Croft E 2008 Feed optimization for five axis CNC machine tools with drive constraints *International Journal of Machine Tools & Manufacture* **48** 733

[17] Sun G and Wright P 2005 Simulation-based cutting parameter selection for ball end milling *Journal of Manufacturing Systems* **24** 352

[18] Sun Y, Jia Z, Ren F and Guo D 2008 Adaptive feedrate scheduling for NC machining along curvilinear paths with improved kinematic and geometric properties *International Journal of Advanced Manufacturing Technology* **36** 60

[19] Toh C K 2005 Design, evaluation and optimization of cutter path strategies when high speed machining hardened mould and die materials *Materials & Design* **26** 517

[20] Weinert K, Enselmann A, Friedhoff J 1997 Milling simulation for process optimization in the field of die and mould manufacturing. *Annals of the CIRP* **46** 325

[21] Choi I H, Yang M Y, Hong W P and Jung T S 2005 Curve interpolation with variable feedrate for surface requirement *Int J Adv Manuf Technol* **25** 325

[22] Lin R and Koren Y 1996 Efficient Tool-Path Planning for Machining Free-Form Surfaces *Journal of Engineering for Industry* **118** 20

[23] Quinstat Y, Sabourin L and Lartigue C 2008 Surface topography in ball end milling process: Description of a 3D surface roughness parameter Journal of Materislas Processing Technology **195** 135

[24] Jung T S, Yang M Y and Lee K J 2005 A new approach to analysing machined surfaces by ball-end milling, part I: Formulation of characteristic lines of cut reminder *Int J Adv Manuf Technol* **25** 833

[25] Chen J S, Huang Y K and Chen M S Feedrate optimization and tool profile modification for the high-efficiency ball-end milling process *International Journal of Machine tools & Manufacture* **45** 1070

[26] Suresh K and Yang D C H 1994 Constant scallop-height machining of free-form surfaces *Journal of Engineering for Industry* **116** 253

Multiscale Analysis of Surface Topography from Single Point Incremental Forming using an Acetal Tool

M Ham[1,2], B M Powers[2], J Loiselle[2]
1. Faculty of Engineering and Applied Sciences, University of Ontario Institute of Technology, Oshawa, ON, L1H 7K4, Canada
2. Worcester Polytechnic Institute, Worcester, MA, 01602, USA

E-mail: marnie.ham@uoit.ca[1]

Abstract. Single point incremental forming (SPIF) is a sheet metal manufacturing process that forms a part by incrementally applying point loads to the material to achieve the desired deformations and final part geometry. This paper investigates the differences in surface topography between a carbide tool and an acetal-tipped tool. Area-scale analysis is performed on the confocal areal surface measurements per ASME B46. The objective of this paper is to determine at which scales surfaces formed by two different tool materials can be differentiated. It is found that the surfaces in contact with the acetal forming tool have greater relative areas at all scales greater than 5 x 10^4 μm² than the surfaces in contact with the carbide tools. The surfaces not in contact with the tools during forming, also referred to as the free surface, are unaffected by the tool material.

1. Introduction

This paper presents information on a new forming tool that has a forming tip made from an acetal resin. Parts are formed with this new tool and a more commonly used carbide tool. The surfaces of the parts are then measured and analyzed using area-scale analysis. This paper compares the scales at which the surfaces produced by the acetal tool are different from the surfaces created by the carbide tool.

Single point incremental forming (SPIF) uses a forming tool that is in contact with one side of a material, the forming tool moves relative to the sheet metal in order to form the part [1]. A die is not required on the other side of the material; this surface is referred to as the free surface. SPIF has the potential to bring complex sheet metal parts to market, without the need for forming dies. An important step in the advancement of SPIF to a stage where the process can be widely used in industry is the development of good surface quality during manufacturing. The main shortcoming of surface quality in SPIF is the greater roughness. The mean values of average roughness (Ra) reported have typically been in the range of 4-12μm [2], while more traditional forming processes have produced surfaces with mean Ra 0.8-3μm [3].

Much of the previous study on surface topography in SPIF has discussed the topography that is easy to visually observe [4-9]. Previous work has focused on ridges created in the part from the toolpath and tool geometry as the tool steps down into the part (step down ridges). These ridges provide discernible surface topography. Generally, the toolpath is described in terms of step down and forming angle. The most common tool geometry is a hemispherically tipped cylinder. The literature on SPIF toolpaths refers to the amount the tool "steps" into the material during each spiral of the toolpath;

commonly described as the step down [10]. The step down is the most frequently studied SPIF process variable in relation to surface topography.

The side of the sheet metal that does not contact the tool during forming is referred to as the free surface [11]. The surface topography of the free surface has seldom been studied in SPIF. The roughening of the free surface during forming is referred to as "orange peel". This roughness is generally described as occurring on the scale of grain size and occurs with movement of grains relative to the surface [12]. Although other work has found the mean length of the roughness to be ten times as long as the grain size [11], high-speed compression testing with the tool contacting one side of the sheet metal, found that roughness increases with increasing strain when the grain size is constant [11]. The mode of deformation and the grain size have also been shown to affect this roughness [11]. In SPIF, it was found that the tool's feed direction (or toolpath) can sometimes be seen on the back side of the part [6]; implying that the tool diameter and step down might influence the surface topography of the free surface.

Area-scale fractal analysis can be used to calculate fractal parameters described in ASME B46.1 [13] or to search for differences [14] or correlations [15] between surfaces as a function of scale. Area-scale fractal analysis first virtually tiles the surface with triangular tiles. The area of each tile is the scale. Then relative area is calculated by taking the sum of the area of all of the tiles and dividing that sum by the nominal area. The relative area is calculated at a series of scales and can be plotted and analyzed at each scale.

F-tests have previously been used as a function of scale to determine at which scales the relative area of two surfaces are significantly different [14]. Area-scale analysis has also been used to find the scale with the greatest correlation [15]. Scale-base correlation has also been established by filtering with a shifting band-passed filter before calculating roughness parameters to test for correlation [16] A Mann-Whitney U-test has been previously employed to determine if the relative length of groups of surfaces are significantly different as a function of scale [17]. The advantage of this approach was that it did not assume that the relative length at each scale was normally distributed. The Mann-Whitney U-Test is more robust due to not having a required distribution of data, such as a normal distribution.

Non-parametric statistics are statistical techniques that do not rely on the data belonging to any particular distribution. Non-parametric statistics are typically used when either a data set is shown to be non-normal or the distribution of the data set is not known. Two of the non-parametric tests of interest in this study are the Kruskal-Wallis test, which is similar to a one-way ANOVA, and the Mann-Whitney U-test, which is analogous to a t-test [18]. The Kruskal-Wallis test is used to see if samples originated from the larger sample or population. The Kruskal-Wallis test only determines whether all the samples belong to the population or if at least one does not. It cannot determine which sample does not belong. The Kruskal-Wallis test is a ranked sum test, if the rankings have no order then they all belong to the same large sample. A post-hoc test, after the Kruskal-Wallis test, is needed to determine which samples do not belong. A typical post-hoc test is the Mann-Whitney U-test. The Mann-Whitney U-test compares two samples and determines if they are both part of a larger sample or population. This is done through ranking the sums and determining whether they are lumped or segregated.

The objective of this paper is to determine a) at which scales the contact and free surfaces formed by two different tool materials can be differentiated and b) to gain insight into the mechanisms that form these surface.

2. Methods

The methodology in this work consists of the forming methods used to create the SPIF test pieces, with either the acetal or carbide tool, and the techniques used to characterize the topography of the formed surfaces. SPIF methods are discussed first.

The sheet material used in these experiments is AA 3003-O 1.6 mm thick. AA3003-O has previously been investigated in SPIF studies due to its high formability [1]. The mechanical properties of AA3003-O include: ultimate tensile strength (UTS) = 110MPa, yield strength (YS) = 40MPa, and

Brinell hardness number (BHN) = 28 and n = 0.242 [19]. The mean grain size of the AA3003-O prior to SPIF was previously found to be 25μm; however, not all of the grains are equiaxial [20].

Figure 1 shows the fixture and backing plates utilized to hold the sheet metal during forming. A continuous spiral toolpath is generated using the computer aided manufacturing (CAM) software Esprit®. A step down of 0.13 mm is used, with a spindle speed of 600 RPM and a feed rate of 9 m/min. This toolpath is shown in figure 1.

A Haas VM3 vertical machining center is used to form truncated pyramids of 76 mm by 76 mm, with a depth of 25 mm and a forming angle of 60-degrees. The truncated pyramid is shown in figure 2. The truncated pyramid shape was selected for this study due to the relatively flat sides of the pyramid and because the optical instrument used for measurements has a short working distance. Thus, the relatively flat specimens can be cut from the sides of the pyramids to facilitate measurement.

All of the specimens are created using one of two 12.5 mm diameter tools to form parts. One tool is a conventional carbide tool and the other new acetal-tipped tool. Figure 3 depicts the two different forming tools. The carbide tool is a solid carbide tool with a ground hemispherical tip. The carbide tool requires lubrication, during forming. In these experiments, the lubrication used is petroleum-based wax. The acetal-tipped tool consists of an acetal tip that is press fit into a steel shank. The acetal resin utilized is DuPont™ Delrin® AF 100, consisting of Teflon™ fibers uniformly dispersed throughout the acetal resin. This acetal resin is selected due to its low friction, wear characteristics, and because it is self-lubricating; thus, it doesn't require additional lubrication.

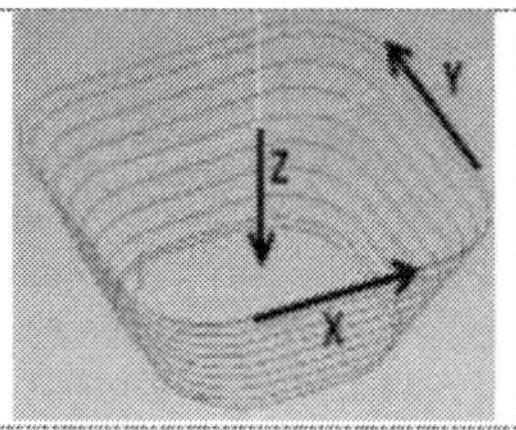

Figure 1. Toolpath generated using Espirit®

Figure 2. SPIF fixture and backing plate with a formed truncated pyramid.

Figure 3. Forming tools: a) acetal-tipped tool b) carbide tool.

Optical surfaces [21] are acquired from each side (contact and free surface) of the eight formed parts (four parts formed with the acetal tool, and four parts formed with the carbide tool). A vertical scanning laser confocal microscope with a 408 nm violet laser light is used for the surface measurements, specifically an Olympus LEXT OLS4000. For all specimens, the microscope is used with a 50X objective with a numerical aperture of 0.95 and a 250 nm sampling interval, nine overlapping surface measurements, each with a measurement region of 250 μm x 250 μm, are stitched together to form a single surface optical surface. For each optical surface, relative area is calculated at a series of approximately 2800 scales, based on the standards ASME B46.1 [13], using the Sfrax software. Each surface is grouped according to the tool used to produce the part and the specific side of the part (contact or free surface). The statistical analysis is used to prove whether or not the differences in relative area at each scale are significant.

The statistical analysis starts with a Kruskal-Wallis test and, if needed, a post-hoc Mann-Whitney U-test is completed. This is done at each scale. The Kruskal-Wallis compares multiple (more than two) groups of relative areas at a scale to determine if they are not independent. The first step in the test is to rank all of the data from smallest relative area to largest relative area, independent of the scale it is from. The next step in the test is to set the hypothesis that states all ranks from all scales are the same; the alternate hypothesis states not all ranks are the same. The ranks are then tested with the Kruskal-Wallis test statistic, equation 1. A p-value (probability of obtaining a test statistic) is obtained. This is the lowest α value at which a scale is determined to be independent. If the p-value is less than the selected α (typically $\alpha = 0.05$), then an ad-hoc test is needed to determine which scales are different. As a post-hoc test, the Mann-Whitney U-test is performed multiple times, one time for each pair of scales that needs to be tested or each pair that are to be compared.

$$H = \frac{12}{N(N+!)}\sum_{i=1}^{k}\frac{{R_i}^2}{n_i} - 3(N+1) \quad (1)$$

The Mann-Whitney U-test is a rank sum test used to compare two sets of relative areas at one scale. This is done by ranking the data in the two set of relatives areas independently of the set from which it originated. The next step in the test is to create a hypothesis that states that the ranks from the two sets of relative areas are the same; the alternate hypothesis states that the ranks are not the same. The ranks are then tested with the Mann-Whitney test statistic, equation 2. A p-value (probability of obtaining a test statistic) is obtained. This is the lowest α value at which a sample is determined to be different from the larger set of relative areas. If the p-value is less than the selected α (typically $\alpha = 0.05$), then the sample is determined to be different.

$$U = n_1 n_2 + \frac{n_i(n_i+1)}{2} - \sum R_i \quad (2)$$

3. Results

Figure 4 shows parts formed by a) the acetal tool and b) the carbide tool. Figure 4 shows images of the side that has the forming tool in contact with the material. The acetal tool in contact with the material produces a more matte appearance than that produced by the carbide tool. The direction of the lay is in-line with the original roll marks of the sheet metal for the parts formed with the acetal tool. The roll marks refer to the surface topography created from cold rolling of the sheet. The lay on the parts formed with the carbide tool follows the toolpath.

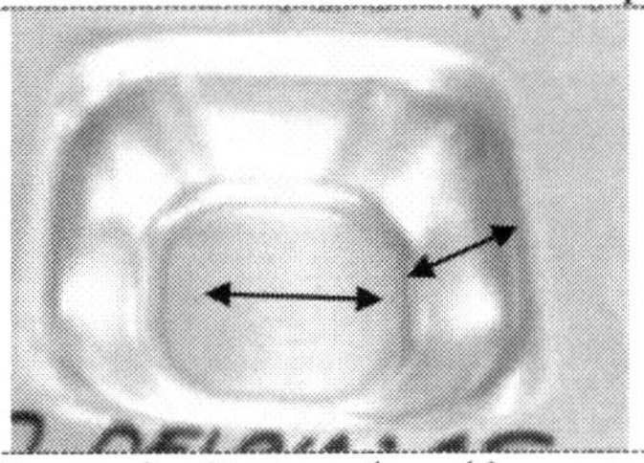

Figure 4a. Part produced by acetal tool; the arrow indicates the orientation of the roll marks.

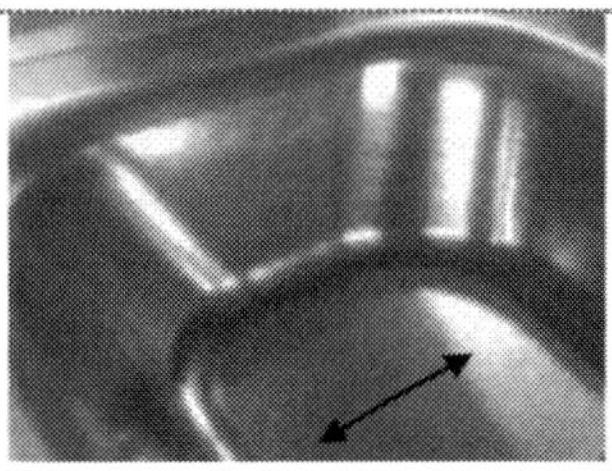

Figure 4b. Part produced by carbide tool; the arrow indicates the orientation of the roll marks.

Two variables (tool material and tool and surface interaction) are changed in order to create four groups of surfaces. The tool material is either acetal or carbide. The tool and surface interaction is "contact" if the tool is in contact with the surface during forming or "free" if the tool is on the other side of the sheet metal during forming. This creates four groups of surfaces referred to the tables and figures as AC for acetal contact, AF for acetal free, CC for carbide contact and CF for carbide free.

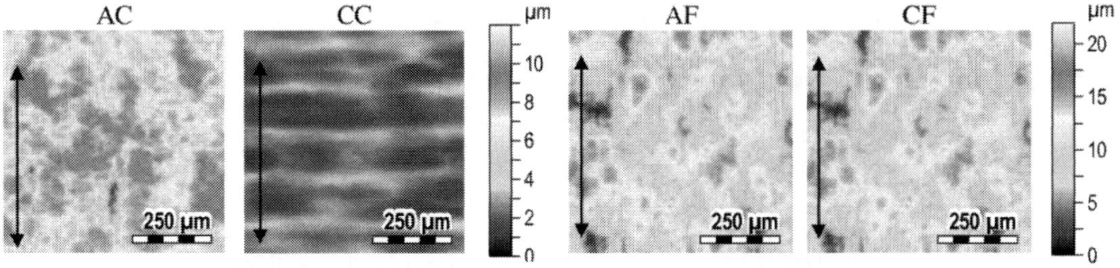

Figure 5. Height maps of selected surfaces from all groups.

Figure 5 shows the height maps of the free surfaces of the parts that are formed by both the carbide and acetal tools. The middle color scale bar is for both the AC and CC height maps and the right color scale bar is for the AF and CF height maps. There is sometimes an apparent texture in the roll mark direction, and the scale bar indicates a much greater roughness than that on the contact surfaces.

Figure 6 is the log-log plot of relative area vs. scale to the relative area for the tool materials and tool-and surface interaction. The four groups resulting from the combinations of the two tool materials and two types of tool-surface contact: carbide tool in contact with surface, carbide tool and free surface (not in contact with tool), acetal tool in contact with surface, acetal tool and free surface (not in contact with tool). Figure 7 is a relative area vs. scale plot for the scales of 0.01 to 1 μm^2. Since this study is an investigation to understand the surface differences created by: a) the different tool materials and b) whether the tool is in contact with the forming surface, it is important to compare the results of the combinations.

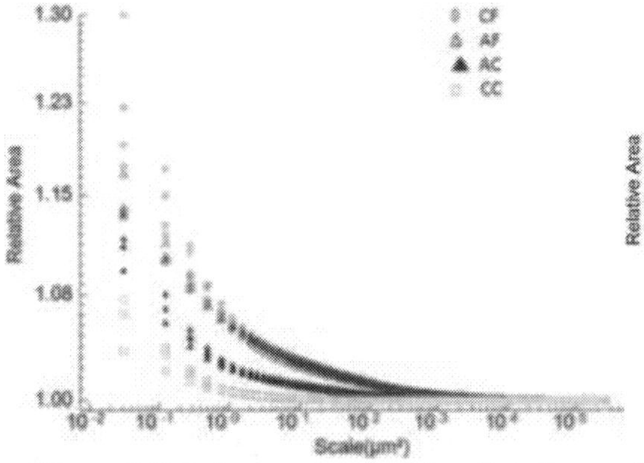

Figure 6. Relative area vs. scale for all surfaces

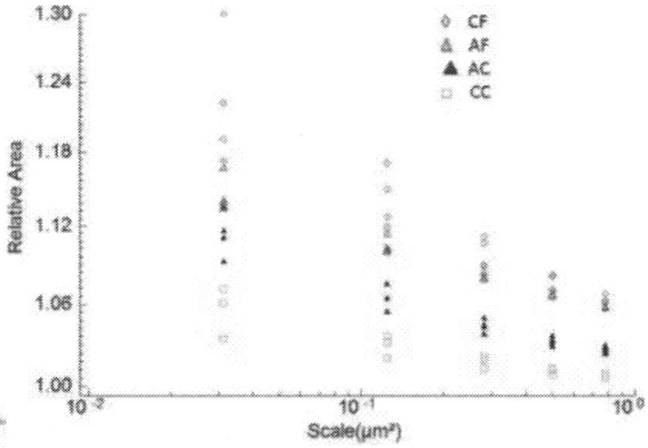

Figure 7. Relative areas for scales of 0.01 to 1 μm^2

The resulting relative areas for the scale of 0.03125 μm^2 are provided in table 1(a) for each of the groups. The relative areas of this scale are provided as a detailed example of the analysis done on figure 6 for interpretation of the results of the groups. The hypothesis to test is that all of the median ranks of relative areas for the four combinations of tool materials and contacts are systematically the same, as shown in equation 3. The alternative hypothesis is that there is a tendency for the relative areas to rank systematically higher or lower for at least one of the combinations when compared to the other combinations, as shown in equation 4. The relative areas in table 1(a) are ranked in table 1(b), irrelevant of the group it comes from.

$$\text{Ho: } \theta_{AC} = \theta_{AF} = \theta_{CC} = \theta_{CF} \quad (3)$$

$$\text{Ha: At least one } \theta_{ij} \text{ is different} \quad (4)$$

Table 1. Relative areas for a scale of 0.03125 μm² **(a)** all groups, **(b)** all groups ranked, **(c)** ranks for acetal groups (AC and AF).

(a)

AC	AF	CC	CF
1.111	1.139	1.035	1.300
1.136	1.167	1.061	1.173
1.093	1.140	1.034	1.190
1.117	1.135	1.072	1.221

(b)

Relative Area	Rank	Group
1.034	1	CC
1.035	2	CC
1.061	3	CC
1.071	4	CC
1.093	5	AC
1.111	6	AC
1.117	7	AC
1.135	8	AF
1.136	9	AC
1.139	10	AF
1.140	11	AF
1.167	12	AF
1.173	13	CF
1.190	14	CF
1.221	15	CF
1.301	16	CF

(c)

Relative Area	Rank	Group
1.093	1	AC
1.111	2	AC
1.117	3	AC
1.135	4	AF
1.137	5	AC
1.139	6	AF
1.140	7	AF
1.167	8	AF

Visual inspection of table 1(b), shows that the ranks are not in random order. This needs to be verified by calculating the Kruskal-Wallis H statistic and comparing this to the H critical statistic.

The $H_{test} = 13.79$ and the $H_{crit} = 7.65$, with an $\alpha = 0.01$ and df = 3. The α value of 0.01 is chosen since a high level of confidence is desired in determining that the relative areas are the same or not. As the $H_{test} > H_{crit}$, the null hypothesis is rejected, which suggests a significant difference in the medians of the relative areas. A post-hoc test is needed to determine which of the groups are different.

The Mann-Whitney U-test is used as the post-hoc test. The Mann-Whitney U-test is another rank sum test, but this time only two groups are compared at once. There are six relevant comparisons to consider: AC-AF, CC-CF, AF-CF, AC-CC, AF+CF-CC, AF+CF-AC. Table 1(c), shows the relative areas for all of the surfaces, using an acetal tool. Table 1(c) is used to demonstrate the Mann-Whitney U-test; the other Mann Whitney U-tests are shown in table 2.

Visual inspection of table 1(c) shows that the ranks are not random order. Calculation of the Mann-Whitney U-test is needed to determine if the absence of randomness is statistically significant, in order to determine that the two groups are different. This is done by calculating the U statistic for each of AC and AF.

$$\begin{gathered} U_{AC} = 4(4) + \frac{4(4+1)}{2} - \sum 1+2+3+5 = 15 \\ U_{AF} = 4(4) + \frac{4(4+1)}{2} - \sum 4+6+7+8 = 1 \\ U_{test} = \min(U_{AC}, U_{AF}) \end{gathered} \tag{5}$$

The result of this calculation is $U_{test} = 1$ and $U_{crit} = 0$ at $\alpha = 0.05$ or $\alpha/2 = 0.025$ for $n_1=4$ and $n_2=4$. As the $U_{test} \geq U_{crit}$ then the null hypothesis is not rejected, this suggests that there is no real difference in the medians of the relative areas. The test statistics can also be evaluated using software and the p-value is calculated. The p-value is the probability of calculating a test statistic that is at least the value of the critical value if the null hypothesis is true. If the p-value is less than your alpha, you reject the null hypothesis. For the above example, the p-value was calculated as 0.06. This p-value is greater than the chosen alpha ($\alpha = 0.05$), as $0.06 > 0.05$ then the null hypothesis in not rejected. Table 2 shows p-values of the Mann Whitney U-tests for all the groups at a scale of 0.03125 μm².

Table 2. Mann Whitney U-test results at a scale of 0.03125 μm²

Group	Comparative Groups	p-value
Acetal	AC-AF	0.0606
Carbide	CC-CF	0.0304
Free	AF-CF	0.0304
Contact	AC-CC	0.0304
	free-CC	0.0085
	free-AC	0.0138

Comparing the results from the Mann-Whitney U-test to the plot in figure 7, it is noted that one of the AC relative areas is mixed in with the AF relative areas. From the Mann-Whitney U-test, it is known that this test comparison resulted in p-values greater than 0.05, with a chosen $\alpha = 0.05$ and thus the null hypothesis is rejected. In other words, when the ranks of the relative area of one group is not greater than the other, it is not possible to determine at an $\alpha = 0.05$ that the comparators are different. For the cases in figure 7, where the ranges of the relative area for each group do not overlap (CC to CF, AF to CF, AC to CC, free surface to CC and free surface to AC), it is found that at an $\alpha = 0.05$, it is possible to determine that these surfaces are different.

Table 3 shows comparisons of different groups in figure 6 to demonstrate which groups are visually evident as distinct, or to show that there is no overlap of points. The plot of relative area vs. scale (figure 6) is a common method of interpreting area-scale results. The results from table 3 demonstrate that the free surface and the contact surface are always distinguishable from each other.

Table 3. Graphical Determination of Comparisons (groups are differentiable - ✓, groups are not differentiable - X)

Scale	Individual	Contact vs Free	CF vs. AF	CC vs. AC
10^{-2} - 10^{-1}	X	✓	✓	✓
10^{-1} - 10^{0}	✓*	✓	✓*	✓
10^{0} - 10^{1}	X	✓	X	✓
10^{1} - 10^{2}	X	✓	X	✓
10^{2} - 10^{3}	X	✓	X	✓
10^{3} - 10^{4}	X	✓	X	✓
10^{4} - 10^{5}	X	✓	X	X
$> 10^{5}$	X	✓	X	X

Note: * at all scales within range but 0.781μm².

At most scales, it is not possible to distinguish the acetal tool from the carbide tool, except at the 0.1 – 0.5 μm² scales. On the contact side, the surface formed by the acetal tool has a greater relative area than the surfaces formed by the carbide tool at all scales less than 5 x 10^4 μm. All of the individual groups could be distinguished from each other only at the 0.1 – 0.5 μm² scales. These are also the only scales at which the relative areas of the free surfaces (AF and CF) can be distinguished.

4. Discussion

The non-parametric statistical methods have been applied to enable significance testing as a function of scale. With only four measurement regions considered for each group, any overlap between relative areas in a group implies that the scale differences in relative area between the two groups is not significant. While no overlap, implies that the differences between the two groups are significant. The effect of tool material is significant in free surfaces are only different at scales range from 0.1-0.5 μm² but is significant in the contact surfaces at all scales greater than 5 x 10^4 μm².

This method has shown that the topography of the free surfaces are at minimum, mostly independently of the material of the tool that forms them. That the tool material has no effect on the surface topography of the free surfaces makes sense, as the step down in this work is smaller than the previous study where the bulge of the tool part was evident on the free surface [6]. The surfaces from these two groups are essentially the same in appearance and relative area. When this information is considered collectively, it appears that the surface topography of the free surface is independent of the material of the forming tool.

The carbide tool appears to be burnishing (smearing the surface texture) and creating a pattern of burnished tool marks (consists of ridges and valleys aligned with toolpath). The acetal surface does not form the distinctive pattern of ridges that is typical of most SPIF parts, including the carbide contact surfaces in this study. The result of this apparent absence of burnishing is a greater roughness and an increased relative area at scales greater than 5 x 10^4 μm². The acetal tool either does not burnish the surface or is only burnishing the surface on some of the peaks. This is likely due to the more conforming nature of the acetal compared with the carbide which reduces the local stresses on the surface due to the increased contact area and thus, lower localized stresses on the surface texture. If burnishing is occurring on the surface when the acetal tool is used, it could be expected to occur on the peaks of the surface texture of the aluminum sheet metal. If this is occurring, than burnishing might be a factor in the surface of the contact side with the acetal tool being smoother than the free surface of the acetal tool.

The surface of the acetal contact surface appears to form in a similar manner to the free surfaces as the appearance of the surface texture is similar although the amplitude of the surface is less. Two possibilities for reduced amplitude are: 1) that either the surface is burnished at the peaks thus reducing the amplitude, or 2) the forces from the tool restrict the out-of-plane displacements of the grains. Thus on the contact side, the acetal tool is forming surface texture with out-of-plane displacements that may be restricted by the forming tool contact or the surface texture may be partially smeared by localized burnishing. The free surfaces are formed in a manner consistent with the "orange peel" effect. That is the mechanism at work in creating these surfaces appears consistent with the slipping of grains in out-of-plane displacements [11].

5. Conclusions

The first objective of this work has been met using non-parametric statistics (Kruskal-Wallis test and Mann-Whitney U-test) to determine at which scales there are significant differences (≥ 95% confidence level) in relative area. It is found that the surfaces in contact with the forming tool have greater relative area when the acetal tool is used than when the carbide tool is used at all scales greater than $5x10^4$ μm². The tool material is not found to have an effect on the surface topography of the free surface.

The second objective of the paper was to gain insight into mechanisms that form these surfaces. The carbide contact surface texture appears to be formed by burnishing. While the acetal contact surface appears to be formed by out-of-plane displacements of grains and possible burnishing, and the free surfaces formed in a manner consistent with "orange peel" effect.

6. Acknowledgements

Natural Sciences and Engineering Research Council Canada (NSERC) for funding. Olympus and the WPI Surface Metrology Laboratory for the use of the LEXT OLS 4000. Digital Surf for the use of the MountainsMap software used to calculate the surface texture parameters.

7. References

[1] Jeswiet J, Micari F, Hirt G, Duflou J, Allwood J, and Bramley A 2005 *CIRP Ann.* **54/2** 130
[2] Durante M, Formisano A, and Langella A 2010 *J Mater Process Technol* **210/14** 1934.
[3] Boothroyd G, Dewhurst P, and Knight W 2002 *Product Design for Manufacture and Assembly* (New York : Marcel Dekker)
[4] Bhattacharya A, Maneesh K, Venkata Reddy N and Cao J 2011 *J Manuf Sci Eng* **133/8** 621
[5] Alves M, Silva M, Alves L, and Martins P 2008 *Proc SPIE Int Conf on Exp Mech* (Nanjing, China)
[6] Hamilton K and Jeswiet J 2010 *CIRP Ann* **59/1** 311
[7] Hagan E and Jeswiet J 2004 *Proc Inst Mech Eng Part B J Eng Manuf* **218/10** 1307
[8] Ham M, Powers BM, Brown CA, Jeswiet J, and Hamilton K 2009 *NAMRC 37* 411
[9] Ham M and Powers BM 2013 *Scanning* (pre-press SCA 21072)
[10] Ham M and Jeswiet J 2006 *CIRP Ann* **55/1** 241
[11] Osakada K and Oyane M 1971 *Bull Soc Mech Eng* **16/68**
[12] Hosford WF and Caddell RM 2011 *Metal Forming-Mechanics and Metallurgy* (Cambridge, MA: Cambridge University Press)
[13] ASME B46 2002 *Surface Texture, Surface Roughness, Waviness and Lay* (New York: ASME)
[14] Moreno MC, Bouchon P, and Brown CA 2010 *Scanning* **32/4** 212
[15] Berglund J, Brown CA, Rosén B-G, and Bay N 2010 *CIRP Ann* **59/1** 577
[16] Berglund J, Agunwamba C, Powers B, Brown CA, and Rosén B-G 2010 *Scanning* **32/4** 244
[17] Powers BM, Ham, M and Wilkinson MG 2010 *Scanning* **32/4** 199
[18] Corder GW and Foreman DI 2009 *Nonparametric Statistics for Non-Statisticians: A Step-by-Step Approach* (New York: Wiley)
[19] Hatch J 1984 *Aluminum: Properties and Physical Metallurgy* (Materials Park, OH ASM International)
[20] Ham M, Powers BM, Loiselle J 2013 *Key Eng Mater* **549** 84
[21] ISO 25178 2010 *ISO Standard Geometrical product specifications (GPS)* (Brussels:ISO).

Global and local mapping of motor blocks liners roughness for the analysis of honing performance

F Cabanettes[1], L Fahlgren[2], T Hoering[3] and B-G Rosén[1]

[1]School of Business and Engineering, Halmstad University, PO Box 823, SE-301 18 Halmstad, Sweden

[2]Volvo Car Corporation, Base Engine Dept., Gothenburg, Sweden
[3]NanoFocus AG, Oberhausen, Germany

E-mail: frederic.cabanettes@hh.se

Abstract. The manufacturing and finishing (honing) of cylinder liners for the automotive industry is a constant challenge in order to reduce friction losses and oil consumption. A better knowledge of surfaces generated during plateau honing is then required for optimization of the process. Despite a well-known and controlled honing process, variations in surface roughness appear at both global (due to honing tool wear) and local (TDC, middle stroke, BDC) scales and need to be mapped and analysed. The following paper proposes to map the global and local variations in roughness by using a confocal 3D measuring equipment able to measure and scan any area of a cylinder liner. Six motor blocks (five liners each) are evaluated with twenty topography measurements per liner. In total, six hundred 3D measurements of size 1x1 mm are performed and roughness parameters are computed. The results show that some parameters do correlate with the honing tool wear specific to each cylinder. Experimental models could be built. Furthermore surface roughness varies significantly over the axial length of the liners due to waviness deviations combined with a lack of flexibility of the honing tool in axial direction.

1. Introduction

The specifications of components in the automotive industry are strongly ruled by customer and legislation demands. Indeed, engineers are aiming for the production of engines with lower fuel consumption while keeping low emissions [1]. The cylinder liner is one of the engine components with real potential for lowering friction: a recent study by Holmberg et. Al. [2] showed that the piston assembly is responsible for 45% of the engine friction losses in passenger cars. One of the solution for reducing friction is to improve and control the surface roughness of the liners [3,4].

The plateau honing super finishing process is commonly used to obtain surfaces with good functionalities for the ring/liner contact. Nowadays surface quality control of the honing shows a lack of information (2D measurement of a non-functional area of the cylinder) to predict the functionality of the liners produced. Furthermore, despite a well-known and controlled honing process, spreading of surface roughness cannot be avoided and topography variations do exist at global scales (due to honing tool wear) and local scales (TDC, middle stroke, BDC). The existing variations need to be mapped and analysed for a better knowledge of the surfaces produced. By a better understanding of the process generating surfaces, optimizations leading to lower friction are possible.

The following paper proposes to map the global and local variations in cylinder liner roughness by using a confocal 3D measurement equipment able to measure and scan any area of a cylinder liner from passenger cars.

2. Materials and methods

2.1. Motor blocks

Six standard motor blocks from the same production line were picked up with two days interval. The blocks are from the Volvo T5 [5], 2.5 litres petrol engine (254 hp) with 5 cylinders (see figure 1). The interval for picking up blocks provides a complete scan of the entire honing tools life which gives also a relevant overview of the surface roughness spreading.

Figure 1. Motor block with 5 cylinders.

The blocks are labelled and the honing tools wear is recorded for each cylinder from each motor (see table 1). In this case, the honing tool wear is defined as the opening diameter of the tool to compensate wear of the honing stones.

Table 1. Honing tool wear (normalized) for each cylinder and block (cell in gray corresponds to a change of tool).

	Cylinder 1 (tool wear)	Cylinder 2 (tool wear)	Cylinder 3 (tool wear)	Cylinder 4 (tool wear)	Cylinder 5 (tool wear)
Motor block 1	63%	49%	0%	18%	74%
Motor block 2	73%	55%	8%	33%	84%
Motor block 3	80%	59%	14%	44%	92%
Motor block 4	82%	62%	16%	48%	95%
Motor block 5	86%	66%	21%	54%	100%
Motor block 6	99%	79%	34%	74%	1%

2.2. Measuring equipment

For this study, an optical 3D surface measuring system is used (see figure 2): the μsurf cylinder [6] is based on the confocal principle and allows automated cylinder liner inspections by positioning the optical objective according to depth (axial direction of the liner) and angle (circumferential direction of the liner). The methodology for acquiring 3D measurements (area: 1 by 1 mm, lateral resolution: 0.95μm, vertical resolution < 10nm) is presented in the section 3.1.

Figure 2. Confocal 3D measuring equipment installed on a Volvo engine block. System fully automated for scanning a cylinder liner in axial and circumferential direction.

2.3. Complementary measuring techniques

In order to complete the study, additional measurements were performed with a Phase Shift Technology MicroXam interferometer [7] and a mechanical stylus Surfascan 3CS [8]. The measuring area is approximately 600μm*800μm (1.089μm by 1.287μm lateral resolution and 0.1nm vertical resolution) for the interferometer. The profiles measured (see figure 4) with the stylus have a length of 98 mm (1μm lateral resolution and 6nm vertical resolution) allowing measuring at once much of the cylinder length (150mm).

The 3D measurements by interferometry are performed on replicas of the cylinder liner surfaces (see figure 3). A good agreement between replicas and original honed surfaces has been previously shown [9].

Figure 3. Mechanical stylus measuring over the length of a cylinder liner.

Figure 4. Replication of liner surfaces for different depth in order to measure with the interferometer.

3. Results and analysis

The six motor blocks presented in section 2.1 are measured using the measuring equipment presented in section 2.2. Each cylinder from each block is measured according to a methodology presented below in section 3.1. Additional measurements according to section 2.3 are performed to confirm trends observed.

3.1. Measuring methodology

Each cylinder from each block is measured according to the following measuring pattern:

- 4 angular positions in the circumferential direction of the liner are chosen as shown by the figure 5 (0°, 90°, 180°, 270°)
- 5 depth positions in the axial direction of the liner are chosen as shown by the figure 6 (10mm, 40mm, 50mm, 60mm, 70mm). More measurements are performed in the middle of the liner since it is a crucial area in terms of functionality. Furthermore, as mentioned before, for each depth 4 angular positions are measured. It is decided to perform the 4 measurements in a zigzag pattern along the reference depth to randomize the results as proposed by Rosén and Garnier [10].

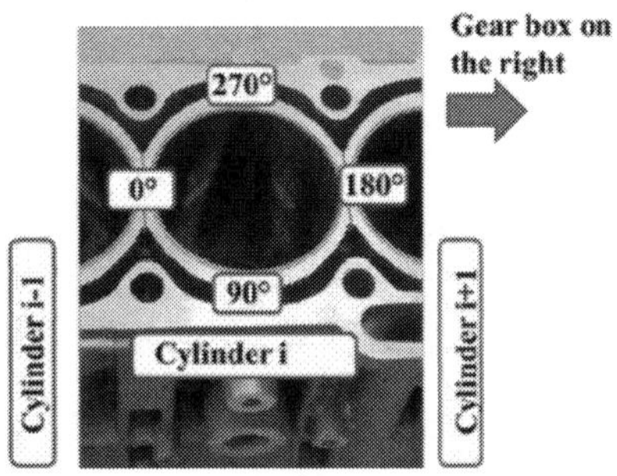

Figure 5. Labelling of cylinders and angles.

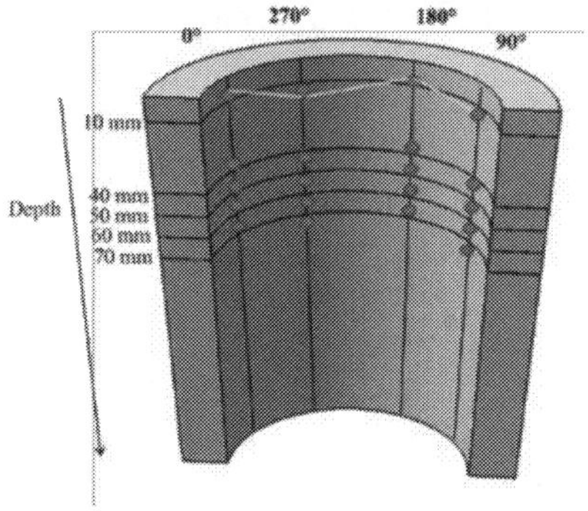

Figure 6. Unfolded cylinder and position of measuring points in a zigzag pattern along each depth position line.

3.2. Characterization

Once the surfaces acquired, the following pre-processing is performed by using MountainsMap® Premium 6.2 [11]: a polynomial filter of order 2 removes the form. Afterward, data with low light intensity are removed and a robust Gaussian filter (250 µm cut-off) is used to isolate the roughness.

A wide range of surface parameters is computed from the pre-processed surface (see figure 7) and includes the main families: amplitude, spatial, hybrid, functional, feature parameters.

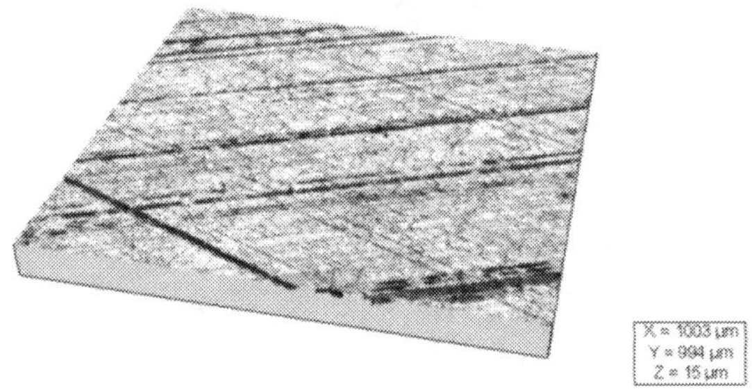

Figure 7. Example of a cylinder liner surface obtained after the pre-processing. The 3D surface is combined with the intensity layer for a more realistic visualization.

3.3. Correlations of surface roughness with honing tool wear (global variations)

Only the most interesting parameters are discussed in the following sections. A total of 600 measurements are processed (6 blocks, five cylinders, twenty positions) giving statistically significant results. The error bars in the following graphics represent the standard deviation to the mean. All results are presented as percentages of the mean value. The figure 8 presents the summit curvature parameter Ssc as a function of honing tool wear (6 motors picked up with different honing tool wear stages) for the different cylinders.

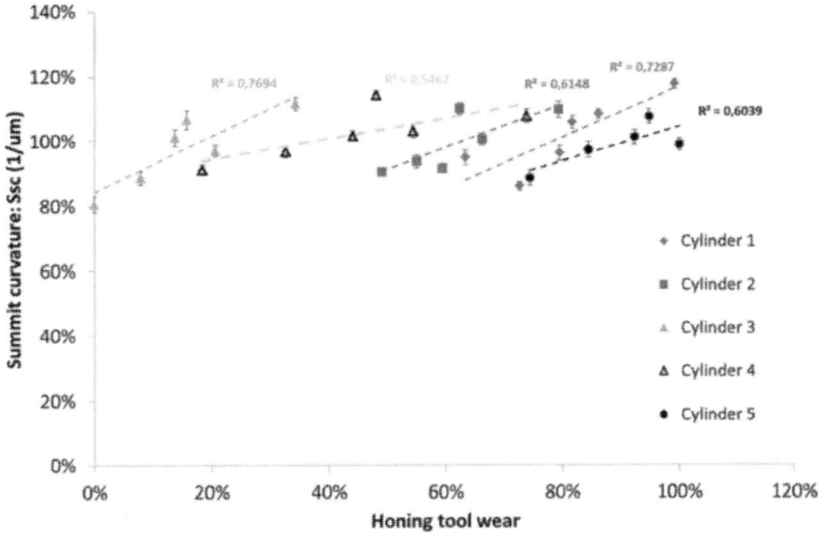

Figure 8. The summit curvature parameter Ssc is increasing with the honing tool wear specific to each cylinder.

No clear dependence can be observed between all the measurements and the honing tool wear. However, if the results are analysed separately, cylinder by cylinder, it is possible to see that Ssc increases with the honing tool wear specific to each cylinder (see trend lines on figure 8). A good correlation does exist between Ssc (as well as Spk and Sk) and the tool wear for a given cylinder (see figure 9) meaning that models could even be built for each cylinder. The possible reasons for finding good correlations only if each cylinder is analysed separately are the following: i) the proposed definition of the honing tool wear (opening diameter of the tool) is not accurate enough and not an absolute value ii) the honing process for each cylinder is complex and is depending on a wide amount of variables (lubrication, forms, heating) making it difficult to find a general model for every cylinder.

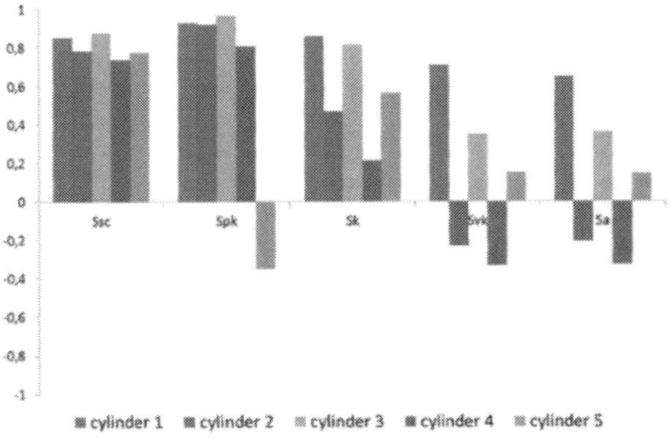

Figure 9. Correlation between different roughness parameters and the honing tool wear specific to each cylinder: Ssc, Spk and Sk increases with the honing tool wear whereas Svk and Sa have no clear dependence with the tool wear.

Globally, it is possible to analyse the results in the following way: as the honing tool wears down: i) the plateau and peaks are rougher (Sk, Spk) ii) asperities gets sharper (Ssc) iii) valleys are not significantly influenced (Svk).

3.4. Surface roughness variations over the liners (local variations)

More local evaluations of the honing process are also performed. The figure 10 presents the variations of the Sk parameter at different depths inside the liners. Similar behaviours are observed for Spk, Svk and Ssc.

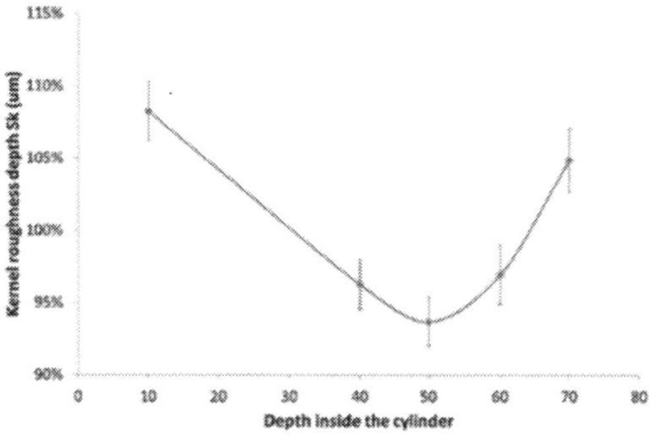

Figure 10. Trend of the roughness depending on depth observed with the confocal system

The same trend is confirmed by the analysis of 60 additional interferometer measurements along the axial direction (see figure 11). The complementary measurements also provide information deeper inside the liner. According to the figure presented below, the roughness varies along the axial direction of the liners due to the existing waviness (valleys of the waviness are less honed compared to the waviness hills). On the waviness hills, the honing tool is probably more in contact with the surfaces which results in a lower roughness. A lack of flexibility of the honing tool in the axial direction seems to be the reason. Techniques presented by Schmid [12] to compensate form deviations are based on similar assumptions. No influence of angular position is observed in the results since the tool is more flexible in circumferential direction.

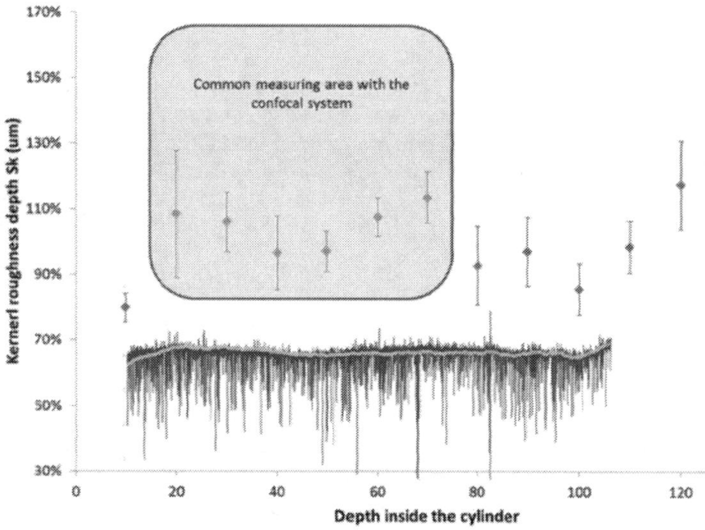

Figure 11. Confirmed trend observed by the interferometer for the cylinder 4 of motor 5. The measured stylus profile is superimposed to show that the Sk roughness values are depending on the waviness profile.

4. Conclusion

In order to improve the performance of the ring liner system, a better knowledge of surfaces generated during plateau honing of the cylinder liners is required. The paper proposed to use a 3D confocal measuring system with automated positioning in order to measure 600 surfaces from motor blocks picked up from production at different honing tool wear stages. The following conclusions were found:

- The most interesting parameters to link manufacturing and characterization are Spk, Sk and Ssc.
- As the honing tool wears down:
  - The plateau and peaks are rougher (Sk, Spk)
  - Asperities gets sharper (Ssc)
  - Valleys are not significantly influenced (Svk)
- Surface roughness varies significantly over the axial length of the liners but not around the circumferential direction
- The variations observed are due to waviness deviations combined with a lack of flexibility of the honing tool in axial direction. In other words, the honing stones cannot conform to the waviness in order to produce a homogeneous surface.

In the future, experimental models predicting the roughness (and the friction) as a function of the tool wear could be created leading to optimization opportunities.

Acknowledgements

The authors wish to thank Volvo Car Corporation for their financial support, their contribution in man hours and material. The authors are also thankful for the collaboration with Digital Surf through Mount Shasta program.

References

[1] http://www.dieselnet.com/standards/eu/hd.php, March 2013
[2] Holmberg K, Andersson P and Erdemir A 2011 Global energy consumption due to friction in passenger cars *Tribology International* **47** 221-34
[3] S. Johansson, P. H. Nilsson, R. Ohlsson, C. Anderberg, B.-G. Rosén, 2008 New cylinder liner surfaces for low oil consumption, *Tribology International* **41** 854-9
[4] Z. Dimkovski, F. Cabanettes, H. Löfgren, C. Anderberg, R. Ohlsson, B.-G. Rosén, 2012 Optimization of cylinder liner surface finish by slide honing *Proceedings of the Institution of Mechanical Engineers, Part B Journal of Engineering Manufacture* **226** 575-84
[5] http://www.volvocars.com/uk/all-cars/volvo-v40/specifications/pages/technical-spec.aspx, March 2013
[6] http://www.nanofocus.de/en/produkte/usurf/usurf-cylinder/, March 2013
[7] http://www.kla-tencor.com/mems-wafer-inspection/microxam-1200.html, March 2013
[8] http://www.jenoptik.com/en_home, March 2013
[9] F. Cabanettes, J. Claret-Tournier, J. Mohlin, P.H. Nilsson, B.-G. Rosén, L. Xiao 2009 The evolution of surface topography of injection cams, *Wear* **266** 570-3
[10] B.-G. Rosén, J. Garnier 2010 Uncertainties and optimized sampling in surface roughness characterization, *Wear* **271** 610-5
[11] http://www.digitalsurf.com, March 2013
[12] J. Schmid 2010 Friction optimization of cylinder surfaces from the perspective of production technology, *MotorTechnische Zeitschrift* **6** 8-12

Topographic modelling of haptic properties of tissue products

B-G Rosen[1,2,3], A Fall[4], S Rosen[3], A Farbrot[4], P Bergström[4]
[1]School of Business and Engineering, Halmstad University, PO Box 823, SE-301 18 Halmstad, Sweden
[2]Chalmers University of Technology, Göteborg, Sweden
[3]Toponova AB, Halmstad, Sweden
[4]SCA Hygiene Products AB, Göteborg, Sweden

E-mail: bg.rosen@hh.se

Abstract. The way a product or material feels when touched, haptics, has been shown to be a property that plays an important role when consumers determine the quality of products For tissue products in constant touch with the skin, softness" becomes a primary quality parameter. In the present work, the relationship between topography and the feeling of the surface has been investigated for commercial tissues with varying degree of texture from the low textured crepe tissue to the highly textured embossed- and air-dried tissue products. A trained sensory panel at was used to grade perceived haptic "roughness".

The technique used to characterize the topography was Digital light projection (DLP) technique, By the use of multivariate statistics, strong correlations between perceived roughness and topography were found with predictability of above 90 percent even though highly textured products were included. Characterization was made using areal ISO 25178-2 topography parameters in combination with non-contacting topography measurement. The best prediction ability was obtained when combining haptic properties with the topography parameters auto-correlation length (Sal), peak material volume (Vmp), core roughness depth (Sk) and the maximum height of the surface (Sz).

1. Introduction

For tissue products touch and tactile sense is the most important factor used to judge the overall consumer acceptance [1, 2]. For different types of tissue products the significance of the tactile sense varies. A tissue product can be judged in many different ways by the tactile sense, but the outcome usually is depicted either as the bulk feel or as the surface feel. The present work has been focused on the surface feel [3]. The surface feel stands for about 70 percent for facial tissue, 50 percent for toilet paper and 30 percent for household paper of the overall consumer acceptance.

The human structure sensitivity is extremely good. The hand and face are the parts of the body with the best structure sensitivity, especially the fingertips and the lips. The tactile sense is better in discriminating between different feature heights than widths. Height has a threshold of around 1 μm compared to widths of around 10 μm. Also, it is much easier to detect small features during movement - either of the sample or of the body part. From static "non-movement" detection only structures greater than 100 μm can be detected. [4]

The cells and units which are involved in the tactile sense of the skin are called mechanoreceptors. The receptors can be divided into three sub-groups; intensity receptors, rate receptors and acceleration receptors. [5] Many nerve endings can originate from the same nerve cell. They are distributed randomly unlike the Meissner corpuscles and the Merkel discs. These nerve endings associate with the sensations of warmth, coolness, pain, tickling and itching. [6]. How a product feels when touched, contributes substantially to the consumers' overall acceptance or preference of paper and fabric products. The overall sense can be divided into several under groups, which all can be characterized using trained human panels [1, 2]. Paper products can be found in a wide range of areas, from newspaper to juice packet, to furniture cardboards, to toilet paper, to napkins, etc. All papers consist of a fiber network and in almost all cases plant fibers are used. There are three main types, writing paper, cardboard and tissue paper. In this study the focus is on tissue products. All paper types are constructed in a similar manner. The main differences are type and the amount of fiber used and which additives used. For tissue products there are five basic production stages: Pulping, stock preparation, sheet formation, drying, and creping. Embossing is a converting process with purpose to binding two sheets together, improve the visual appearance, increase bulk, increase bulk softness, improve haptics and improve abrasive properties (paper to paper or against some other surface). Embossing usually increase the thickness of the paper several 100 percents in commercialized products.

2. Methods and materials

2.1 Digital Light Projection

A digital light projection (DLP) system GFM[1] MicroCAD [7] with a horizontal- and vertical resolutions (x,y,z) of 16um, 16.2um and 2.6um respectively was used to measure the areal (3D) topography. MountainsMap Premium 4.1 software from DigitalSurf[2] was used for all topography analysis.

2.2. Tissue samples

From each product three different sheets of paper and two different places per sheet were measured, with a measuring area of 26mm*20mm (see figure 1). The samples were placed under a glass plate with a weight of 256g and area of 10*10cm to avoid wrinkles and to keep the sample still during the measurement. Two sets of 6 commercial tissue samples (the calibration and the verification sets) where used to firstly initially identify the significant properties influencing the haptic "roughness" on the first set of six samples and secondly to verify the results on the 2'nd set, the verification set. The samples had a large parameter variation of drying method, no. of ply's (sheet layers), thickness, embossing and fiber type. The finer fibre structure visual in the SEM pictures refer to the "fuzziness" of the tissue is not considered in this study but is besides haptic "roughness" also an important feature of the tissue products.

2.3. Topography characterization

23 Field parameters [8] (Sa, Sq, St, Sz, Ssk, Sku, Sk, Spk, Svk, Sr1, Sr2 , Sdq, Ssc, Sdr, Sds, Str, Sal, Sdc, Smr, Vmp, Vmc , Vvc , Vvv) from the ISO 25178-2:2012 standard on "areal topography" parameter definitions was used to extract a set of significant height- (Sz –maximum height), spatial- (Sal-auto-correlation length), functional- (Sk-core roughness depth, Vmp-material volume of peaks) topography parameters.

The trained sensory panel was used to grade the products perceived haptic "roughness" in a haptic scale rank from 0-100.

Statistical Methods

Multi linear regression (MLR), complemented with principal component analyze (PCA) and partial least square regression (PLS) where used to construct the models explaining the influence of the surface properties to the perceived haptic "roughness" [9, 10].

[1] GFMesstechnik, Berlin, Germany, http://www.gfm3d.com

[2] Digital Surf, Besançon, France, http://www.digitalsurf.fr

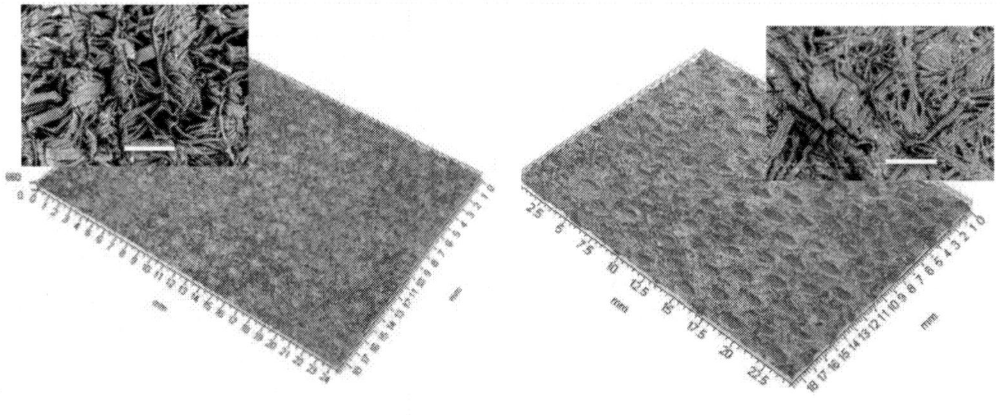

Figure 1. Optically DLP measured 24*19mm examples of two of the tissue producs used in this study. Sample I1 (left) is a facial tissue product with a very smooth surface while I6 (right) is an embossed product with larger surface structures. The smaller pictures are 800*600um SEM measurements displaying the detailed fibre structure.

3. Results

3.1 Material Ratio and material distribution

Material ratio for different depths in a surface can be plotted in "Abbot-Firestone-" or "bearing area" diagram as a "finger print" of the material distribution in a product (figure 2). The Sk parameter in ISO 25178-2:2012 tries to extract the rate of increase of material ratio in the "top- or bearing areas" (see arrows in figure 2) of the surface and is in this study used to quantify the material distribution in the upper region of the surface expressed as the height in um of the top region where the human finger and senses meet the paper.

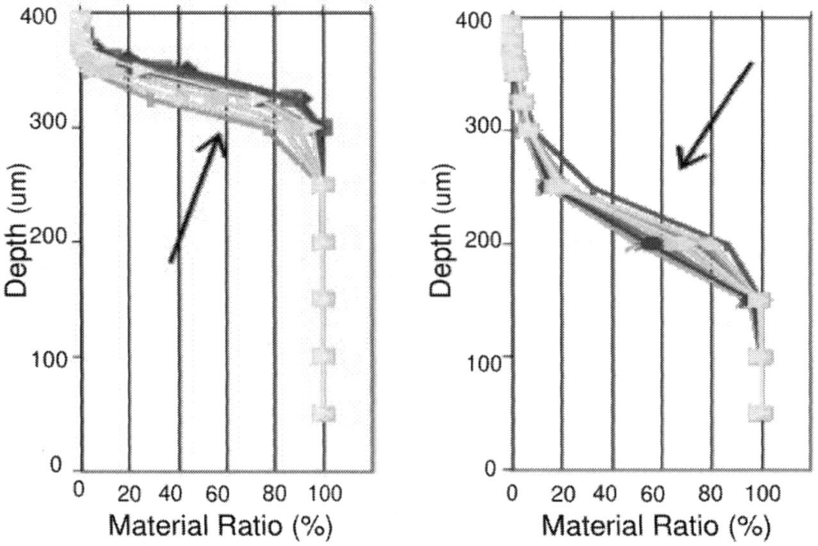

Figure 2. The Material Ratio Curves measured for product I1 (left) and I6 (right) of the calibration set. The steep part of the slope is significantly steeper (high Sk value) for I6 than for I1 (lower Sk-value) indicating a concentration of material distribution in the upper region of the surfaces.

3.2 Calibration set analyze

All the topography parameters were analyzed by studying mean values for measurements of the sheets in combination with the human panel sensory observations of the "perceived roughness". In figure 3 (left) below, it can be seen how a complete set of ISO 25178-2:2012 parameters contribute to explain the "perceived roughness" forming the circular pattern. By extracting a minimum set of parameters with relatively high "loading" values for the two principal components in the diagrams (PC1 and PC2) like in figure 3 (right) a numerical model describing the "perceived roughness" as a function of the chosen parameter set can be formulated like in equation (1) below.

Numerical regression models based on only topography parameters ended up with low R2-values A "product type" parameter, DL, distinguishing between multiply and singleply product types (DL: -1 or1), had to be added in order to achieve models with higher R^2-values than 0.77. DL together with the topography parameters Sk, Vmp, Sz and Sal, create models explaining the haptic "roughness" with a high prediction ability (R^2-values around 0.98).

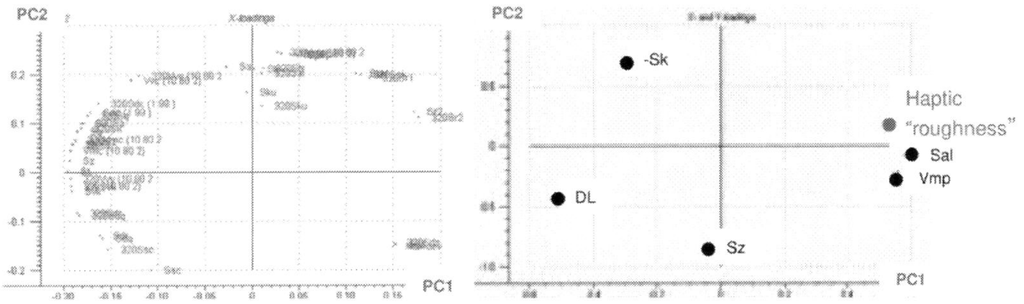

Figure 3. Left: Principal component loading plot using the complete set of topography parameters. Right: A loading plot with the selected parameters DL, -Sk, Sz, Vmp, Sal with PCs explaining in total 96 percent of the variability in haptic "roughness".

When the mechanical parameters tension stiffness index and tension stretch index were added into the numerical regression models the R^2-value increased insignificantly but the robustness in prediction improved by removing unwanted trends in the pattern of residuals.

A positive regression coefficient (figure 4 left) implies that with increasing value for the parameter that the surface will feel "rougher", the negative regression of the –Sk parameter in figure 4 below indicate a positive influence on haptic "roughness" of increasing Sk.

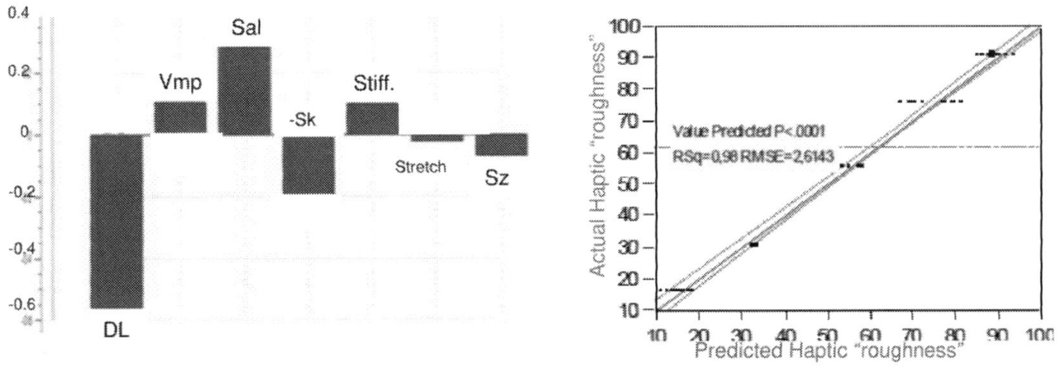

Figure 4. Left: Normalized regression coefficients for the PLS modelling with mechanical parameters included. Right: MLR model of the calibration set using the DL-, and the topography parameters here resulting in a R^2 of 0.98.

3.3 Validation Set analysis

The validation set evaluation was based on the parameters DL, Vmp, Sal, Sk, Stiff, Stretch and Sz, obtained from the calibration set evaluation.

If only topography parameters were used a regression model with R^2-value of 0.77 was obtained for the validation set. When DL and mechanical properties where included the R^2-value increased to 0.97. These strong correlations where achieved when topography parameters Sz, Vmp, Sk, Sal, were completed with the mechanical properties Stiffness index and Stretching index along with DL and resulted in the numerical model in eq. 1 below. Constants A to H where given by the PLS evaluation using the validation set surface measurement values in combination with the human panels' evaluation of the haptic "roughness".

"perceived haptic roughness" = *A+B*-**DL**+C****Vmp****+D****Sal****+E*(**Sk**)+F* **Stiff**+ G*-**Stretch** H*-**Sz*** (1)

From equation 1, DL, Sz, and Stretch had negative regression coefficient values, which imply that an increase in the parameter value results in a decrease in perceived haptic "roughness". The coefficients for Vmp, Sk and Sal were positive, hence positive correlated with the haptic "roughness".

3.4 Optimization example of Haptic Property

With help of results from the above regression it is possible to optimize the roughness feel for a product. Product 4 in the validation set is toilet paper product with an embossed pattern.

The mean haptic "roughness" for the front side was 67 and 82 for the back side on the 0 – 100 scale. The haptic "roughness" for product 4 is about the same as the mean of all the products in the validation test and to increase customer satisfaction and to lower the haptic "roughness" eq.1 above and table 1 below present a number of options for the product developers at SCA. Comparing topography values for product 4 compared to the "mean for all products", the Sal parameter stands out as a candidate for design changes.

A decrease of Sal would according to eq. 1 be one effective way of improve the haptic feeling for the product and decreasing the "rough" feeling. A decreased Sal would in practice result in a reduction of long wave-length components, i.e the repeating pattern around 1mm size (based on the Sal 822um and 762um in table 1).

Table 1. *Top:* Max-, min-, and mean parameter values for all measurements. *Down:* Top: Max-, min-, and mean parameter values for product 4 for both the back side and front side measurements.

		Haptic Roughness	min. Sk	Sal	Vmp	Sz	"Stiffness"	Stretch
			(um)	*(um)*	*(um3/um2)*	*(um)*	*(Nm/g)*	*(%)*
All Products	*Max*	90	361	1512	807	2360	0,31	14
	Min	15	22	0,52	1	125	0,05	9
	Mean	60	47	170	170	445	0,15	12
Product 4	*Front side*	63	44	822	0,74	449	0,06	13,2
	Back side	82	58	762	9,38	433	0,06	13,2

4. Conclusions

- Topography plays a significant role in explaining the perceived haptic "roughness" of tissue products.
- Topography combined with mechanical properties can predict haptic "roughness" with more than 90 percents accuracy.
- The areal topography parameters *autocorrelation length* (Sal)-most important, *maximum height* (Sz), *core height* (Sk) and *peak volume* (Vmp) all show significant impact on haptic "roughness".
- Models for optimisation of haptic "roughness" can be designed using PCA, PLS, and MVA methods.
- Non-contact areal metrology using structured light capture the significant topography responsible for the haptic "feeling of "roughness of tissue products.

5. Future

The models of the haptic "roughness" will be further developed by the investigation of the contact pressure and material properties influence as well as the extension towards using areal topography metrology to quantify the "fuzziness" of the tissue products and relations to the haptic "roughness" and the customers total "feeling" for the tissue products.

Acknowledgement

The Authors also want to thank Linda Davnert, David Lindqvist and Martin Stenberg for their valuable contribution to the results.

References

[1] Civille D 1990 Development of Terminology to Describe the Handfeel Properties of Paper and Fabrics *J. Sensory Studies* **5** 19-32
[2] Meilgaard M C Civille G and Carr B T 2007 *Sensory Evaluation Techniques* London CRC Press, 4th edition
[3] Hollmark H 1983 Evaluation of Tissue Paper Softness, *Tappi J.* Feb.
[4] Berland C 2006 *Skin Sensitivity to Texture* SCA Report F3589
[5] Lamb Determination of Softness of Tissue Products München 3rd PTS Tissue Symposium
[6] Tortora G J and Reynolds Grabowski S 2004 Introduction to the Human Body USA John Wiley & sons
[7] Chen L C and Liao C C 2005 Calibration of 3D surface profilometry using digital fringe projection *Meas. Sci. Technol.* **16** 1554–1566
[8] Jiang X Scott P J, Whitehouse D J and Blunt L 2007 Paradigm shifts in surface metrology Part II The current shift *Proc. R. Soc.* **A 463/2085** 2071-2099
[9] Altman D G 1994 *Practical Statistics for Medical Research* London Chapman & Hall 1st edition
[10] Esbensen K H 1994 Multivariate Data Analysis – in practice Oslo CAMO Software AS 5th edition

The Use of Feature Parameters to Asses Barrier Properties of ALD coatings for Flexible PV Substrates

Liam Blunt[1], David Robbins[2], Leigh Fleming[1] and Mohamed Elrawemi[1]

[1]University of Huddersfield, West Yorkshire. Huddersfield, HD1 3DH, UK.
[2]Centre for Process Innovation Limited, County Durham. Sedgefield, TS21 3FG, UK.

E-mail: l.a.blunt@hud.ac.uk

Abstract. This paper reports on the recent work carried out as part of the EU funded NanoMend project. The project seeks to develop integrated process inspection, cleaning, repair and control systems for nano-scale thin films on large area substrates.

In the present study flexible photovoltaic films have been the substrate of interest. Flexible PV films are the subject of significant development at present and the latest films have efficiencies at or beyond the level of Si based rigid PV modules. These flexible devices are fabricated on polymer film by the repeated deposition, and patterning, of thin layer materials using roll-to-roll processes, where the whole film is approximately 3um thick prior to encapsulation. Whilst flexible films offer significant advantages in terms of mass and the possibility of building integration (BIPV) they are at present susceptible to long term environmental degradation as a result of water vapor transmission through the barrier layers to the CIGS (Copper Indium Gallium Selenide $CuIn_xGa_{(1-x)}Se_2$) PV cells thus causing electrical shorts and efficiency drops. Environmental protection of the GIGS cell is provided by a thin (40nm) barrier coating of Al_2O_3. The highly conformal aluminium oxide barrier layer is produced by atomic layer deposition (ALD) where, the ultra-thin Al_2O_3 layer is deposited onto polymer thin films before these films encapsulate the PV cell. The surface of the starting polymer film must be of very high quality in order to avoid creating defects in the device layers. Since these defects reduce manufacturing yield, in order to prevent them, a further thin polymer coating (planarization layer) is generally applied to the polymer film prior to deposition.

The presence of surface irregularities on the uncoated film can create defects within the nanometre-scale, aluminium oxide, barrier layer and these are measured and characterised. This paper begins by reporting the results of early stage measurements conducted to characterise the uncoated and coated polymer film surface topography using feature parameter analysis. The measurements are carried out using a Taylor Hobson Coherence Correlation Interferometer an optical microscope and SEM. Feature parameter analysis allows the efficient separation of small insignificant defects from large defects. The presence of both large and insignificant defects is then correlated with the water vapour transmission rate as measured on representative sets of films using at standard MOCON test. The paper finishes by drawing conclusions based on analysis of WVTR and defect size, where it is postulated that small numbers of large defects play a significant role in higher levels of WVTR.

1. Introduction

Flexible Photovoltaic cells are one of the newest developments in the renewable energy field. Currently the most efficient cells are those based on Copper Indium Gallium Selenide (CIGS) materials, up to 19%. These flexible cells are however highly susceptible to environmental degradation. The most critical problem with PV modules is the transmition of water vapour through the barrier films on the PV functional layers. This water vapour transmission is caused by the presence of tiny defects in the barrier coating and results in decreased cell efficiency and decreased cell lifespan. One of the most effective and reliable methods of protecting these cells is to apply a barrier coating of Al_2O_3 to the encapsulation material. The highly conformal Al_2O_3 barrier layer is produced by the atomic layer deposition (ALD) technique. The surface of the encapsulation substrate polymer (PET) film must be of very high quality; in order to achieve this high quality the substrate film is further planarised. However, even this ALD barrier is not at present effective enough; water vapour can still permeate through the barrier due to the presence of micro and nano-scale size defects on the barrier films, thus causing electrical shorts, efficiency drops and, ultimately, failure. This paper reports the results of measurements conducted to characterise the uncoated and barrier coated polymer film surface topography using segmentation feature parameter analysis. The presence of defects is then correlated with the water vapour transmission rate as measured on representative sets of films using a standard MOCON test. The overall process for manufacturing flexible PV's is by roll to roll technology. The results in this paper provide the basis for the development of roll to roll in process metrology devices for defect detection

1.2 Flexible PV Module

Flexible solar modules comprise of four functional layer groupings as shown in Fig [1]. The main focus of the investigation in this paper is the encapsulation/barrier layer. The barrier layer is typically formed from a planerised Polyethylene Naphthalate (PEN) sheet,with the Al_2O_3 barrier coating (40 nm thick) being produced by an ALD process. Despite the excellent barrier properties provided by this material, all of the published data indicates there is some remaining permeation, even when the barrier coating is reasonably thick ($\geq$ 50nm) [1]. This outstanding transmission of water vapor is considered to be due to the presence of small defects on the film.

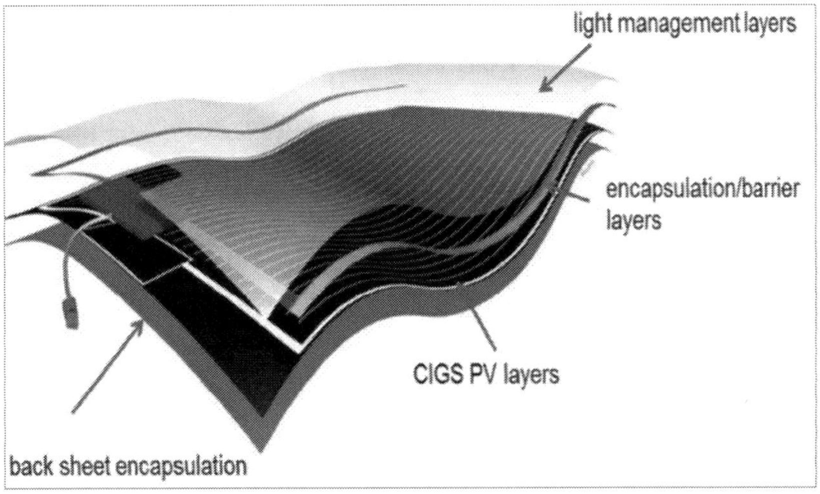

Figure 1: Schematic of the flexible PV Module (Courtesy of Flisom, Switzerland)

2. Water Vapour Transmission Rate Assesment:

All coated samples were measured for water vapor transmission rate (WVTR) using standard MOCON test instrumentation prior to the surface measurement. With this method, the test specimen is held within the instrument such that it separates into two sides of a test chamber. One side, the "wet side", is exposed to the gas or vapor to be measured. On the detector or the "dry side", the sample is subjected to zero relative humidity. The dry side is purged with a carrier gas which carries away any transmitted water vapour to a infrared sensor which records the transmission rate. The stady state rate was recorded along with the time to stable transmission. The area exposed to the water vapour was approximately $80mm^2$

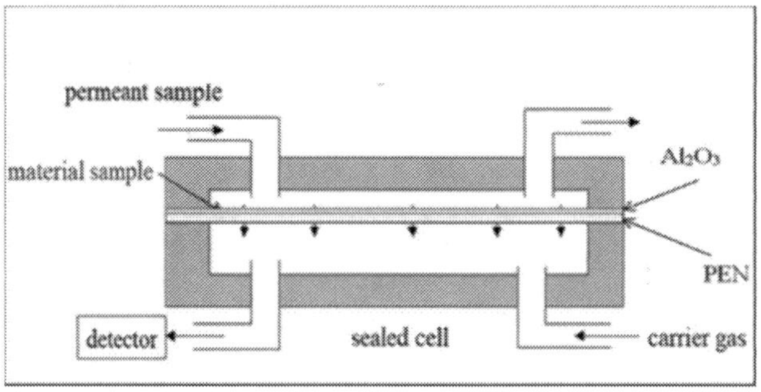

Figure 2: WVTR Measurement using MOCON method, exposed area $80mm^2$[2]

2.1. *Measurement Methodology*

Four samples of ALD coated substrate with 40 nm Al_2O_3 as well as a non-coated substrate were assessed. The WVTR results were kept blind to determine the presence and the size of the features that had a potential influence on the WVTR value. Initially an optical microscope was used for visual inspection of the coated samples. Different types of defects were noticed for each sample and varied from one sample to another in terms of type, width, and height as shown in Fig [3] and Fig [4].

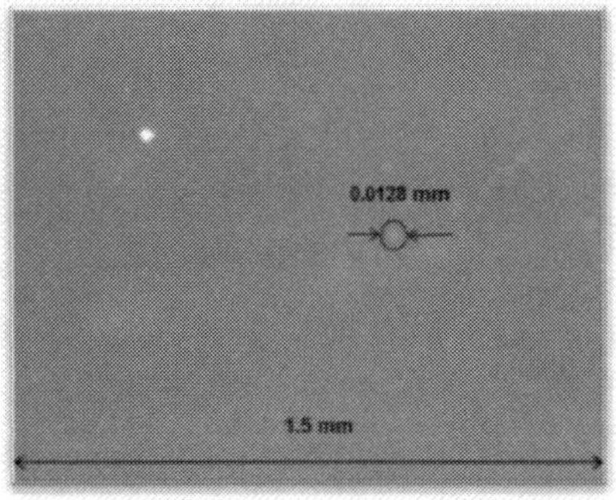

Figure 3: Scale of small defects x 200 Mag

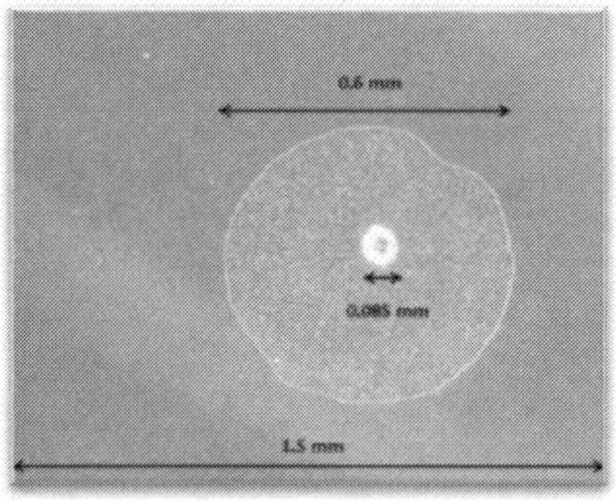

Figure 4: Scale of large defects x 200Mag

2.2. *White Light Scanning Optical Interferometry*

A Taylor Hobson Coherence Correlation Interferometry (CCI) was used to provide a rapid means of characterizing a number of important features; including surface roughness, defects density and other surface variations all of which were considered to have a major influence on the barrier properties of the Al_2O_3 layer and hence module efficiency. The instrument was calibrated according to the manufacture specifications prior to use. The issue of the measurement area is an important point to be considered, in the case 20× objective lens magnification was used this allows the instrument to measure a sample area of approximately $1mm^2$. This lens gave adequate spatial resolution to detect defects at both small and large sizes. However, it was not feasible for measurement of the entire surface, over which the WVTR is assessed ($80mm^2$). Consequently, 700 measurements, equating to 14% of the total surface area of all the specimens was measured.

3. Results

The WVTR (water vapor transmission rates) for each sample are shown in Table [1]. The MOCON test has shown that three samples have a relatively low WVTR and one sample has higher WVTR value (2705).

Table [1] Water vapour Transmission Rate values

Sample No	WVTR (g/m²/24 hrs.)	Stabalisation Time
Sample 2701	Low	11 days
Sample 2702	Low	11 days
Sample 2705	***High***	***5 days***
Sample 2706	Low	5 days

The surface topography analysis of the Al_2O_3 barrier layer showed that many defects were observed on each sample; these defects varied from one sample to another, and were classified depending on the type and prominence. Fig [5] and Fig [6] show two of these typical defects which are peaks and holes.

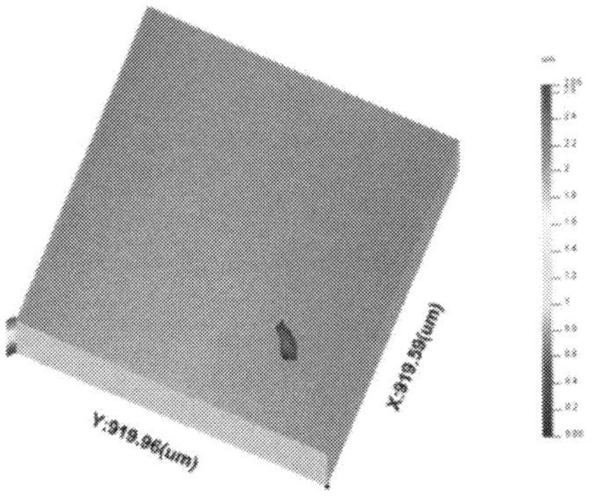

Figure 5: "Hole" type defect

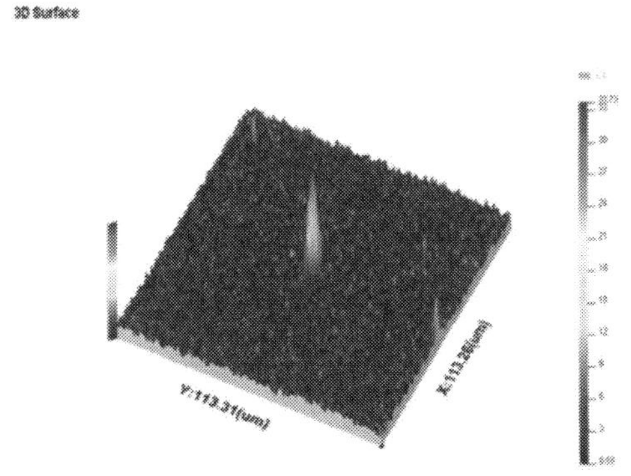

Figure 6: "Peak" type defect

3.1. Areal Surface topography assessment

Feature Parameters as defined in ISO 25178-pt2 [3,4] were used to adapt a scheme for setting appropriate boundaries in order to employ a suitable Wolf pruning method. The "toolbox" method was adapted to automatically detect significant hills and dales features. In this procedure in particular a parameter *Sfd* was used (where *Sfd* = the number of significant hills + significant dales) significance was defined as any peak/pit greater than 20% of the total peak to valley roughness (Sz). Using this default significance value for the defects the results showed no clear correlation with the WVTR results. Therefore, was conducted with additional prune conditions, those defects which satisfied the criteria of only large defects (6σ (Sq=0.8nm) height & 180μm^2) were measured and recorded for effective discrimination of significant and non-significant defects.

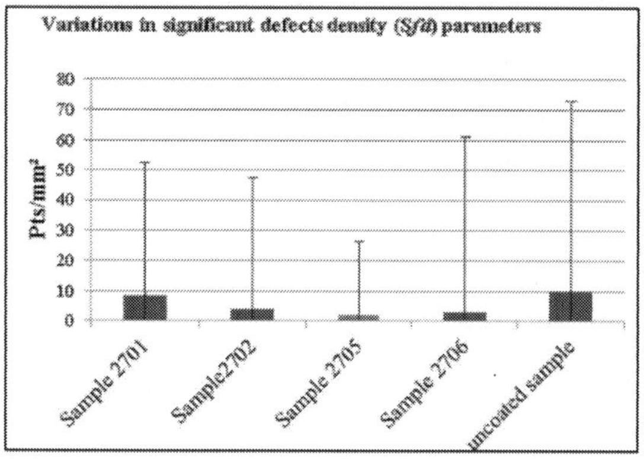

Figure 7: Variation in defects density initial 20% height Wolf prune

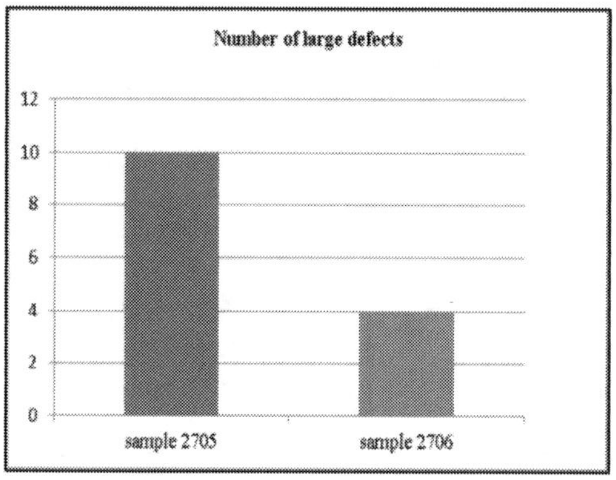

Figure 8: Significant defect count with additional prune criteria

The present findings would appear to suggest that small numbers of large defects (as detected by the optimal prune criteria) are the dominant factor in determining WVTR. Any detection system should focus on these scales of defects as a determinant of function.

4. Conclusions:

The use of feature parameter analysis allows separation of small insignificant defects from large significant defects automatically. The numerical approach has provided a clear evidence for the correlation of surface roughness, defect density, and the transmission of water vapour through the barrier coating layer. Hence, from the previous analysis, it can be concluded that the total permeation rate corresponding to small numbers of larger defects is much greater compared to the total permeation rate corresponding to large numbers of small pinhole-type defects over the same area of substrate. This result provides novel information to enable automatic detection and correction of potential restrictors to PV module performance in the line processes. Work is continuing to check repeatabiliy of these tests and produce "cleaner" substrates.

Acknowledgement

The authors would like to thank the EC for funding this work (NanoMend NMP4 LA-2011-280581)

Refrences

[1] Carcia, P. McLean, R. Hegedus, S. (2010). Encapsulation of Cu (InGa) Se_2 solar cell with Al_2O_3 thin-film moisture barrier grown by atomic layer deposition. Solar Energy Materials and Solar Cells. 94 (12), p2375-2378.

[2] Duncan, B. Urquhart, J and Roberts, S. (2005). Review of Measurement and Modelling of Permeation and Diffusion in Polymers. NPL Report DEPCR 012.

[3] Jiang, X., Scott, P. J., Whitehouse, D. J. and Blunt, L. (2007). 'Paradigm shifts in surface metrology.' Part II. Proceedings of the Royal Society A. 463 (2), pp. 2071-2099.

[4] ISO 25178-2:2012(E), Terms, definitions and surface texture parameters. 1-47(2012).

Contribution of human skin topography to the characterization of dynamic skin tension during senescence: morpho-mechanical approach

H. Zahouani[1], M. Djaghloul[1], R. Vargiolu[1], S. Mezghani[2], M. EL Mansori[2]

[1]University of Lyon, Laboratory of Tribology and Dynamic of Systems
UMR-CNRS 5513. ENISE – ECL 36 Avenue Guy de Collongue, 69131 Ecully
France
[2]LMPF-EA4106, Arts et Metiers Paris tech, France

E-mail: hassan.zahouani@ec-lyon.fr

Abstract. The structuring of the dermis with a network of collagen and elastic fibres gives a three-dimensional structure to the skin network with directions perpendicular and parallel to the skin surface. This three-dimensional morphology prints on the surface of the stratum corneum a three dimensional network of lines which express the mechanical tension of the skin at rest. To evaluate the changes of skin morphology, we used a three-dimensional confocal microscopy and characterization of skin imaging of volar forearm microrelief. We have accurately characterize the role of skin line network during chronological aging with the identification of depth scales on the network of lines ($z \leq 60\mu m$) and the network of lines covering Langer's lines ($z > 60$ microns). During aging has been highlighted lower rows for elastic fibres, the decrease weakened the tension and results in enlargement of the plates of the microrelief, which gives us a geometric pertinent indicator to quantify the loss of skin tension and assess the stage of aging. The study of 120 Caucasian women shows that ageing in the volar forearm zone results in changes in the morphology of the line network organisation. The decrease in secondary lines ($z \leq 60$ µm) is counterbalanced by an increase in the depth of the primary lines ($z > 60$ µm) and an accentuation of the anisotropy index.

1. Introduction

The longevity of life is a recent unsurpassed demographic phenomenon in the history of the humanity and impact of which will be considerable. The analysis of ageing of populations and its social consequences requests an interdisciplinary approach. By gathering capacities of research in these different domains, it becomes possible to treat in a transverse way the big questions linked to ageing and to longevity, by understanding better all the factors that drive to the fragility of the old tissues (origin of diseases, of dysfunctions and handicaps).

Old age is inevitable and a natural period of life characterized by a decrease in physical function, loss of social role as an adult, changes in physical appearance and routing to a gradual decrease in capacity. Aging is part of a continuing evolution in the course of human development, rigorously following embryogenesis, puberty and maturation. Throughout this process, the organs develop into effect after a specific time. The "program" of aging is probably mediated by the endocrine hormones, neurotransmitters acting on certain target organs. Finally the cell is genetically programmed and the programming could be compared to a turtle horloge. A turtle is programmed for 100 years, a monkey for 20 to 25 years and 120 years for humans. For against this life expectancy is compromised by damage to genetic or acquired by a malfunctioning biological or enzymatic cell. The first signs that we perceive aging are changes in the body that they depend on internal changes. It is important to note that the rate of aging varies with individuals. Thus the change of hair color, which gradually become gray and then white, and the appearance of wrinkles are a reflection of the inevitable passing of time. Skin problems, which are more common in the elderly than in younger individuals, are not only related to the fact that the skin is exposed to a long remained more or less cumulative dose of ultraviolet rays, but also to change in structures the skin itself (connective tissue, collagen, elastic fibres, etc..).

The skin is a set of grouped cells in the form of a flexible and resistant fabric, composed of several layers, and covering the whole body. The skin consists of three distinct parts: the stratum corneum, the epidermis (epi = above) whose main role is to protect the body, and the dermis, only one with vessels allowing nutrients (nutrients carried by the blood) to diffuse to the epidermis. In an adult, the skin weighs between 3.5 kilograms and 4.5 kilograms. Its total area up to two square meters. It is highly vascularized and also has a large number of glands producing sweat (sweat glands), sebaceous glands (sebum secreting = fatty substance that protects the skin) and nerve receptors for tactile sensations and pressure. This cover (integument) very flexible but strong helps protect the body against external aggression (infections, temperature variations, etc. ...). The thickness of the skin varies between 1.5 and 4 mm depending on the region of the body in question. The epidermis covering the soles and palms is thicker than the rest of the body. Below the dermis lies the hypodermis, also called superficial fascia. It consists of adipose tissue (fat) tissue and looser than the dermis. The dermis has a role to adapt to the movements of structures located below it (muscles, tendons, and fascia) but also protect the body shots, thanks to its constitution fat. It is at this level that is fat stores of the body, which accumulate in humans in the abdomen, and women on the thighs and breasts. 1) The epidermis contains different varieties of cells, including keratinocytes (which are the most numerous), and melanocytes.

1.1. Skin Lines Network

The skin surface shows a specific topography depending on the anatomical site, age and sex. In general, the skin morphology presents a 3D network of lines, who by his organization expresses all the multidirectional tensions of elastic fibres and the collagen beams. Micro-lines, primary lines, fine wrinkle and wrinkle represent, in fact, the special organization of collagen bundles and elastic fibres in the superficial dermis, and there is a relationship between the morphology of skin lines and elastic network. Different functions can be attributed to the lines network. The first function is the retention and drainage canals of the sebum and sweat. They collect preferentially and retain for long time the substances applied to the skin: they are thus preferential sites of percutaneous absorption. This reservoir function, allow the applied topical products to be stored on the skin surface and then eventually to diffuse in its different layers. The second function is mechanic, during ageing the depth, width, density and orientation of skin lines change. Some lines become more marked; they evolve progressively in marked anisotropy connected to the decrease of the elasticity of the collagen fibres.

1.2. Skin Tension and Anisotropy of Skin Lines Network

When analysing the mechanics of skin in vivo, a significant property is its natural tension. Discovered by Dupuytren [1], and mapped by Langer [2], the non-uniform skin tension lines exist. Langer has identified these lines by puncturing the skin with a circular device, figure (1). The wounds then assume an elliptical shape and by joining the major axes of the ellipses a system of lines can be drawn, Some authors propose other methods to obtain these lines, such as wrinkling of the skin by Borges[3].

Skin resistance to traction predominates in the Langer's lines direction and varies with body site. On all body sites, the skin tension is greater in the direction of Langer's lines, figure (2).
This phenomenon is the source of the Young's modulus anisotropy[4], whose distribution angle shows a maximum in the Langer's lines axis [5],. This result favour a similar orientation of the elastic fibres involved in the skin. Assuming that the fibres are independent, it has been calculated that, on the calf, 76% were in the direction of Langer lines and 5.1% perpendicular [6]. Of course the distribution concerns only the elastic fibres, which are parallel to the skin surface. Observation of the dermis with scanning electron microscopy confirms this data[7]. In retracted skin the collagen bundles look tortuous, with no special direction, and sinuous elastic fibres are fixed to them in several places, especially in their concave portion.

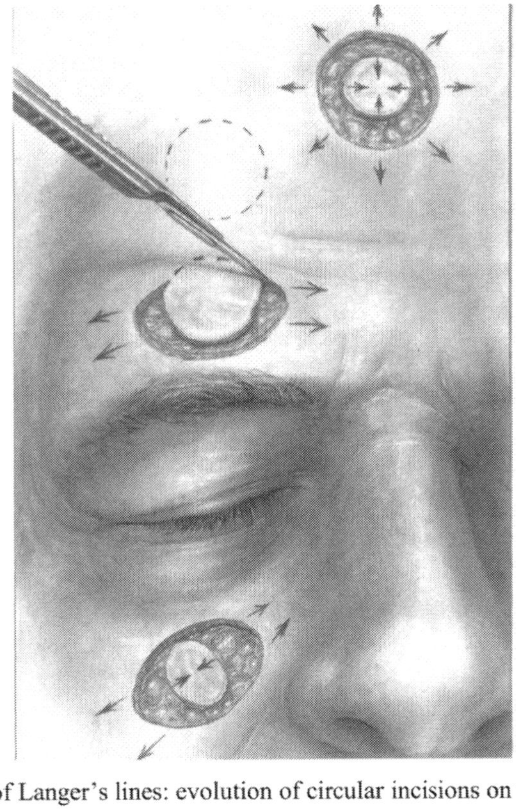

Figure 1. Detection of Langer's lines: evolution of circular incisions on the face by Waldorf [4].

In non-retracted skin, the thinnest collagen bundles as well as the elastic fibers are straightened in the direction of the Langer's lines and almost parallel; the thickest bundles remain tortuous and oriented in all directions, but their shape seems to be modified by the traction from the oriented bundles and fibres. Contrary to common belief in the past, Langer's lines do not reflect anisotropy of the collagen density but anisotropy of the reticular dermis collagen bundles' direction and elastic fibres' tension.

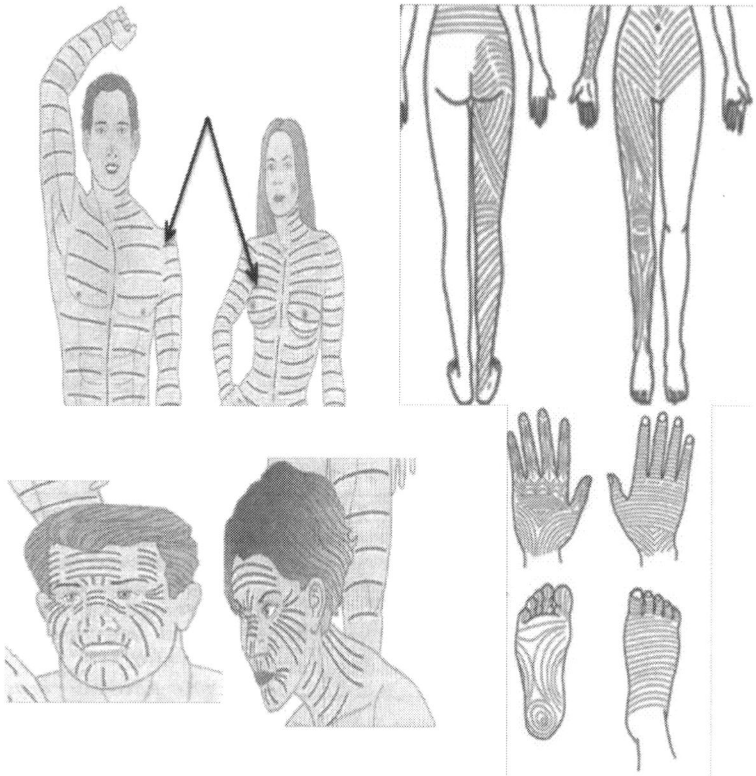

Figure 2. Schematic representation of Langer's lines on different areas of the body.

1.3. Identification of Langer's Lines

The morphology of lines network reflects the intrinsic tension of the skin, which is distinct from the additional tension induced by increases in volume of the underlying tissues, such as muscle contraction, edema, or particular posture which stretches the skin. The identification of Langer's lines must therefore, be made on relaxed skin.

Several methods of identification of Langer's lines were developed, as evidenced by the following works of literature:

- Stark's Method [8]

According to the definition, the direction of the maximum tension can be found by stretching the skin in several directions with an equal force, the direction of minimum elongation is that of Langer's lines. Stark developed a simple device comparable to a compass: two branches with claws at their extremities part spontaneously by 30 mm under the action of a 14.2g/mm spring. He could measure the elongation of the skin into eight directions quickly, each measurement requiring only 1.5 seconds.

- Borges' Method[3]

This similar method is less accurate, but even quicker, and consists of creasing the skin between the thumb and the index in all directions until the furrows are regular and parallel. They follow Langer's lines. In the other directions, they are impeded by the skin tension that makes them irregular.

- Barbenel's Method [9]

The measurement of the extensibility of the skin using the suction method is valid only if slip page of the skin into the suction chamber is presented. If, inversely, this movement is facilitated, the most extensible direction of the skin will appear easily, that is to say perpendicularly to Langer's lines. If the contour of the chamber is drawn in during suction, when the chamber is removed, an oval outline is observed instead of a circle and its main axis corresponds to Langer's lines.

- Skin micro- topography : Zahouani's method [10,11,12]

Apart from the palms and soles, the skin mirorelief is made of plateaus separated by valleys. The latter are roughly parallel and oriented in different directions, and this layout is characteristic of each body area. The direction of the deepest valleys matches Langer's lines. There may be one or two other preferential directions, indicating an ordered no orthogonal mechanical anisotropy. This method has an advantage over the others as it is insensitive to extrinsic skin tensions. Its physiological interpretation is simple. The cutis is normally retracted (skin tension) and extensible, whereas the epidermis has none of these properties. Therefore, the epidermal creasing responsible for the micro relief appears to be a transformation of tension allowing the creases to be flattened by stretching them. The superficial dermis, an intermediary zone between epidermis and cutis, is precisely the place where the skin relief begins. This mechanical transduction is one of its functions.
It is this approach that we developed to study the chronological aging of human skin. The method uses the wealth of three-dimensional imaging of the skin and the possibility of linking a signature printed on the surface of the stratum corneum with the organization in volume of the different skin layers.

2. Morphological evolution of skin lines network of caucasian french women during aging

Several studies have demonstrated changes in the network of lines with age, leading to deepening of certain lines and the disappearance of others [13-22]. However, quantitative and detailed descriptions of the modifications of skin lines with age are rarely reported in the literature since most of the published results are based on standard parameters which give an overall description of the topography of any surface, without specificity for the skin morphology. For this reason, we have adapted a 3D confocal microscope working with a high vertical and lateral resolution, which enables precise characterisation of the skin lines network. Negative skin replicas were taken with silicone rubber (Silfo®, Flexico Ltd, England) from 120 Caucasian French women equally divided into six age groups (20-29 years, 30-39 years, 40-49 years, 50-59 years, 60-69 years and 70-80 years). Replicas were taken from the women's left volar forearm at the same pre-determined area, after a 30-minute rest period in an environmentally controlled room (temperature: 21 ± 2°C and relative humidity: $50 \pm 5\%$).
Resolution refers to the smallest distance the confocal microscope can accurately measure. It can be considered in terms of lateral or and vertical resolution. The vertical resolution value of the confocal microscope is about 0.01 µm with a vertical range of 1000 µm. .The lateral resolution depends on the quality of displacement in the plane x, y. In the case of skin aging, the lateral resolution which was used is 1 µm in the directions x, y.
Figure (3) shows the evolution of skin volar forearm morphology of Caucasian women aged 20 to 80 years.

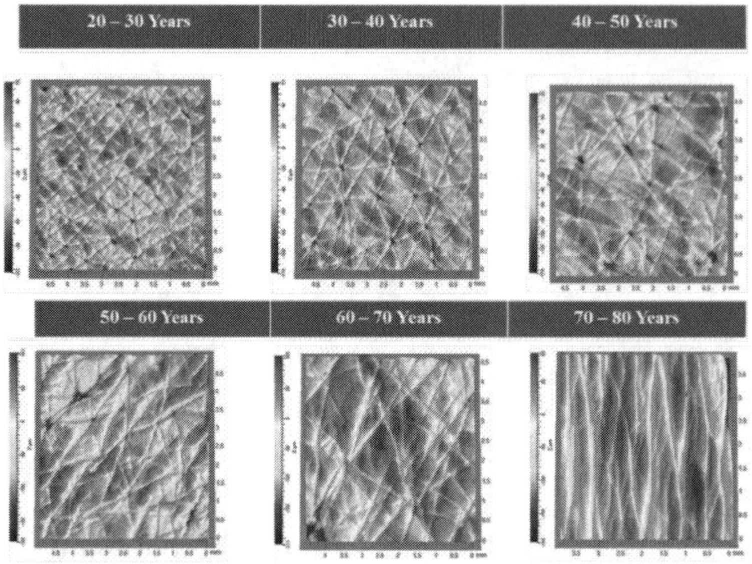

Figure 3. Morphology of skin relief of Caucasian women during aging (volar forearm site).

The images reflect the three-dimensional character of skin lines network. The scale height is expressed here by a colour scale. The deepest lines are expressed by the Colour Blue, heights that are at the top of plates expressed through into colour red – black. The full scale (pick to valley) can reach 500 µm

2.1. Multi-Scale Analysis of Skin Lines Network Morphology

Microscopic observations have shown that skin morphology contains a network of lines whose organisation reflects the multidirectional tensions of elastic and collagen fibres in the superficial dermis (11). Hashimoto (12) gave a precise four-level classification of the line network scales: (I) The primary lines are clearly marked and are between 20 and 100 µm deep, (II) The secondary lines are more discrete and correspond to a depth of 5 - 40 µm, and are perpendicular to the primary lines, (III) The tertiary lines correspond to the corneocyte border (about 0.5 µm), (IV) The quaternary lines correspond to the morphology of each corneocyte (about 0.05 µm). Tertiary and quaternary lines cannot be seen without magnification.

To study the transformation of the 3D skin line network during aging, it is necessary to identify all its local motifs. A motif of skin line is defined by the association of two peaks separated by the hollow of a valley, the height is determined by the difference between the highest peak and the hollow of the valley.[23-24], figure (4) :

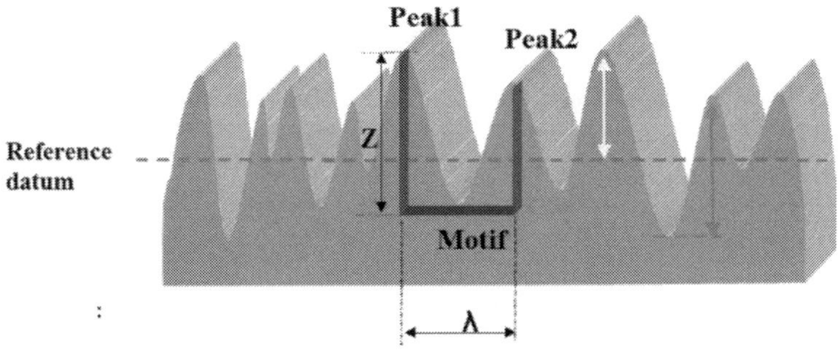

Figure 4. Definition of tension lines patterns.

The width of the motif (λ) is given by the distance between both peaks. The direction of the motif which coincides with the main direction of the line is defined in the orthogonal direction at the maximal variation of the local gradient, figure (3).

This quantitative approach enables us to follow the evolution of the different families of lines in relation to age. The depth Z depends on the skin site and two classes of depth $Z1 \leq 60$ µm and $Z2 > 60$ µm were adopted for the volar forearm aging. This choice was fixed after sampling every 10 µm as class depth. The results showed a marked decrease in density of the family lines of depth less than 60 µm and a net increase in density of the family of lines at depths exceeding 60 µm:

- $Z1 \leq 60$ µm, related the tension effect of elastic fibers network
- $Z2 > 60$ µm, related to Langer's line

Analysis of the morphology of skin tension lines of Caucasian women aged between 20 and 80 years, shows significantly decrease of the density of lines of depth <60 microns and an augmentation of deep lines beyond 60µm. This important result presented in figure (5) shows the mechanical role of elastic fibres in maintaining skin tension and firmness of young skin. This elastic relaxation mechanism of the reduction of elastic fibers according to age is the basis of the phenomenon of the appearance of wrinkles.

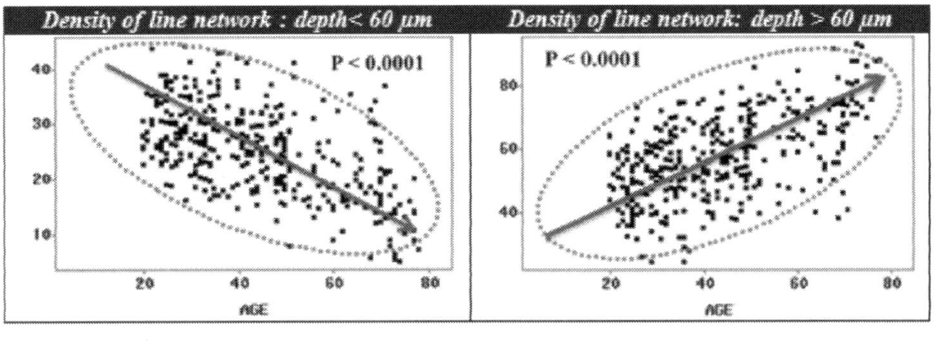

Figure 5. Evolution of the scale lines of tension during aging.

2.2. Changing the Orientation of Lines and Anisotropy during Aging

For better identification of the shift of skin lines network, we privileged to work on the rest state of the skin of the forearm by characterizing the image of the skin relief at different stages of aging. The methodology is to identify the local orientation of a pattern of lines and quantify the overall direction of the network of lines over the 0 ° direction taken as the axis of the body from head to foot [25, 26, 27]. Two conditions are adopted: i) the direction must coincide with the local normal of the plan formed by the three features of the motif, figure (6), ii) if the first peak is point A, the hollow point B and the second peak point C, the normal vector N is approximately collinear to the vector operation:

$$\vec{N} = A\vec{B} \wedge B\vec{C} \tag{1}$$

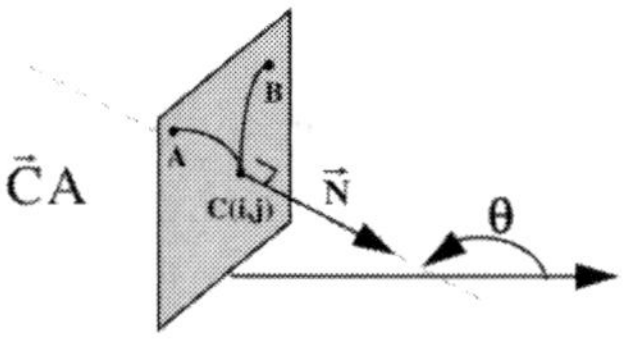

Figure 6. Detection of the skin line direction.

The direction of the motif must coincide with the direction of the valley which is composed of a succession of motifs, it is necessary to know if the hollow C(i,j) belongs to the move of the neighbouring points, figure (7). To achieve this, the intersection points between half a straight line stemming from C(i,j) and the nearest points are noted. If there is an even number of these intersection points, then the point is outside the direction of the skin line, otherwise the hollow is in the direction of the skin line. The number of iterations of this procedure is equal to the number of valley points of the detected motifs, and the three elements of the motif Z, λ and θ are memorised for each motif; these three parameters will be the fundamental components to build the 3D morphological tree of skin lines network.

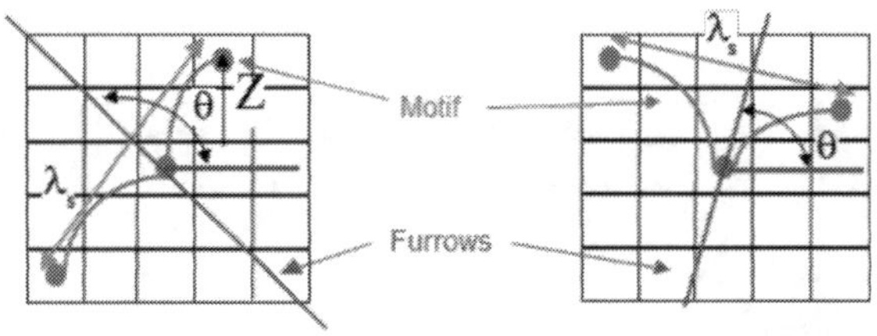

Figure 7. Orientation of local motif and skin line tension.

This approach quantifies the morphology of skin lines according to their orientation and their depth. The orientation distribution of the skin lines was quantified as a compass rose. This graphic representation plots in 20° intervals the density of lines between 0° and 180°, with the body axis used as the principal axis of orientation, figure (8). Thus, the density of lines in a triangular part corresponds to the percentage of line patterns which have this orientation.

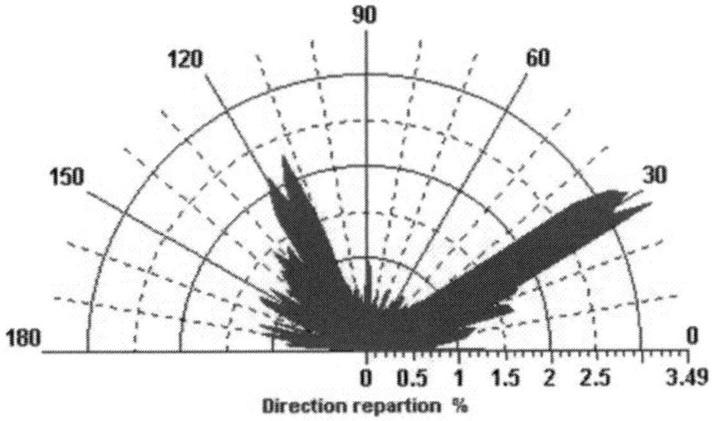

Figure 8. Anisotropy of kin lines network of the forearm of a subject 40 years.

2.3. Dynamic Rotation of Skin Line Network during Aging

To monitor the rotational dynamics of skin lines during aging, the overall results for the densities of line orientation from 20 to 80 years were collected for a comprehensive representation of the dynamic change of direction in function of age. We have chosen to represent this change for those aged under 60 and over 60 years. Figure (9) shows the rotational dynamics and the significant decrease in tension lines in the directions between 90 and 180 ° and the establishment of a marked anisotropy between the directions 20 and 60 degrees. This result demonstrates the relationship between the voltage loss of elastin fibres and lower voltage lines printed on the plates of the relief and the depth is less than 60 microns.

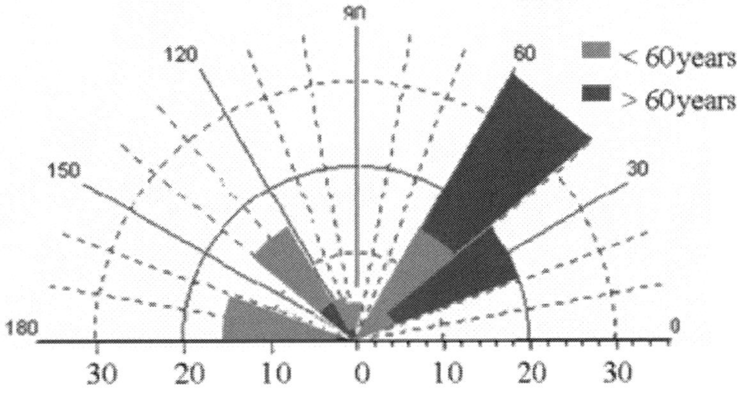

Figure 9. Aging effect on the dynamic rotation of skin line network of 120 Caucasian women.

2.4. Anisotropy Index of Skin Lines Network during Aging

The parameter resulting from the information on the density of lines according to their depth and orientation is the anisotropy. To assess the degree of the skin lines anisotropy during ageing, we introduced an anisotropy index from direction roses. A completely anisotropic surface gives a rose oriented into only one angular sector. Conversely, a perfectly isotropic surface leads to a circular direction rose. In consequence, if N is the number of angular sectors between 0 and π, the anisotropy index (A.I) can be defined as [26,27]

$$AI = \frac{1}{2}\frac{\sum_{i=0}^{N-1}|R_i - S/N|}{S - S/N}*100 \quad (2)$$

where R_i is the rose value corresponding to angular sector i and [26]

$$S = \sum_{i=0}^{N-1} R_i \quad (3)$$

S/N should be the R_i value for all i in the case of a perfectly isotropic surface and the factor 1/2 derives from the fact that an R_i value greater than S/N must be exactly compensated by lower values. As a result of the increase of the density of lines deeper than 60 µm and the diminution of the density of lines < 60 µm, results in the increase of the anisotropy index significantly during aging, figure (10).

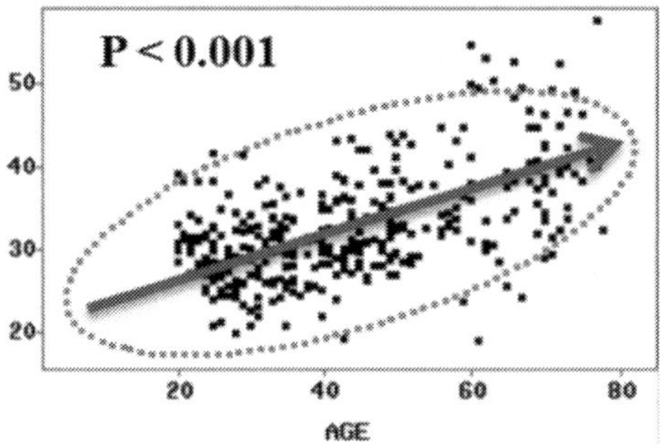

Figure 10. Aging effect on the skin line anisotropy index of 120 Caucasian women.

2.5. 3D Reconstruction of Skin Line Network: Volumetric Anisotropy and Tree of Skin Tension Network [27, 28, 29, 30]

The dermis is the layer of living skin. It is a supporting connective tissue rich in fiber which gives the skin elasticity and strength. The dermis contains the appendices of the skin. Histologically, the dermis can be divided into two layers: the papillary and reticular layer. Dermis and epidermis are closely meshed into each other through many outgrowths of the surface ripples of the dermis called papillae. The papillary dermis contains many nerve endings (thermo-receptors, tactile receptors). The reticular dermis consists of a network of collagen bundles (thick, wavy, perpendicular to the basal membrane) more visible because more dense within the reticular dermis. The network of elastic fibres which underlies the undulations of collagen fibres bundles and around the latter is anchored to their concavity. The reticular dermis is the strongest part of the dermis. His mobility results from unfolding the undulating collagen fibres bundles of allowing their extension and their return to their original position by the action of elastic fibres.

This structuring of the dermis with a network of collagen and elastic fibers gives a three-dimensional structure to the skin network with directions perpendicular and parallel to the skin surface. This three-dimensional morphology prints on the surface of the stratum corneum a three dimensional network of lines which express the mechanical tension of the skin at rest.

The approach developed specifically to the skin morphology allows the identification of the lines network anisotropy at different scales of depth and orientation. For each plane at a certain depth of the skin surface, are determined three parameters of the point belonging to the line of tension: the density of depth z, the width of the line and the rose of directions between 0 and 180° [27, 28, 29]. The figure (11A) represents the identification of the network of skin tension lines in different directions about a 25 year old. The depth of the skin lines is illustrated by the range of colours from blue to red. One can distinguish the family of skin lines printed on the plates: secondary lines (colours of green, yellow, and red correspond to a variation of depths between -17 microns and 50 microns). The family of skin tension lines in the main colour blue are in a scale between -17 and -84 microns. The identification of the orientation of skin lines in different directions is illustrated by the figures: 11B, 11C, 11D, 11E. The figure (12), illustrates the anisotropy distribution versus the depth of skin network families.

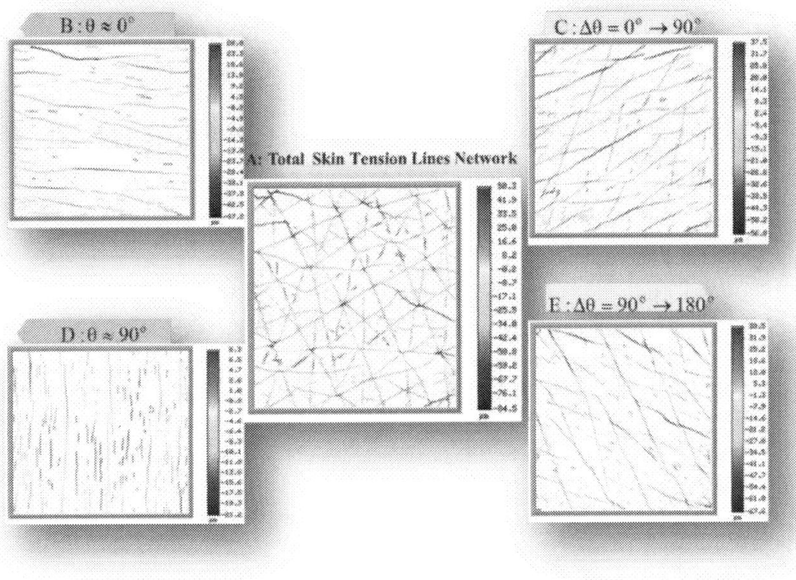

Figure 11. Network of skin tension lines.

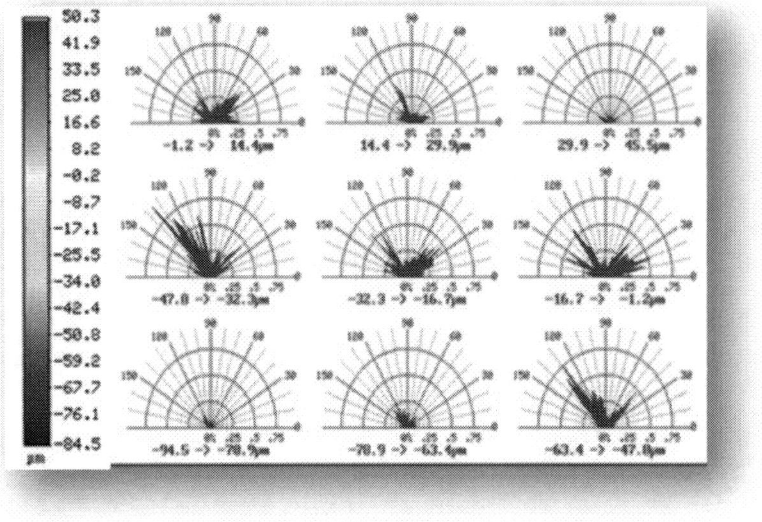

Figure.12. Reconstruction volumetric skin tension lines anisotropy.

2.6. Morphological Tree of Skin Line Network [28, 29, 30, 11]

Appropriate and quantitative representation of skin line network has been developed. It allows reconstructing all network lines as a morphological tree. Each trunk of tree represents the density of lines in a given direction and for a given depth, figure (13)

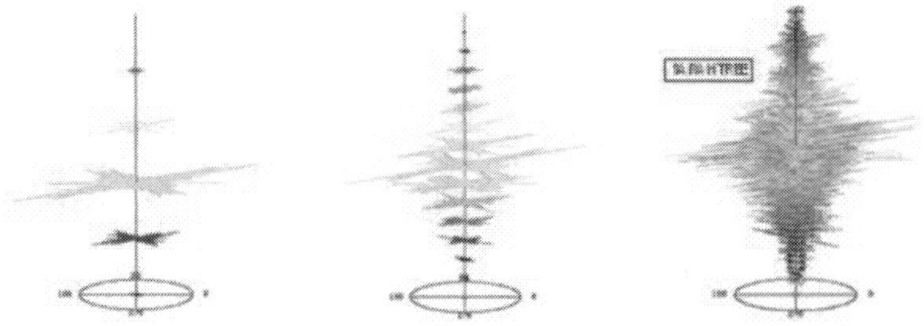

Figure 13. 3D Reconstruction: Tree of skin tension lines network.

One family of lines is described as a branch of the tree, in relation to its depth and direction. With this original method, it is possible for the first time to quantify the different scales of skin line accurately and to follow the morphological changes of the surface in relation to age. This multi-morphological decomposition of line network can be used to assess mechanical tension of elastic fibres and collagen bundle during ageing. In the other hand, this approach can be used as a preventive test for certain diseases of the elasticity of the skin. The examples of figure.(14) shows the use of this approach in the analysis of the transformation of the 3D line network during ageing [11,26].

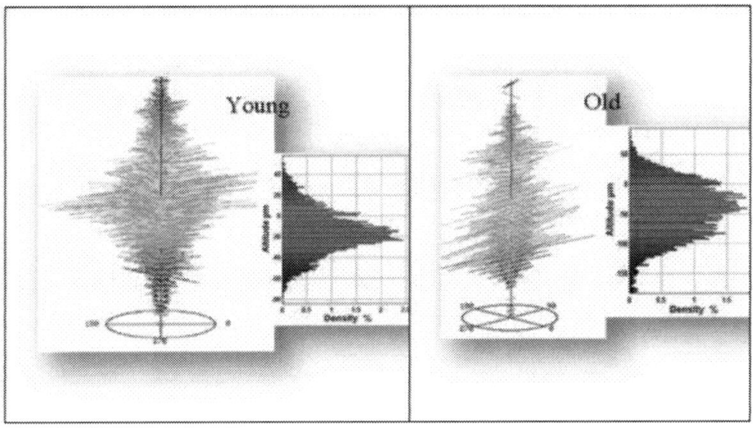

Figure 14. Aging effect on the Tree of skin tension lines network [11, 26, 27, 28].

3. Conclusion

Changes of dermal collagen and elastin content are characteristic for skin aging as well as for pathological skin conditions. The formation of lines and wrinkles in light-exposed areas throughout the body, such as the face, throat and hands is a well-known sign of skin aging. Lines and wrinkles are influenced by both intrinsic and extrinsic factors. The numerous intrinsic factors are age, gender, genetic disposition and race. In exposed and unprotected zones of the body such as on the skin of the hands and the face, extrinsic factors such as UV light, weather and climatic influences, nutrition, tobacco and alcohol abuse, effect the formation of lines and wrinkles.

With increasing age, the physiology and appearance of the human skin will change. Alterations in structure, loss in tightness, smoothness and a decrease in the skin's functional capacity are phenomena which may be attributed to the aging mechanism. An increase in dryness and thus roughness as well as a loss in elasticity and even pigmentation are also a sign of increasing skin aging. Wrinkles on flaccid skin develop with growing age. There is a decline in the subcutaneous fatty tissue. Today, relatively little is known about the exact biomechanical processes of skin aging.

Primarily, changes in the skin's appearance are a result of a general aging process of the connective tissue of the subcutis. This leads to an atrophy of the epidermis which adjoins the papillary layer and to an irregular decrease in the elasticity of the elastic nets which are structures accompanying the collagen fibres in the connective tissue. As a result of the changed amount and chemical composition of the basic substance of the connective tissue, a loss of liquids is the result, which consequently leads to a decrease in glycosaminoglycans, the basic structures of the connective and supporting tissues. As a consequence, the youthful turgor, i.e. the skin's tension, is lost. Melanocytes disintegrate or lose close contact to epidermal cells and finally lead to a spotted pigmentation of the skin.

To evaluate these changes, we used the characterization of skin imaging of microrelief. The study of 120 Caucasian women shows that ageing in the volar forearm zone results in changes in the morphology of the line network organisation. The decrease in secondary lines ($z < 60$ μm) is counterbalanced by an increase in the depth of the primary lines ($z > 60$ μm) and an accentuation of the anisotropy index. This phenomenon is known to be accelerated by actinic radiation (extrinsic photoageing) which increases the degradation of the elastic fibres.

References

[1] G. Dupuytren. (1834), Traité théorique et pratique des blessures par armes de guerre,

[2] K. Langer (1978). On the anatomy and physiology of the skin. 1: the cleavability of the cutis (English transaction by T Gibson). *Br J Plast Sur* 31: 3 - 8.

[3] A.F Borges (1989), Relaxed skin tension lines. *Dermatol Clin* 7:169-177

[4] J.C Waldorf, G. Perdikis and S.P Terkonda (2002°, Planning incisions. *Operative Techniques in General Surgery*, 4(3): 199-206, September 2002

[5] PFF Wijn, AJM Brakkee, JP Kuiper, AJH Vendrik (1981), The alinear viscoelastic properties of human skin in vivo related to sex and age. In: Marks R, Payne PA (eds), Bioengineering and the skin, MTP Press, Lancaster, pp 135-146)

[6] JFM Manschot, PFF Wijn, AJM Brakkee (1982), the angular distribution function of the elastic fibres in the skin as estimated from in vivo measurements. In: R Huiskes, DH Van Campen, de JR Wijn (eds) Biomechanics: principles and applications. Vol 1: developments in biomechanics. M . Nijhoff, The Hague, pp 411-418

[7] GE Pierard, CM Lapière (1987) Microanatomyof the dermis in relation to relaxed skin tension lines and Langer's lines. *Am J Dermatopathol* 9:219-224

[8] H.L Stark (1977) Directional variations in the extensibility of human skin. *Br J Plast Sur*. 30: 105 – 114

[9] JC Barbenel (1995) Identification of Langer's lines. In Serup J, Jemec GBE (eds) Handbook of non invasive methods and the skin. CRC Press, Boca Raton, pp 341 – 344

[10] H. Zahouani (2002) Méthodes de caractérisation de la surface cutanée. Encyclopédie Médico-Chirurgicale. Editions Scientifiques et Médicales Elsevier SAS Paris 50-140-H-10.

[11] H. Zahouani, R. Vargiolu (2005) Skin Line Morphology: Tree and Branches. Editions Springer ISBN 3-540-01771-2. Measuring the Skin. sous la direction du Professeur P. AGACHE. pp 40-59

[12] H. Zahouani, (2006), Skin tension lines network during ageing", Handbook of Non-Invasive Methods & the Skin, Ed. J. Serup, G. B. E. Jemec, G. L. Grove, CRC Taylor & Francis, 2nd Ed., ISBN 0-8493-1437-2, (1048 pages), pp 191-204.

[13] P. Corcuff, J. de Rigal, S. Makki, JL. Leveque, P Agache (1983). Skin relief and aging. *J Soc Cosmet Chem* 1983; 34:177-90.

[14] J. Mignot, H. Zahouani, D. Rondot, PH. Nardin (1987). Morphological study of human skin relief. *Bioeng Skin*; 3:177-196.

[15] P. Corcuff, JL. Leveque, GL. Grove, AM Kligman (1987). The impact of ageing on the microrelief of periorbital and leg skin. *J Soc Chem*; 82:145-152.

[16] P. Corcuff, O. de Lacharière, JL. Leveque (1991). Extension induced changes in the micro-relief of the human volar forearm: variation with age. *J Gerontol Med Sci*; 46:223-227.

[17] H. Zahouani, J. Asserin, A. Mavon, D. Blanc, P. Agache (1996). Morphological and spectral rose identification of the anisotropic skin micro-relief, furrows and wrinkles. *Skin Res Technol*; 2:201.

[18] JL Grove, MJ Grove, JJ Leyden and al (1991). Skin replica analysis of photodamaged skin after therapy with tretinoin emollient cream. *J Am Acad Dermatol* 1991; 25-231-237.

[19] S Makki, P Agache, J Mignot, H Zahouani (1984). Stastistical analysis and three dimensional representation of human skin surface. *J Soc Cosmet Chem*; 35:311-325.

[20] P. Corcuff, F. Chateny, JL. Leveque (1984). A fully automated system to study skin surface patterns. *Int J Cosmet Sci*; 6:167-176

[21] H. Zahouani, M. Chuard, J. Mignot, S. Makki, P. Agache (1985). Etude tridimensionnelle du relief cutané *ITBM*; 6:447-460.

[22] JM. Lagarde, C. Rouverais, D. Black, S. Diridillou, Y. Gall (2001). Skin topography measurement by interference fringe projection: a technical validation. *Skin Res Technol*; 7:112-121.

[23] GE. Pierard, JF. Hermans, CH. Lapiere (1974). Stéréologie de l'interface dermo-epidermique. *Dermatologica*; 149:266-273.

[24] L. Hashimoto (1974). New methods for surface ultrastructure. *Int J Dermatol*; 13:357-381

[25] H. Zahouani (1998), Spectral and 3D motifs identification of anisotropic topographical components. Analysis and filtering of anisotropic patterns by mophological rose approach. *Int J of Machine Tools & Manufacture*; 38:615-623.

[26] H. Zahouani, R, Vargiolu, (2008), "Skin morphology and volume: methods of evaluation", Injection Treatments in Cosmetic Surgery, Ed. B. Ascher, M. Landau, B. Rossi, Informa Health Care, Series in Cosmetic and Laser Therapy, New York, ISBN 9780415386517, (480 pages), pp 13-33.

[27] H. Zahouani, R. Vargiolu, (1998). 3D Morphological tree representation of the skin relief. A new approach of skin imaging characterization. XXth Congress. International Federation of the Societies of Cosmetic Chemists, CANNES, France september 14 -18, 1998.Paper N° 30, pp 69 - 80.

[28] H. Zahouani, R. Vargiolu, (2000). Mesures du relief cutané et des rides `` Collection Explorations Fonctionnelles Humaines. Physiologie de la Peau et Explorations Fonctionnelles Cutanées. Sous la direction du Professeur Pierre AGACHE. ISBN : 2-7430-0360X., pp 41-57

[29] H. Zahouani, (2007), Tension des fibres élastiques et lignes cutanées : application au vieillissement et à la cicatrisation", Les secrets de l'anti-âge, Ed. J.L Lévy, Groupe Liaisons SA ISBN 978-2-7184-1161-3, pp 67-79.

[30] J.C. Guimberteau *, J. Sentucq-Rigall, B. Panconi, R. Boileau,P. Mouton, J. Bakhach, (2005), Introduction à la connaissance du glissement des structures sous-cutanées humaines. *Annales de chirurgie plastique esthétique* 50 .19–34.

Surface design methodology – challenge the steel

M Bergman[*1,2], B-G Rosen[1,2], L Eriksson[3], C Anderberg[4]

[*]E-mail: martin.bergman@hh.se

Abstract. The way a product or material is experienced by its user could be different depending on the scenario. It is also well known that different materials and surfaces are used for different purposes. When optimizing materials and surface roughness for a certain something with the intention to improve a product, it is important to obtain not only the physical requirements, but also the user experience and expectations. Laws and requirements of the materials and the surface function, but also the conservative way of thinking about materials and colours characterize the design of medical equipment. The purpose of this paper is to link the technical- and customer requirements of current materials and surface textures in medical environments. By focusing on parts of the theory of Kansei Engineering, improvements of the companys products are possible. The idea is to find correlations between desired experience or "feeling" for a product, -customer requirements, functional requirements, and product geometrical properties -design parameters, to be implemented on new improved products. To be able to find new materials with the same (or better) technical requirements but a higher level of user stimulation, the current material (stainless steel) and its surface (brushed textures) was used as a reference. The usage of focus groups of experts at the manufacturer lead to a selection of twelve possible new materials for investigation in the project. In collaboration with the topical company for this project, three new materials that fulfil the requirements -easy to clean and anti-bacterial came to be in focus for further investigation in regard to a new design of a washer-disinfector for medical equipment using the Kansei based Clean ability approach CAA.

1. Introduction

It is well known that a proper material selection is important when designing a new product; a minor mistake can cause terrible consequences to its user. An outcome could only be a success if the person who matters understands the message [1] and [2]. Nevertheless, material selection is not only about ensuring safety in a construction or optimising weight in a car for instance. Zoom into the material beyond what we can see with th enaked eye, and the micro structure will expose a landscape which affects us as users more than we can understand.

Professor Mitsuo Nagamachi (Hiroshima International University) had a vision about improving products on a more detailed level than before. Hence, he developed the method Kansei Engineering (KE) in the 1970's which has its roots from the Japanese concept of Kansei, ("intuitive mental action of the person who feels some sort of impression from an external stimulus") [3]. According to Professor Nagamachi the Kansei concept includes; "a feeling about a certain something that likely will

1 University of Halmstad, SE-301 18 Halmstad, Sweden.
2 Chalmers University of Technology, SE-412 96 Göteborg, Sweden
3 Jönköping University, P.O.Box 1026, 551 11 Jönköping, Sweden
4 Getinge Infection Control AB, Research & Development Dept, P.O Box 69, SE-310 44

improve one's quality of life". KE can also be defined as a customer-oriented approach to product development [4], [5], [6] and [7].

By using the framework of KE as an approach and focusing on finding correlations between the functions; customer requirements, function requirements, design requirements and process requirements; a higher level of user quality could be obtained.

The perceived quality is related to the experience of the product, and the experience is directly connected to the interaction and stimulation of the senses [8]. The customer describes their experience of a product with descriptive words such as; elegant, plastic or stylish and so on [9]. Yet, what they describe is their feeling, nothing else! By focusing on the material selection and surface processing (which is controlled by us, described with numbers/parameters) the experience of the product could vary, and by that also the customers feeling [9]. Now, is it possible to link those parameters to the customer's feelings and experience (and put a number onto it), but at the same time fulfil the technical requirements of the product?

2. Method

The project is basically implemented by the framework of KE, although it is modified to fit into the topic. The adapted method has been used systematically in the research work as a reference to obtain qualitative data but also to be able to develop a tailor made toolbox for designers where special tools for product optimization are possible. The approach is briefly described in this paper, the reader is recommended to read "Surface appearance and impression" [9] for a deeper understanding of the method.

The research approach used in this project handles 6 different phases/steps, initially though, the domain and the context has to be defined;

1. Pilot Study – In this phase it is important to define the questions what, who, where, how, why, and when. But also to implement a market analysis.
2. Describe the Experience, *"feeling"*– In this phase focus is lying on the description of the experience, which is possible by firstly collecting adequate describing words, *adjectives*, which the user expresses when interacting with the domain.
3. Define Key Product Properties – In this phase it is important to know the product and its features. It is about finding properties that affect the user, yet it is important that the properties are measurable and adjustable.
4. Connect the Experience,-the adjectives, and Product physical properties – By gathering typical users into a focus group whose mission is to analyse and evaluate the domain; it is possible to obtain information about the experience. In this phase focus is lying on finding connection between the describing adjectives and the key product properties.
5. Validity Check Point – When the correlation is established, it is important to verify potential other domains tested in the project. This is possible by measuring and comparing the key product properties, either the competitive domain fails the test or it will proceed to the next level.
6. Synthesis and Modelling the Domain – When the experience is evaluated towards a number of different product properties and the correlation is verified, the development of a new domain is possible using the research work as a solid ground to reach a higher level of quality.

However, the main issue in this project mainly handles the "clean ability" of the material and surface. Focus is both lying on finding a novel approach, which facilitates the confirmation of what a clean surface is or not but also finding correlations between the clean ability, and bacterial growth prevention, i.e. "Clean Ability Approach" (CAA). The CAA is about analysing the material and surface, which is directly linked to the product properties of the domain [10]. However, verification of an analysed material is difficult without a reference. Hence, the topical company´s trademark surface, the stainless steel with brushed texture, will be used as a comparison reference.

Journal of Physics: Conference Series **483** (2014) 012013 doi:10.1088/1742-6596/483/1/012013

3. Result

The result is based on a literature study and measurements of the company´s reference surface, but also external materials. The analysed theoretical material is handling medical- and food contact surfaces, but also surfaces of different coatings. The thirteen (13) selected materials which were tested in a focus group were; *Anodized aluminium, Bronze, Aluminium/plastic sandwich material, Stainless steel (reference), Copper/aluminium, Tinted glass, Acrylic plastic, Zinc, White plastic, Active glass (coated), Spray painted aluminium, Fibre cement* and *Glass*.
The limitation with a wide range of tested materials was positive in terms of provoking different experiences by the users in an initial phase. The function requirements on the other hand, obviously affected the material selection as well. However, three materials came to be selected for further investigation; *glass, spray painted aluminium* and *acrylic plastic*, which all passed the focus group and the function requirements.

3.1. Pilot Study and Describe the Experience

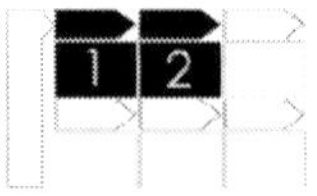

Now, the domain and the context were already defined by the company. The domain is a washer-disinfector for medical equipment, focusing on the material of the front of the product, figure 2. The context is the medical environment the product is located in. Nevertheless, by defining the target audience, possibilities of finding user centred experiences were possible, and by that also correlation between the experience and the selected material samples. By converting the experience to describing words such as adjectives, it is easier to understand and analyse the outcome of the focus group. Words that continuously arose and were discussed in regard to a visionary product were; *robust, resistant, clean, warm, sleek* and *elegant*.

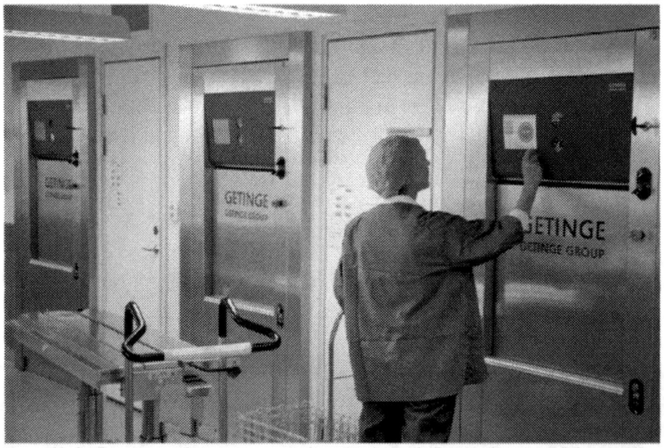

Figure 1. The washer-disinfector units are complex industrial products with medical demands on clean ability and bacterial resistance among several other demands like scratch resistance and chemical resistance. The current material in most of the product is today stainless steel.

3.2. Define Key Product Properties

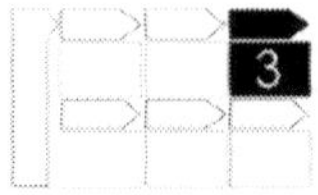

The functional requirements (table 1) and the physical design features responsible for the functions are in focus in phase 3, *-Correlation between expectations of the experience adjectives and the functional requirements.* The CAA is here a subset of all the functional requirements of the products. In this phase measurable functions and physical product properties are in focus; what is possible to measure is possible to modify, and by that providing a possibility of correlation between the experience adjectives and the CAA functional properties [10].

Table 1. The relation between functional requirements and judged texture related design parameters for CAA (crosses).

			Physical product properties, Design parameters **DP's**						
			Panel qualitatively judged properties						
			Surface texture	Chemical resistance	Color suitability	Darkness of color	Gloss	Heat conductance	Hardness
Functional requirements, **FR's**	Other	Shock resistance							
		Scratch resistance							
	Clean Ability Approach, CAA	Cleanability	x						
		Bacterial resistance	x						

The selection of functional demands and product properties were made by consulting the company R&D department in order to limit the study towards surface associated functions (see table 1). By judging the different functional requirements, the FR's and connecting them to product physical properties, the Design parameters DP's, the focus group considered CAA components to be: *bacterial resistance,* and *clean ability* as directly connected to surface texture.

3.3. Connect the Experience adjectives and Product Properties

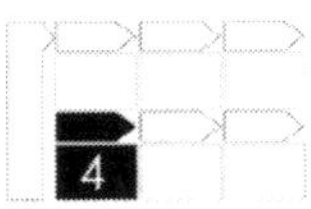

To be able to gather information about a certain product and its material, a survey is a powerful way to go forward [11]. This may lead towards the result that new products may be designed to create desirable, enjoyable and meaningful experiences. The main idea with the focus group (figure 3) was to obtain their thoughts about the current design of the product [12], focusing on the materials and the surfaces. By observing the focus group and obtaining words, adjectives, that describe their experience of the materials, correlation between the materials/surfaces and the experience was possible to find (table 2.). In this phase the functional requirements were on hold to give the focus group unbiased possibility to explore new materials purely in relation to the adjectives of experience.

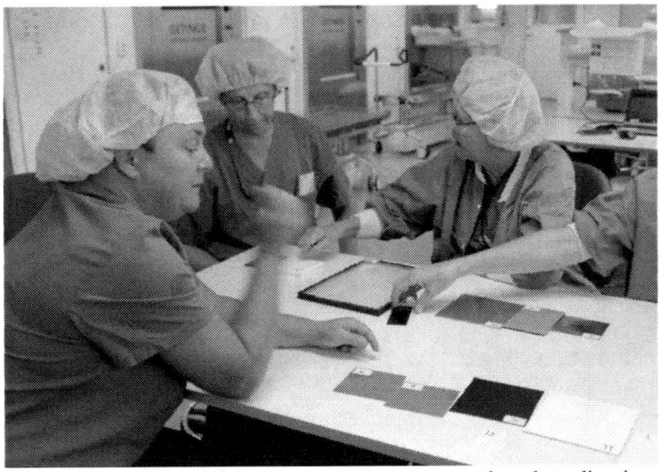

Figure 2. The focus group consisting of company experts connecting the adjectives describing the "feeling" and experience of the surfaces of the 12 judged surfaces, resulting in table 2 below.

The words from phase 2, *robust, warm, sleek, elegant, resistant,* and *clean,* were now in focus. The valuation was simple, although significant. Using sight and tactile senses the focus group participants evaluated and compared to the reference stainless steel the surfaces towards the describing adjectives, either better or worse in comparison. For instance; correlation between the material *fibre cement* and

Journal of Physics: Conference Series **483** (2014) 012013 doi:10.1088/1742-6596/483/1/012013

the describing word *clean* was not very high in comparison with the lower texture amplitude in the reference stainless surface.

The result of the 4'th phase was that the focus group decided to choose three (3) surfaces for the next phase. The three surfaces chosen were *Acrylic plastic, Spray painted Aluminium,* and *Glass* (see table 2). All the three surfaces were considered as fulfilling the emotional response related to not only the CAA adjectives *resistant* and *clean* but also the other 4 adjectives found in phase 2.

3.4. Validity Check Point

The validity check provides the opportunity of a check point where selected materials are measured and analysed. The material samples which advanced through the focus group were measured using a MicroXAM phase shifting- and coherence scanning interferometry instrument[5] and analysed in Mountains Map Premium 6[6] to establish quantitative measures of the surface texture (see table 3). Initial focus was concentrated on the surface texture arithmetical mean height, ***Sa***, according to ISO 25178:2011. *Sa* is considered as the surface texture property closely connected in internal company standards to the legislative demands on *bacterial resistance*, and *clean ability* i.e. connecting *Sa* and CAA. The CAA is considered "OK" when the Sa value for a given surface is <0.8μm [10]. NOTE: *Originally the "0.8μm rule" was determined for the profile ISO 4287:1997 Ra parameter, in this study replaced by the areal defined Sa parameter.*

Now, the correlation in between the experience and "feeling" (psychological requirements) and the function requirements (physical requirements) have to be established as well. It is obvious that the describing adjective words (*robust, resistant, clean, warm, sleek* and *elegant*) are possible to link to the function requirements. For instance, the adjective *clean* is directly connected to the demands; *clean ability* as discussed earlier but also secondly related to *chemical resistance*, and *scratch resistance*

Table 3. The relation between functional requirements and focus group judged as well as physically measured design parameters for CAA.

			Physical product properties, Design parameters **DP's**						
			Focus group qualitatively judged properties				Measured Properties		
			Surface texture	Chemical resistance	...	...	**Surface texture**	Haptic friction	Hardness
Functional requirements, **FR's**	Other	Shock resistance							
		Scratch resistance							
	Clean Ability Approach, **CAA**	Cleanability	x				**Sa<0.8um**		
		Bacterial resistance	x				**Sa<0.8um**		

while the material used will deteriorate in function if it is affected negatively by chemical agents and mechanical wear by surface scratching.

To further indicate the complexity, the functional demand on *scratch resistance* is regarded as linked to the adjectives *robust* and *elegant* by the focus group. Hence, it is obviously correlation not only between one describing word and a functional demand or property but also between several of the words and different demands and properties. It is therefore important to be aware of the complexity of reaching a certain experience or "feeling" for a product, without losing another experience when selecting between different product properties.

[5] ADE Phase Shift MicroXAM Optical interferometric profiler characteristics [online]. http://www.tcd.ie/CMA/misc/MicroXam.pdf

[6] Digital Surf, Besançon, France, http://www.digitalsurf.fr

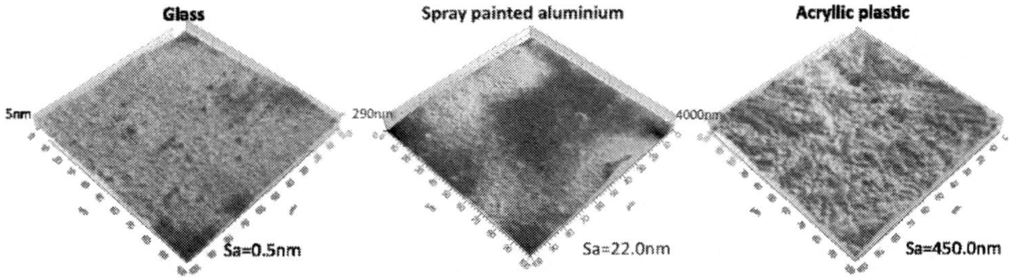

Figure 3. Coherence scanning interferometer 0.101x0.101mm measurements of the three candidate materials for replacing the stainless steel for washer-disinfector units. Arithmetic mean height (*Sa*) of the surfaces varies from below one nanometer to about 0.5μm –a factor of 900. All surfaces pass the current regulations for clean ability and bacterial resistance demands.

From figure 3 it is clear that all the three selected surfaces meet the requirements of CAA, *clean ability*, and *bacterial resistance*, as defined by the criteria Sa<0.8μm but two questions arise when considering this. Firstly, two of the surfaces have a Sa<<800nm and what does this mean for the function. Is an almost atomically flat scratch free surface like the *Glass* and *Spray painted aluminium* more beneficial? Secondly, is the *Sa* criterion valid for both functional CAA requirements? *Sa* is only one sub set of "field parameters" defined in ISO 25178:2 describing the mean amplitude of the surface texture. Spatial- or hybrid (combinations of vertical- and horizontal geometrical properties like slope and material distribution) geometrical information provides additional significant surface texture information.

By committing additional functional tests on CAA response by bacteria cultivation and cleaning tests in combination with more advanced surface texture characterisation more detailed models for the understanding of the function in relation to texture can be made. Optimizing models would be possible to design for detailed control of *clean ability* and *bacterial* growth and answers of the first question above could be given.

The fact that many surface texture parameters are closely correlated is also important to consider when answering the second question. Different physical explanations of CAA function could be masked by correlations between texture parameters describing different significant geometrical features. The *Sa* parameter correlates to 15 of the 20 ISO parameters in table 4 below. The possibility is high that some of the 15 parameters other than *Sa* have a more close relation to the physical reasons behind *clean ability* and *bacterial growth*.

Journal of Physics: Conference Series **483** (2014) 012013 doi:10.1088/1742-6596/483/1/012013

Linear Correlation, R, between Surface texture parameters according to ISO 25178:2

Measured properties / Areal surface texture	Sq	Sp	Sv	Sz	Sa	Smr	Smc	Sxp	Sdq	Sdr	Vm	Vv	Vmp	Vmc	Vvc	Vvv	S10z	S5p	S5v	Sda	Sdv
Sv	0,94	0,87	1,00																		
Sz	0,96	0,91	1,00	1,00																	
Sa	0,99	0,95	0,90	0,92	1,00																
Smr	-0,66	-0,82	-0,60	-0,65	-0,65	1,00															
Smc	0,98	0,95	0,86	0,89	1,00	-0,66	1,00														
Sxp	0,98	0,90	0,96	0,96	0,95	-0,62	0,92	1,00													
Sdq	0,94	0,87	1,00	0,99	0,90	-0,59	0,87	0,95	1,00												
Sdr	0,93	0,85	0,97	0,97	0,91	-0,49	0,87	0,94	0,98	1,00											
Vm	0,61	0,81	0,54	0,60	0,62	-0,95	0,63	0,54	0,54	0,44	1,00										
Vv	0,98	0,95	0,86	0,89	1,00	-0,66	1,00	0,92	0,87	0,87	0,63	1,00									
Vmp	0,61	0,81	0,54	0,60	0,62	-0,95	0,63	0,54	0,54	0,44	1,00	0,63	1,00								
Vmc	0,97	0,95	0,86	0,89	0,99	-0,64	0,99	0,91	0,87	0,88	0,62	1,00	0,62	1,00							
Vvc	0,95	0,94	0,80	0,84	0,98	-0,66	0,99	0,87	0,81	0,82	0,65	0,99	0,65	0,99	1,00						
Vvv	0,97	0,88	0,96	0,97	0,94	-0,58	0,91	1,00	0,96	0,96	0,50	0,91	0,50	0,90	0,85	1,00					
Spd	-0,36	-0,39	-0,35	-0,36	-0,35	0,37	-0,35	-0,35	-0,33	-0,30	-0,40	-0,36	-0,40	-0,35	-0,35	-0,34					
Spc	0,61	0,75	0,57	0,61	0,61	-0,92	0,62	0,57	0,57	0,47	0,84	0,61	0,84	0,62	0,62	0,53					
S10z	0,88	0,80	0,97	0,96	0,83	-0,50	0,79	0,90	0,97	0,95	0,47	0,79	0,47	0,79	0,73	0,91	1,00				
S5p	0,95	0,84	0,93	0,93	0,93	-0,45	0,90	0,95	0,94	0,97	0,40	0,90	0,40	0,90	0,86	0,97	0,92	1,00			
S5v	0,80	0,74	0,94	0,92	0,75	-0,49	0,70	0,83	0,94	0,91	0,48	0,70	0,48	0,71	0,64	0,84	0,99	0,85	1,00		
Sdv	0,93	0,83	0,99	0,98	0,89	-0,50	0,85	0,94	0,99	0,99	0,45	0,85	0,45	0,85	0,80	0,96	0,97	0,96	0,94	0,64	1,00
Shv	0,78	0,76	0,91	0,90	0,71	-0,60	0,66	0,83	0,88	0,82	0,58	0,66	0,58	0,65	0,59	0,82	0,91	0,75	0,93	0,75	0,87
	Sq	Sp	Sv	Sz	Sa	Smr	Smc	Sxp	Sdq	Sdr	Vm	Vv	Vmp	Vmc	Vvc	Vvv	S10z	S5p	S5v	Sda	Sdv
	µm	µm	µm	µm	µm	%	µm	µm	-	%	µm	µm	µm	µm	µm	µm	µm	µm	µm	µm	µm
	Areal surface texture																				
	Measured properties																				

Table 4. Correlation table expressing the linear correlation, R, between Areal surface texture parameters measured using optical interferometry. Figures in bold indicate a correlation higher or equal to 0.8 ($R^2>=0.64$).

4. Conclusions

- Kansei engineering is a strong product developing method generating information about the customer expectations and "feelings" of a sterilisation unit, its functional requirements and the connection to geometrical texture properties.
- Clean ability and bacterial growth are the main surface related requirements of the sterilisation unit.
- *Glass*, *Spray painted aluminium* and *Acrylic* plastic are all strong possible candidates to replace stainless steel for sterilisation units.
- The arithmetic amplitude of the surface texture, *Sa*, is closely historically related to clean ability and bacterial growth factors of the washer-disinfector unit.
- *Sa* is highly correlated to at least 13 of 20 ISO 25178:2 surface texture parameters.
- There exists a potential to closely study surface texture parameters to isolate in more detail other geometrical properties than mean amplitude to explain and model clean ability and bacterial growth of washer-disinfector units.

5. Future

Phase 6, Synthesis and Modelling the Domain Discussion needs to be preceded by a rigorous study to deepen the knowledge about the CAA requirements of a clean and bacterial free surface. Practical tests will be carried out to enable detailed modelling of surface texture as a variable creating desired "feelings" and user experiences from the introduction of new materials and surface textures in the washer-disinfector unit domain.

6. References

[1] Krippendorf K, (2006). The semantic turn – a new foundation for design.Taylor & Francis Group, LLC.

[2] Wikström, L. (2002). Produktens budskap – Metoder för värdering av produkters semantiska funktioner ur ett användarperspektiv. Instutionen för produkt- och produktionsutveckling, Chalmers tekniska högskola. Göteborg, Sweden

[3] Nagamachi M and Lokman A M, (2011). Innovations of Kansei Engineering. Taylor & Francis Group, LLC.

[4] Nagamachi M, (1997). Kansei Engineering and Comfort. International Journal of Industrial Ergonomics 19: 79-80.

[5] Nagamachi M, (2002). Kansei engineering as a powerful consumer oriented technology for product development. Applied ergonomics, vol 33, 3: 289-294.

[6] Frisk M, and Järlskog H, (2002). Handbok i Kansei Engineering. Linköping University IKP.

[7] Hedberg Ö M, (2004). Kansei Engineering som stöd för en designprocess. En explorativ studie. Umeå University.

[8] Wolfe J M, Kluender K R and Levi D M, (2012). Sensation & Perception, 3rd edition. USA: Sinauer Associates, Inc.

[9] Bergman M, Rosén B-G, Eriksson L, (2012). Surface appearance and impression. *International conference on Kansei Engineering and Emotion Research*, Department of Industrial Design, National Cheng Kung University.

[10] Factors affecting fouling and cleanability of open food contact surfaces - A report on findings from the EU Integrated Project PathogenCombat, http://www.pathogencombat.com/Publications/~/media/Adtomic/Unique%20Achievements/Leaflets/Factors%20affecting%20fouling%20and%20cleanability%20of%20open%20food%20contact%20surface.ashx [Acc. 2013-02-26]

[11] Warell A, (2008). Modelling perceptual product experience - Towards a cohesive framework of presentation and representation in design. Institute of Design for Industry and Environment, Massey University, Wellington, New Zealand.

[12] Wibeck V, (2011). Fokusgrupper – om fokuserade gruppintervjuer som undersökningsmetod. Studentlitteratur AB, Lund Sweden.

New method and uncertainty estimation for plate dimensions and surface measurements

Salah H. R. Ali[1] and Jariya Buajarern[2]

[1]*Engineering and Surface Metrology Dept, Precision Engineering Division, National Institute for Standards (NIS), PO Box 136, Giza (12211), Egypt*
[2]*Dimensional Metrology Department, National Institute of Metrology (NIMT), 3/4-5 Moo 3, Klong 5, Klong Luang, Pathumthani 12120, Thailand*

[1]E-mail: SalahAli20@yahoo.com; Mobile: 00201120121545
[2]E-mail: jariya@nimt.or.th

Abstract. Dimensional and surface quality for tile plate manufacturing control is facing difficult engineering challenges. One of these challenges being that plates in large-scale mass production contain geometrically uneven surfaces. There is a traditional measurement method used to assess the tile plate dimensions and surface quality based on standard specifications: ISO-10545-2: 1995, EOS-3168-2: 2007 and TIS 2398-2:2008. A new measurement method of the dimensions and surface quality for ceramic oblong large-scale tile plate has been developed compared to the traditional method. The strategy of the proposed method is based on CMM straightness measurement strategy instead of the centre point in the traditional method. Expanded uncertainties budgets in the measurements of each method have been estimated in detail. The capability of accurate estimations of real actual results of centre of curvature (CC), centre of edge (CE), warpage (W) and edge crack defects parameters has been achieved according to standards. Moreover, the obtained results showed not only a more accurate method but also improved the quality of tile plate products significantly.

Keywords: CMM, straightness, tile plate, dimensions and surface quality; uncertainty.

1. Introduction

Advanced metrology techniques are became used in various fields of science and production engineering applications. There are two new basic approaches, the contact techniques and noncontact measuring techniques. These techniques include scanning electron microscopy (SEM), transmutation electron microscopy (TEM), computed tomography (CT), interferometric method and coordinate measuring machines (CMM) use in dimensional metrology [1]. Coordinate dimensional metrology is the key to global quality assurance systems of industrial engineering products. Quality control measurements performed to ensure quality of hard industrial products like ceramic tile plates help detection of some defected products due to various technical and human reasons [2]. Ceramic tile plates are subject to international and national standards assessed tests such as colour analysis, surface abrasion resistance, thermal expansion, moisture expansion, friction coefficient and wear, breaking

[1] To whom any correspondence should be addressed.

strength; dimensions and surface quality to reveal the durability defect. Furthermore, dimensional and surface quality verification is the main standard test and also the purpose of this investigation. However, the amazing progress in the production technology has allowed a significant decrease of the deviation in the dimensional and surface quality. Accordingly, the measurement methods of dimension and geometrical surface of products should be reviewed using advanced measurement technique such as CMM. It is a custom measurement tool using a point laser for almost plane surfaces of objects [3]. Therefore, there is limited number of studies in this area. While, some publications are carried out using image processing, vision inspection system, morphological techniques and texture analysis [4-9]. Quality inspection of ceramic tile using statistical techniques and neural networks approach is also studied [10]. Most previous research work have been done to assess surface functionalization of industrial ceramic tile plates as surface micromechanical an microstructural using SEM and TEM micrographs to improve the properties of scratch and wear resistance [11-13]. Assessing the environmental impact on production of ceramic tile is also studied [13-14]. While, the American National Institute of Standards and Technology (NIST) developed a sustainability resource guide that compiled the information on the environmental impact of various products, including ceramic tiles [15]. On the other hand, the ability to measure precisely and consistently is of fundamental importance in modern technology. The reliability and safety of new devices and systems are the basis of equal global manufacturing requirements. It must be noted that the detection of defect in dimensions and surface quality is an important area of tiles inspection include measurements, calculation and characterization that are often not available in factories. The dimensions and surface quality test of the ceramic tiles represented in the curvatures of the diagonals, edges, warpages and cracks are essential components of the inspection and evaluation in order to achieve the main purposes such as safe surface use [16-19]. One of the most important functions of the NMI's, the dissemination of the reliable reference data set which can be used as a guide for evaluating the measurement skill and reliability of test results in industry as well as in commercial production and design process. Thus, geometrical form measurements such as straightness form of products surfaces are basic requirements in dimensional metrology for precise manufacturing engineering.

The main of this work is to introduce a new developed measurement method, assist and compare the validity of two different methods (traditional and developed) based on fast tactile 2D metrology. The new measurement method is developed to support the quality assurance in the manufacturing of tiles. Moreover, the possibilities of accurate inspection system to reveal four major problems namely, centre of curvature (CC), centre of edge (CE), warpage (W) parameters and edge crack defects have become assure achieved. The budgets of expanded uncertainty with the measurement results of each method have been estimated accurately. The capability of dimensions and surface quality measurement of tiles has been discussed in details as well. Thus, the proposed measurement method could be able to assist the tiles of different scales to produce not only better accurate method compared to other methods [6-9] but also permit improved manufacturing quality of ceramic tile products.

2. CMM Accuracy Verification

Stationary CMM in the engineering dimensional metrology laboratory is fixed and verified by the manufacture (ZEISS Co., Germany). CMM machine consists essentially of a probe supported on X, Y, and Z coordinates and capable to determine spatial coordinates on a workpice surface [20]. The procedures for measurement using CMM includes: verification of the probing system, defining datum(s) on the tiles surfaces, performing measurement(s), computing the required relative dimensions from measurements made previously and assessing the tiles performance with the specifications. Intermediate verification guarantee by measuring standards reference sphere and comparing the measured value with the specified measurement uncertainty. The environmental conditions of testing room have been adjusted in the rage of standard specifications [20]. Laboratory room temperature was set at 20±1 °C and relative humidity at 50±3%, which are suitable for the

CMM operation. Table 1 comprises the specifications of the CMM machine set-up and strategy of measurements.

Table 1. CMM set-up and measurement strategy

CMM strategy parameters	Specifications
Master probe radius	3.9996 mm
Reference sphere radius	14.9942 mm with S =0.0001 mm
Used long probe radius	4.0000 mm with S = 0.0002 mm
Machine travelling speed	20 mm/s
Probe scanning speed	2 mm/s
STR scanning points	step width = 3 mm
Fitting technique	LSQ

The maximum permissible error (MPE_E) of CMM machine can be moderated according to the ISO 10360-4 using the following equation, where L is the measured length in mm [21].

$$MPE_E = \pm [0.9\ \mu m + (L/350)]\ \mu m \quad (1)$$

3. Traditional measurement method

The traditional method has standard procedure steps using dial gauges in the tile measurement. While, in this research will apply the standard procedures using CMM as alternative technique to avoid the manual errors and increase the speed in measurements. The curvatures of the diagonal, edge, and warpage for the ceramic tiles are measured using suitable measurement strategy of CMM based on the ISO: 10545-2: 1995 [16], EOS: 3168-2: 2007 [17] and TIS 2398-2: 2008 [19]. This traditional method is designed as explained in Fig. 1.

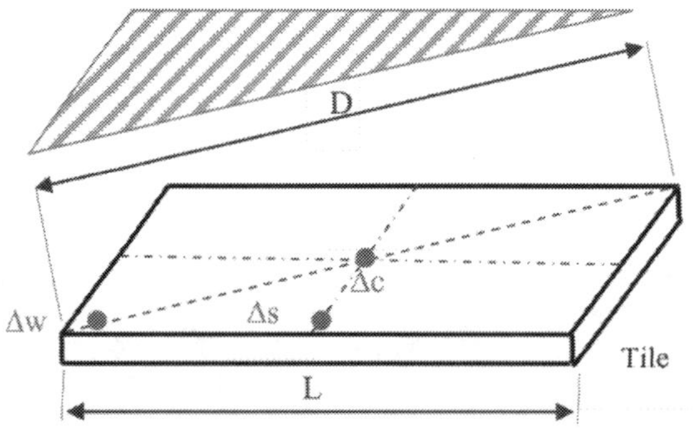

Figure1. The position schematic of the centre points Δs, Δc; Δw and 3 points plane.

D is the length of diagonal for oblong tile. L is the length of the edge for oblong tile. ΔC is the measured vertical difference distance between the centre point of diagonal "D" to the plane of tile. ΔS is the measured vertical difference distance between the centre point of edge "L" to the plane of tile. ΔW is the measured vertical difference distance between the centre point of warpage to the plane of tile. CC is the departure of the centre of one edge of a tile from the plane in which three of the four corner lie. EC is the departure of the centre of a tile from the plane in which three of the four corners lie. W is the departure of the fourth corner of the tile from the plane in which the other corner lie. As shown in Fig. 1, the plane presented in green colour is defined by measuring three points from the four points of the tile surface.

Journal of Physics: Conference Series **483** (2014) 012014 doi:10.1088/1742-6596/483/1/012014

3.1. Experimental results of traditional method

The CMM machine can identify and measure the appropriate plane for three corners and the centre point for each of the diagonal, edge and warpage are identified as obtainable in the ISO-10545-2, 1995 [16] and EOS-3168-2, 2007 [17]. It is then potential to deduce the values of each of ΔC, ΔS and ΔW. Then, can specifying the maximum and average values of measurements for all specimens as illustrated in Tables 2-4.

3.1.1. The first specimen

The measurement results of the traditional method are scheduled in Table 2 in addition to measure the values of both diagonal length "D" and edge length "L" of tiles subject of the study in order to calculate the values of CC, EC and W. The results of length measurements for both D and L are 615.2523 mm and 525.9162 mm respectively. The collection of chosen data can use for the construction of database at NIS and NMIT considering the budget to the industry and available data sources can chosen for building local dataset.

Table 2. The classification results of tile first specimen

Test no.	Measured values ΔC, mm	ΔS, mm	ΔW, mm
1	0.6244	0.6447	0.0012
2	0.6254	0.6462	0.0030
3	0.6262	0.6473	0.0024
4	0.6257	0.6467	0.0022
5	0.6262	0.6473	0.0027
Average	0.6252	0.6464	0.0023
Maximum	0.6262	0.6473	0.0030

Specific calculated parameters, % CC=ΔC/D	EC=ΔS/L	W=ΔW/D
0.11	0.13	0.05

3.1.2. The second specimen

The results of measurements of the second specimen using traditional method are scheduled in Table 3. The measured values for both D and L lengths are of 614.1352 mm and 525.2716 mm respectively.

Table 3. The classification results of second specimen

Test no.	Measured values ΔC, mm	ΔS, mm	ΔW, mm
1	0.8716	0.6038	0.0390
2	0.8787	0.6116	0.0381
3	0.8684	0.6010	0.0257
4	0.8694	0.6017	0.0261
5	0.8696	0.6017	0.0259
Average	0.8715	0.6040	0.0310
Maximum	0.8787	0.6116	0.0390

Specific calculated parameters, % CC=ΔC/D	EC=ΔS/L	W=ΔW/D
0.15	0.12	0.007

3.1.3. The third specimen

The measurement results of this specimen using traditional method are tabulated in Table 4. The results for both D and L lengths measurements are of 613.4891 mm and 524.4620 mm respectively for the third sample. This assessment depends on the estimation of the percentage resulting from dividing ΔC, ΔS and ΔW on the length for each of the D or L to give relative specific values for each: CC, EC and W parameters as recorded above in the results. The results predicted that there is relatively little variation between their values for all measured specimens. Thus, in these cases the specimens have technical acceptance in term of dimensions and surface quality using the traditional method.

Table 4. The classification results of tile third specimen

Test no.	Measured values ΔC, mm	ΔS, mm	ΔW, mm
1	0.6608	0.5723	0.0031
2	0.6511	0.5608	0.0220
3	0.6522	0.5600	0.0225
4	0.6529	0.5606	0.0213
5	0.6529	0.5605	0.0218
Average	0.6540	0.5628	0.0181
Maximum	0.6608	0.5723	0.0225

Specific calculated parameters, %		
CC=ΔC/D	EC=ΔS/L	W=ΔW/D
0.11	0.11	0.04

3.2. Estimation of uncertainty budget

The uncertainties associated with the measurements results are estimated according to GUM [22]. The statistical analyses of *type A* uncertainty of centre point's measurements for the diagonal, edges, and width warpages of the tiles specimens are calculated according to the standard traditional method, the results are presented in Table 5. Where S_D is the experimental standard deviation of five repeated measurements, and u_c is the combined uncertainty due to measurement repeatability ($u_c=S_D/\sqrt{n}$), where n is the number of repeated tests for each target measurement [23].

Table 5. The statistical results of tiles specimens

Specimen	Values	ΔC, mm	ΔS, mm	ΔW, mm
First	S_D	0.0007	0.0011	0.0007
	u_c	0.0003	0.0005	0.0003
Second	S_D	0.0042	0.0044	0.0069
	u_c	0.0019	0.0020	0.0031
Third	S_D	0.0030	0.0053	0.0084
	u_c	0.0017	0.0024	0.0038

The *type B* uncertainty consists of the accuracy of CMM results, MPE, MPE_P of machine and speed of scanning probe respectively. The expanded uncertainty is $U_{exp}=k.u_c$ [23], where k is the coverage factor, which equals to 2 corresponding to 95% confidence level in accordance with the ISO guide to the expression of uncertainty in measurements [23]. Table 6 illustrates the budget of estimated values of the expanded uncertainties for measurements in details. The measurement uncertainties of the first, second and third specimens are 0.98, 3.20 and 3.80μm respectively.

Table 6. Estimation of uncertainty budget in point deviation for tiles measurements

Test Specimen	Sources of uncertainty	Uncertainty, u_c ΔC, μm	ΔS, μm	ΔW, μm	Assumed distribution	Standard uncertainty ΔC, μm	ΔS, μm	ΔW, μm	C
First	Repeatability	0.3	0.5	0.3	Normal	0.15	0.25	0.15	1
	Resolution		0.05		Rectangular		0.029		1
	MPE		0.9+ (L /350)		Normal(based on k=3)		0.30		1
	MPE_P		0.5		Rectangular		0.29		1

Combined Standards Uncertainty, $u_c = \pm 0.49$ µm
Expanded Uncertainty $U_{exp} = \pm 0.98$ µm

Cont. of Table 6

Second	Repeatability	1.9	2.0	3.1	Normal	0.95	1.0	1.55	1
	Resolution		0.05		Rectangular		0.029		1
	MPE		0.9+ (L /350)		Normal(based on k=3)		0.30		1
	MPE_P		0.5		Rectangular		0.29		1
	Combined Standards Uncertainty, $u_c = \pm 1.6$ µm								
	Expanded Uncertainty $U_{exp} = \pm 3.20$ µm								
Third	Repeatability	1.7	2.4	3.8	Normal	0.85	1.20	1.90	1
	Resolution		0.05		Rectangular		0.029		1
	MPE		0.9+ (L /350)		Normal(based on k=3)		0.30		1
	MPE_P		0.5		Rectangular		0.29		1
	Combined Standards Uncertainty, $u_c = \pm 1.9$ µm								
	Expanded Uncertainty $U_{exp} = \pm 3.80$ µm								

where u_c is the repeatability in measurement and C is the sensitivity of measurements.

4. New Measurement Method

The new method adopted to select the actual highest values of ΔC, ΔS and ΔW using straightness (STR) form measurements of all diagonals, edges and widths lengths, so that we can verify the credibility and accuracy of the result performance. The proposed measurement method has been designed in more details as shown in Fig. 2.

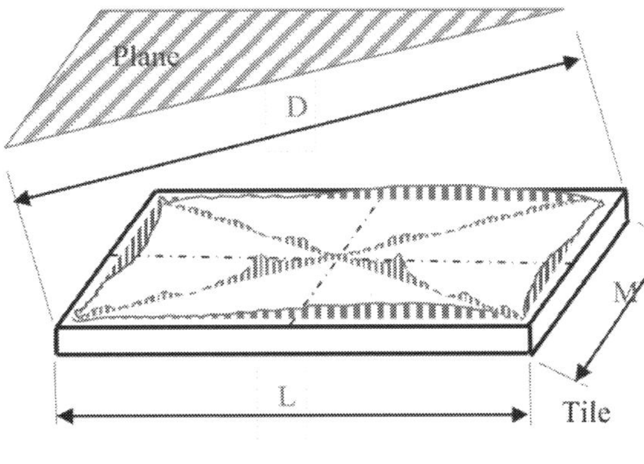

Figure 2. The positions of STR measurement for D, D_1, L, L_1, M; M_1 and 3 points plane

Where, D and D_1 are the diagonals of oblong tile. L and L_1 are the two length edges of oblong tile. M and M_1 are the two widths of oblong tile. ΔC is the achieve value of maximum centre curvature of oblong tile. ΔS is the achieved value of maximum edge curvature of oblong tile. ΔW is the achieve value of maximum warpage of oblong tile. Calculated value of CC parameter represents the maximum curvature expressed as a percentage of the length of diagonal "D" of tile. EC parameter indicates the maximum edge curvature expressed as a percentage of the edge length "L" of oblong tile. While the W parameter represents the maximum warpage expressed as a percentage of the length of diagonal "D" for oblong tile.

4.1. Experimental results of new measurement method

It possible to identify and measure the appropriate plane of three points using CMM machine made as indicated in ISO 10545-2. However, the new here is that there was potential to the replacement of the selected point for each of the diagonal, edge and warpage by measuring the straightness forms of each of the diagonal, edge and width for warpage extension lines as presented in Figs 3-5. It is then possible to obtain the actual values of ΔC, ΔS and ΔW. The maximum and the averagee values of straightness measurement for each specimen are presented in Tables 7-9.

4.1.1. The first specimen

Figure 3 shows the sample behaviour of straightness profiles of one from five measured results along the length of the diagonals D and D_1, length of the edges L and L_1; length of the width M and M_1 of the first ceramic tiles specimen. The results of new measurement method showed that the maximum values for the STR is not at the centre as shown in Fig. 3. While the total results of the first specimen using new measurement method are presented in Table 7 in order to calculate the relative specific values of CC, EC and W parameters.

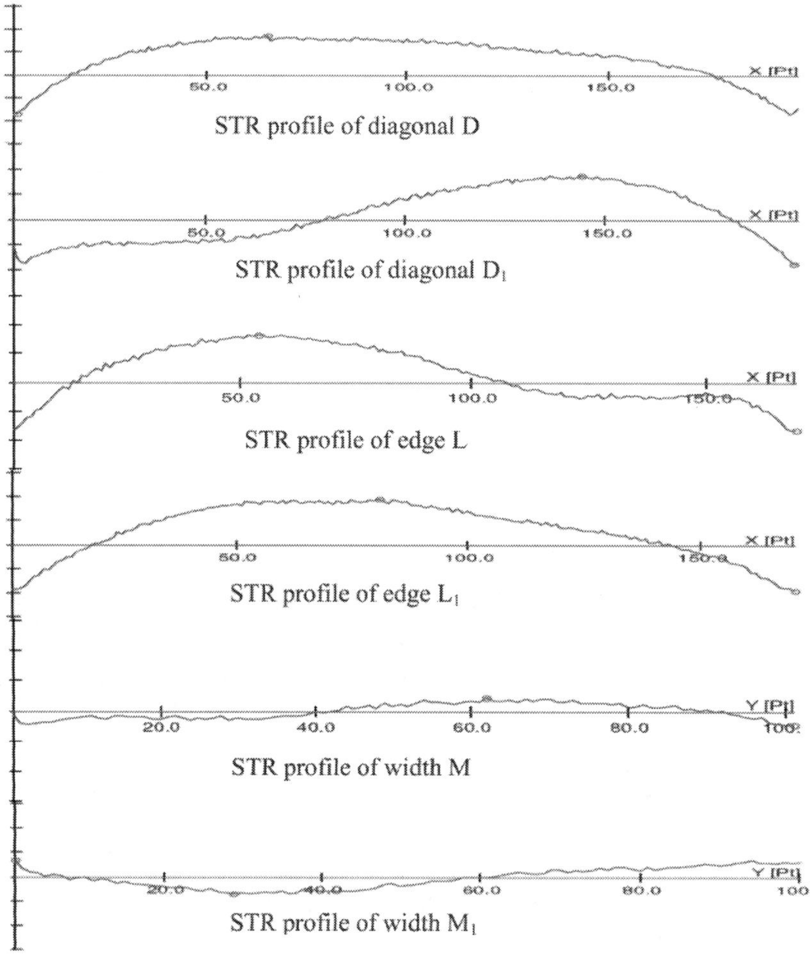

Figure 3. Sample result of STR for D&D1, L&L1; M&M1 lines for tile first specimen using new method

Table 7. The classification results of tile first specimen using new measurement method in mm

Test no.	Measured values of ΔC		Measured values of ΔS		Measured values of ΔW	
	@ D	@ D_1	@ L	@ L_1	@ M	@ M_1
1	0.7321	0.8357	0.7713	0.8275	0.2141	0.3453
2	0.7318	0.8359	0.7715	0.8272	0.2140	0.3456
3	0.7322	0.8361	0.7715	0.8270	0.2139	0.3458
4	0.7319	0.8359	0.7714	0.8267	0.2138	0.3462
5	0.7320	0.8356	0.7706	0.8261	0.2137	0.3461
Average	0.7320	0.8358	0.7713	0.8269	0.2139	0.3458
Maximum	0.8361		0.8275		0.3462	

Specific calculated parameters, %		
CC=ΔC/D	EC=ΔS/L	W=ΔW/D
0.14	0.16	0.06

4.1.2. The second specimen

Fig. 4 shows the behaviour of straightness form samples of the measured result along the length of the diagonals D and D_1, length of the edges L and L_1; length of the width M and M_1 of this specimen from ceramic tiles. The sample profiles show that the maximum values of the STR form are normally not in the middle of lines. While, the total result of the second specimen using the new measurement method that presented in Table 8 including the calculated percentage specific values of parameters CC, EC and W.

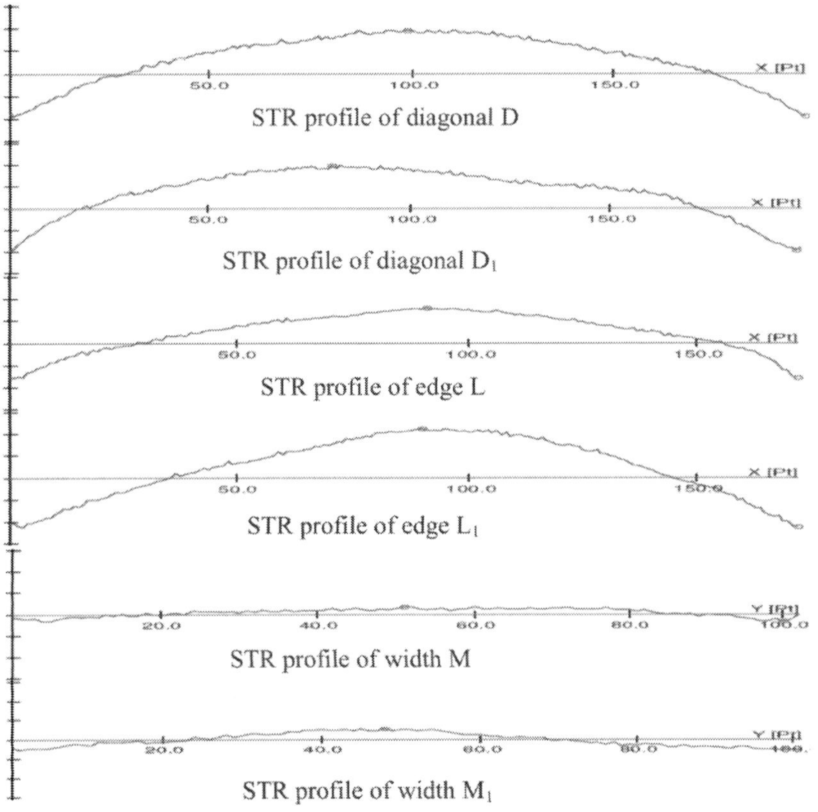

Figure 4. Result of STR form for D&D1, L&L1; M&M1 lines of second specimen in the new method

Journal of Physics: Conference Series **483** (2014) 012014 doi:10.1088/1742-6596/483/1/012014

Table 8. The classification results of tile second sample using new measurement method in mm

Test no.	Measured values of ΔC		Measured values of ΔS		Measured values of ΔW	
	@ D	**@ D_1**	**@ L**	**@ L_1**	**@ M**	**@ M_1**
1	0.7814	0.8375	0.6773	0.9625	0.1453	0.2481
2	0.7812	0.8377	0.6767	0.9625	0.1451	0.2478
3	0.7812	0.8377	0.6771	0.9626	0.1454	0.2480
4	0.7808	0.8377	0.6772	0.9623	0.1453	0.2478
5	0.7805	0.8374	0.6768	0.9625	0.1453	0.2477
Average	0.7810	0.8376	0.6770	0.9625	0.1453	0.2479
Maximum	0.8377		0.9626		0.2481	

Specific calculated parameters, %		
CC=ΔC/D	**EC=ΔS/L**	**W=ΔW/D**
0.14	0.18	0.04

4.1.3. The third specimen

Figure 5 represents straightness profiles of the measured results along the length of the diagonals D and D_1, length of the edges L and L_1; length of the width M and M_1 of this specimen from ceramic tiles. The results of the new measurement method show that the maximum values for the STR form are not in the centre of the line. The results of the third specimen using the new measurement method are presented in Table 9 to be used for calculation the relative specific values of CC, EC and W parameters.

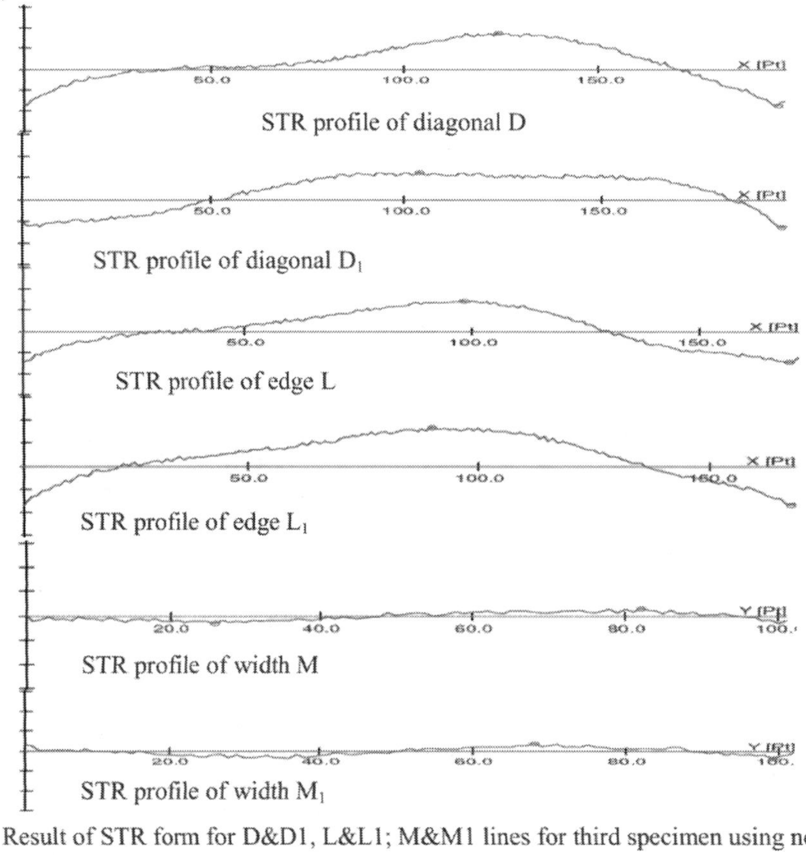

Figure 5. Result of STR form for D&D1, L&L1; M&M1 lines for third specimen using new method

Table 9. The classification results of tile third sample using new measurement method in mm

Test no.	Measured values of ΔC, mm		Measured values of ΔS, mm		Measured values of ΔW, mm	
	@ D	@ D_1	@ L	@ L_1	@ M	@ M_1
1	0.7413	0.6448	0.6227	0.6896	0.1307	0.1626
2	0.7403	0.6450	0.6224	0.6895	0.1307	0.1629
3	0.7399	0.6445	0.6212	0.6894	0.1306	0.1627
4	0.7400	0.6448	0.6212	0.6892	0.1309	0.1630
5	0.7392	0.6449	0.6208	0.6883	0.1309	0.1629
Average	0.7401	0.6448	0.6217	0.6892	0.1308	0.1628
Maximum	0.7413		0.6896		0.1630	

Specific calculated parameters, %		
CC=ΔC/D	EC=ΔS/L	W=ΔW/D
0.12	0.13	0.03

It has become clear that the rate ΔC, ΔS and ΔW to the length of D or L that gives a final relative percentage for CC, EC and W parameters are more accurate than those obtained from the traditional method. While the performance of characterization of results indicate that there is also relatively little variation between these values, but in all cases, the tile specimens are technically acceptable in terms of dimensions and surface quality using the new method. Moreover, the straightness measurement apiaries continuous clear lines without cut, this ensures that there is no edge crack defects in measured tiles, Figs 3-5. These confirm can not be achieved when using conventional measurement method without using the new method.

4.2. Uncertainty budget of the new method

The statistical analyses part of uncertainty *(type A)* of centre point's measurements for the diagonal, edges, and width warpages of the tiles specimens are evaluated and presented in Table 10. Table 11 gives the estimated values of the expanded uncertainties of the proposed method. The estimated values of expanded uncertainty (*Uexp*) of first, second and third measured samples using the new method are 0.86, 0.86 and 0.92 μm respectively.

Table 10. The statistical results of tiles specimens using new method

Specimen	Values	ΔC, mm	ΔS, mm	ΔW, mm
First	S_D	0.0002	0.0005	0.0004
	u_c	0.0001	0.0002	0.0002
Second	S_D	0.0004	0.0001	0.0001
	u_c	0.0002	0.0001	0.00005
Third	S_D	0.0008	0.0008	0.0001
	u_c	0.0003	0.0004	0.0001

Table 11. Estimation of uncertainty budget in straightness deviation for tiles measurement

Test Specimen	Sources of Uncertainty	Uncertainty, u_c			Assumed Distribution	Standard Uncertainty			*C*
		ΔC, µm	ΔS, µm	ΔW, µm		ΔC, µm	ΔS, µm	ΔW, µm	
First	Repeatability	0.10	0.20	0.20	Normal	0.05	**0.10**	0.10	1
	Resolution		0.05		Rectangular		0.029		1
	MPE		0.9+ (L/350)		Normal (based on k=3		0.30		1
	MPE_P		0.5		Rectangular		0.29		1
	Combined Standards Uncertainty, u_c= ±0.43 µm								
	Expanded Uncertainty U_{exp}= ±0.86 µm								
Second	Repeatability	0.20	0.10	0.05	Normal	**0.10**	0.05	0.03	1
	Resolution		0.05		Rectangular		0.029		1
	MPE		0.9+ (L/350)		Normal (based on k=3		0.30		1
	MPE_P		0.5		Rectangular		0.29		1
	Combined Standards Uncertainty, u_c= ±0.43 µm								
	Expanded Uncertainty U_{exp}= ±0.86 µm								
Third	Repeatability	0.30	0.40	0.10	Normal	0.15	**0.20**	0.05	1
	Resolution		0.05		Rectangular		0.029		1
	MPE		0.9+ (L/350)		Normal (based on k=3		0.30		1
	MPE_P		0.5		Rectangular		0.29		1
	Combined Standards Uncertainty, u_c= ±0.46 µm								
	Expanded Uncertainty U_{exp}= ±0.92 µm								

5. Comparison between traditional and new methods

Despite the success of the acceptance for tile specimens after the implementation traditional and proposed methods, but it still shows that there is clear difference in the values of the results of the specific parameters CC, EC and W. This is because the amount of difference of the measured ΔW parameter was very large, it reaches a value more than ten times, while the achieved different in the measured values of the ΔC and ΔS parameters was up to 10% in the new method compared to the old method. Therefore, the values of the ratios CC and EC and W parameters in the new measurement method has achieved a significant increase than the traditional method in measurement. Fig. 6 clearly shows that the maximum value of ΔS is equal to 0.6116 mm using the traditional method rather than the value of 0.9626 mm using the new method. Thus, the new measurement method showed the largest real value to the highest point and this was not achievable when using the old method. This means, applying the new method using straightness measurements, the actual characteristics of surface can be known.

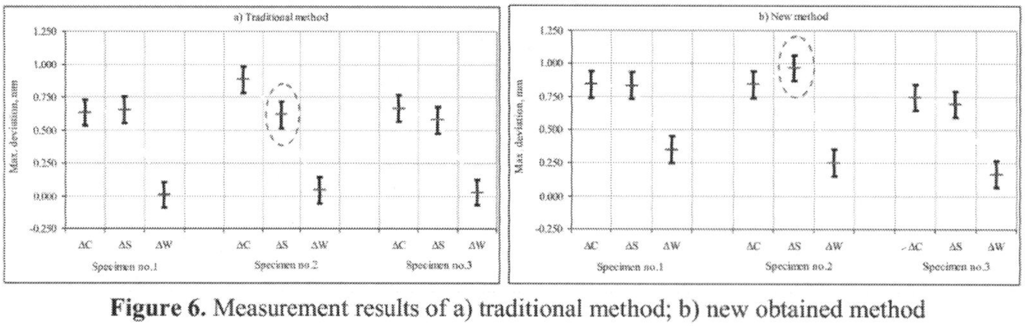

Figure 6. Measurement results of a) traditional method; b) new obtained method

Figure 7 indicates the deviation range in the percentage values of specific parameters between traditional and new measurement methods. It was found that the average percentage values of CC, EC and W parameters are the biggest when we use the new measurement method. This positive difference resulted due to the used proposed method in measuring values of ΔC, ΔS and ΔW and able to reflect the accurately status of ceramic tile surfaces. In additions to the ability of the proposed method is more accurate than the traditional one and the reason is due to the reach of the highest point on the lengths of diagonal, edge and warpage after the use of more precise measurements. Moreover, it can clear that the CMM machine has highly proved in the measurement of oblong large-scale ceramic tiles than the vision inspection methods [6-9].

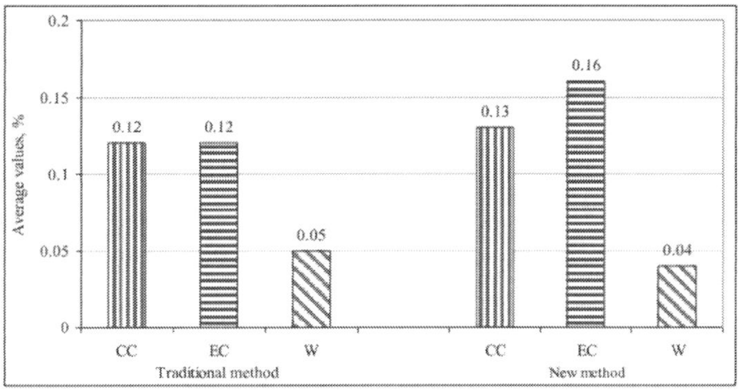

Figure 7. Results of important specific parameters using traditional method and new method

On the other hand, using analytical comparison between the new and traditional measurement methods in terms of average values of the parameters and the uncertainty associated with measuring the repeated five times of three samples of tiles, it can be seen some observations. It was found that the accuracy of the new measurement method to old method up to more than 12% in the measurements of first tile sample, up to more than 73% in the second tile and up to about 76% in the measurements of the third tile respectively. The observed maximum value of expanded uncertainty of the traditional method was larger 4 times than the value derived using the new measurement method, Fig. 8. This means that the value of the associated uncertainty in the results of the new measurement method outweighs the traditional way in all circumstances.

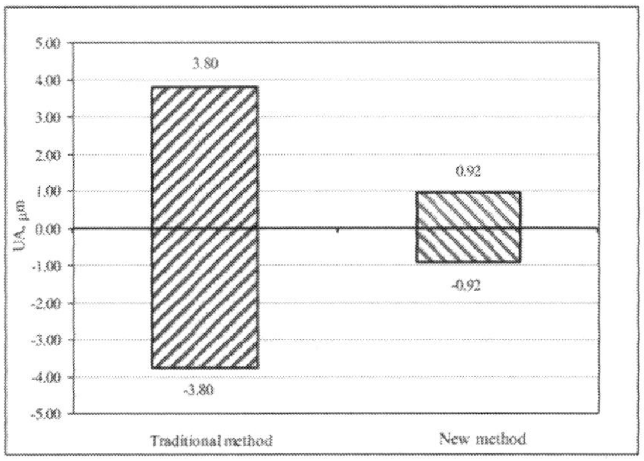

Figure 8. Expanded uncertainty for traditional and new obtained methods

Eventually, the estimation of specific parameters in this work reflects confidence in the high credibility of the new measurement method as shown obviously in Figs 7and 8. Thus, it can say that the method of new measurement is a closer to reality and very much more accurate than the traditional measurement method. In addition, a comparison to illustrate the extent of this improvement in the accuracy of measurement using the new method compared to the traditional way for each tile separately. Moreover, the new measurement method are able to prove the capability of computer numerical control devices as a high efficacy of coordinate measuring machines for evaluation the tile dimensions and surface quality with high accuracy.

6. Concussion

This paper contains analyses and discusses comparison between the traditional method (ISO-10545-2:1995, EOS-3168-2:2007 and TIS 2398-2:2008) and the proposed measurement method that are accepted for quantifying dimensions and surface quality of large-scale ceramic tiles. A number of experimental measurements have been carried out using new strategically method on CMM. The discussions highlighted many differences in the measured values obtained by the two measurement methods. This is very important considering that the suggested measurement method is suitable for use to define the maximum centre curvature (CC), the maximum edge curvature (EC), the maximum warpage (W) and the edge crack defects specifications. Based on the experimental measurements, a relationship between the parameters obtained by the CC, EC and W has been achieved. The presented results can be useful to define the changes in the dimensions and surface of large-scale ceramic tiles specifications. This is due to the confirmed of a new measurement test method as follows:

1- The capability of the new method in the straightness form quality of large-scale ceramic tiles, CC, CE, W and crack defect can be characterized in high accuracy. The actual values of tile dimensions and surface can be measured.
2- Not always the maximum curvature point is at the centre of diagonal, edge or width length as defined in the traditional method.
3- The CMM machine has excellent capability to measure the dimensions and surface quality of tiles with more creditability and better accuracy than the vision inspection methods.
4- The new method procedure is suitable for ceramic, porcelain or any type of tiles and plates measurement.
5- The measurement uncertainty estimated proved the accuracy and preciseness of the proposed measurement method compared to traditional method.

Consequently, the proposed method has excellent capability to assist any scale of tiles and plates, not only has better accuracy and precision, but also permit improvement of the plate quality productions.

7. References

[1] Salah H.R. Ali, 2012. *Advanced Nanomeasuring Techniques for Surface Characterization. Int. Scholarly Research Network, ISRN Optics Journal, Vol. 2012, pp. 1-23, Article ID 859353.*

[2] H. Elbehiery, A. Hefnawy and M. Elewa, 2005. *Surface Defects Detection for Ceramic Tiles using Image Processing and Morphological Techniques, World Academy of Science, Engineering and Technology*

[3] G. Salemi, V. Achilli, M. Ferrarese and G Boatto, 2008. *High Resolution Morphometric Reconstruction of Multimaterial Tiles of an Ancient Mosaic, The Int. Archives of the Photogrammetry, Remote Sensing and Spatial Information Sciences, Vol. XXXVII, Part b5, pp.303-307.*

[4] L. Blunt and L. T. Brown, 2005. *Surface Texture Analysis of Ceramic Tiles Treated with Anti-Slip, Report Ref: CPT 220305.*

[5] M. Mansoory, H. Tajik, G. Mohamadi, and M. Pashna, Dec. 2008. *Edge Defect Detection in Ceramic Tile Based on Boundary Analysis using Fuzzy Thresholding and Radon Transform, Int. Symposium in Signal Processing and Information Technology (ISSPIT), IEEE, pp.58-62.*

[6] Zeljko Hocenski, K. Sobol and R. Mijakovi, 2010. *LED Panel Illumination Design of a Control System for Visual Inspection of Ceramic Tiles, Int. Symposium in Industrial Electronics (ISIE), IEEE, pp.1663-1667.*

[7] J. He, L. Shi, J. Xiao, J.Cheng and Y. Zhu, 2010. *Size Detection of Firebricks based on Machine Vision Technology, Int. Conference on Measuring Technology and Mechatronics Automation (ICMTMA), pp.394-397.*

[8] Ehsan Golkar, Ahmed Patel, Leila Yazdi and Anton Satria Prabuwono, 2011. *Ceramic Tile Border Defect Detection Algorithms in Automated Visual Inspection System, Journal of American Science, Vol.7 (6), pp.542-550.*

[9] H. M. Elbehiery, A. A. Hefnawy and M. T. Elewa, 2005. *Visual Inspection for Fired Ceramic Tile's Surface Defects using Wavelet Analysis, Int. Journal on Graphics, Vision, and Image Processing (GVIP), ICGST, pp.67-74.*

[10] Zeljko Hocenski and Emmanuel Karlo Nyarko, 2002. *Ceramic Tiles Quality Inspection using Statistical Methods and Neural Networks Approach, The World Scientific and Eng. Academy and Society, (WSEAS), pp. 921-926.*

[11] Federica Bondioli, Tiziano Manfredini, Michele Giorgi and Graziano Vignali, January 2010. *Functionalization of ceramic tile surface by soluble salts addition: Part I. Titanium and silver addition, Journal of the European Ceramic Society, Vol. 30, Issue 1, pp. 11-16.*

[12] Federica Bondioli, Martina Dinelli, Roberto Giovanardi and Michele Giorgi, July 2010. *Functionalization of ceramic tile surface by soluble salts addition: Part II. Titanium and silver addition, Journal of the European Ceramic Society, Vol. 30, Issue 9, , pp. 1873-1878.*

[13] Teresa P. Silva, Maria-Ondina Figueiredo, Maria-Alexandra Barreiros and Maria-Isabel Prudêncio, 2013. *Decorative 18thCentury Blue-and-White Portuguese Tile Panels: A Type-Case of Environmental Degradation, Hindawi Publishing Corporation, Journal of Materials, Volume 2013, Article ID 972018, pp.1-6.*

[14] Nachawit Tikul and Panya Srichandr, 2010. *Assessing the environmental impact of ceramic tile production in Thailand, Journal of the Ceramic Society of Japan, Vol.10, pp. 887-894.*

[15] Mary A. Curran, Jonathan G. Overly, P. Hofstetter, R. Muller and Barbara C. Lippiatt, 2002. *Building for Environmental and Economic Sustainability Peer Review Report. National Institute of Standards and Technology (NIST), NISTIR 6865, BEES 2.0, USA.*

[16] ISO-10545, Part 2: 1995. *Determination of Dimensions and Surface Qualities of Ceramic Tile.*

[17] EOS-3168, Part 2: 2007. *Determination of Dimensions and Surface Qualities of Ceramic Tiles, Egyptian Organization of Standards and Quality.*

[18] ASTM-C485: 2009. *Standard Test Method for Measuring Warpage of Ceramic Tile.*

[19] TIS 2398, Part 2: 2008. *Determination of Dimensions and Surface Qualities of Ceramic Tiles, Thai Industrial Standards Institute, Thailand.*

[20] David Flack, July 2001, UK. *Measurement Good Practice Guide No. 42, CMM Verification, National Physical Laboratory, ISSN: 1368-6550.*

[21] ISO-10360, Part 4: 2000. *Geometrical Product Specifications (GPS) - Acceptance and Reverification Tests for Coordinate Measuring Machines (CMM), CMMs used is Scanning Measuring Mode.*

[22] ISO/IEC-98, Part 3, 2008. *Uncertainty of measurement -- Part 3: Guide to the expression of uncertainty in measurement (GUM:1995).*

[23] Keith Birch, March 2003. *Measurement Good Practice No. 36: Estimating Uncertainties in Testing, an Intermediate Guide to Estimating and Reporting Uncertainty of Measurement in Testing, British Measurement and Testing Association, NPL, UK.*

Journal of Physics: Conference Series **483** (2014) 012015 doi:10.1088/1742-6596/483/1/012015

Metrological characterization of optical confocal sensors measurements (20 and 350 travel ranges)

H Nouira[1], N El-Hayek[1,3], X Yuan[1], N Anwer[2] and J Salgado[1]

[1] Laboratoire Commun de Métrologie (LNE-CNAM), Laboratoire National de Métrologie et d'Essais (LNE), 1 Rue Gaston Boissier, 75015 Paris, France,
[2] Ecole Normale Supérieure de Cachan, The University Research Laboratory in Automated Production, 61 avenue du Président Wilson, 94235 Cachan, France.
[3] Arts et Métiers de Lille (ENSAM), Laboratory of information sciences and systems (LSIS), 8 Boulevard Louis XIV, 59046 Lille, France,
E-mail: hichem.nouira@lne.fr

Abstract. Confocal sensors are usually used in dimensional metrology applications, like roughness, form, thickness and surface profile measurements. With the progress of technologies, metrological applications require measurements with nanometer-level of accuracy by using ultra-high precision machines, which should present a minimum and stable metrology loop. The loop is equipped with sensors with nanometer-level of resolution and linear residual. The study presented here, is mainly focused on the characterization of Confocal sensors in order to identify their performance practically. Such information is useful to establish a correction model in the digital signal processing (DSP) software. In this context, LNE developed an ultra-high-precision machine, dedicated to the roughness measurement with an uncertainty of a few nanometres (< 30 nm) by using a tactile sensor. In order to match this machine to Confocal sensors, an experiment has been recently developed to characterize the behaviour of two commercial Confocal sensors with the measuring range of 20 µm and 350 µm. The experiment permits the evaluation of the major error sources: axial and radial motion errors as-well-as the deviation/tilt of the sensors.

1. Introduction

Confocal chromatic probes are widely used in mechanical industry and prove to operate at fast rates to measure any error of form, roughness, profile and thickness, as well as they are distinctively transparent to surface deformation.

The working principle of confocal chromatic probes for surface profiling is reported in [1-3]. We usually deal with a white light source (LED), in which the spectral components are focused at different distances along the optical axis. The reflected light is collected and analyzed with a spectrometer (diffraction grating).

The large use of confocal probes creates the need for a serious study of their limitations and performances. The calibration of such probes is usually achieved by comparing the confocal probe's measurement with a second measurement serving as a reference. Boltryk et al [4-5] conducted experiments and compared optical measuring techniques on a cylindrical artefact with sinusoidal grooves. Confocal and triangulation laser probes are tested along with a stylus profilometer as a reference for comparison. Results reveal that confocal probe measurements are slower but more accurate than triangulation probes.

This paper summarizes the characterization of two confocal probes with a measuring range of 20 and 350 µm. A test bench was specifically designed to ensure that measurements exhibit a nanometric level of accuracy. The linear residuals of the confocal chromatic probes are investigated with different materials and colour paints. The effect of the inclination, roughness, speed and error related to the artefact geometry are also detailed.

2. LNE ultra-high precision profilometer for 3D measurement

The 3D apparatus performs three independent motions (x, y and z) generated by three independent high-precision mechanical guiding systems equipped with encoders. The horizontal (x and y) guiding systems support the table made of zerodur material, and is able to move along both the x- and y-axes in dynamic or static operation. The vertical guiding system moves the structure supporting the measuring probe along the z-axis. Each x- and y-motion is controlled by a laser interferometer in order to locate the part at the nanometre level of accuracy. The z-motion is controlled by a differential laser interferometer to shorten the metrology loop [6]. The supporting structure is made of massive granite materials and support all the guiding systems. The metrological structure, represented with blue colour in Figure 1(a), is made of Invar material and is less sensitive to the environment and to structure thermal deviation. Any vertical variation of the main granite structure (1) by both thermal drift and its own mass slightly influences the metrological structure with few nanometres. This variation is indirectly compensated by the differential laser interferometer.

The architecture of the machine perfectly respects the Abbe principle [7] along all the axes. The measuring single scanning probe is collinear with the differential laser beam (2) and represents the "measuring Abbe Z-axis". The horizontal x- and y-laser interferometer beams and the touching element of the tactile probe are designed to be in the same XY-plane and axis (Abbe x- and y-axes). These Abbe axes never change because the measuring probe and the x- and y-laser interferometers are fixed on the same Invar metrological structure. By that, any vertical motion of the metrological structure induces a similar vertical motion of the x- and y-laser interferometers and the measuring probe. Each laser interferometer is fixed on an intermediate mechanical structure to ensure two motions and one rotation of the laser head in order to adjust the alignment of the laser beam on each respective mirror.

The entire x- and y-working range is equal to 50 mm. The z-working range can reach 100 mm. In-situ calibration of the measuring probe can be achieved by moving the metrological structure which contains the tactile probe along the z-axis with perfect respect of the Abbe principle.

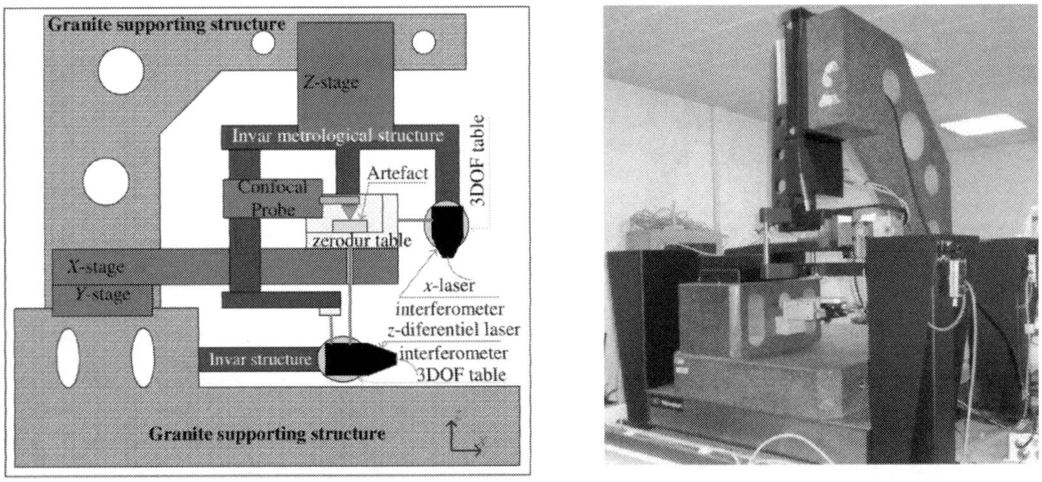

Figure 1. (a): Architecture of the ultra-high precision profilometer, (b): Photograph of the apparatus

The performance of the LNE profilometer depends on both the behaviour of the measuring probe and the quality of the mechanical guiding systems, which is reduced by using the reference laser interferometers. The current configuration of the machine allows measurements of the target with a tactile single scanning probe. For a working range of the tactile probe of 150 µm the evaluated uncertainty is estimated to 30 nm.

In some applications such as the metrological characterization of optical aspherical lenses with an accuracy of few tenths of nanometres, it is interesting to use an optical scanning probe instead of a tactile probe in order to avoid any deformation or scratching of the part's surface coating. In this way, two optical confocal scanning probes are selected for the application and need to be qualified. According to the behaviour of the such probes, corrections have to be taken into account in the software.

3. Error sources in displacement measuring confocal probes

The main degrees of freedom that impact the measurement and are worth being evaluated are:

- translation (T_x) along the x-axis: it corresponds to the axial motion of the confocal probe with respect to a flat artefact. The evaluation of T_x allows the characterization of both the linearity and the sensitivity of the probe.
- translation (T_y) along the y-axis: this degree of freedom corresponds to the transverse motion of the confocal probe axis from the generator of the cylindrical artefact. The evaluation of T_y gives an idea about the influence of the transverse motion error on the measurement.
- rotation (R_y) along the y-axis: this degree of freedom corresponds to the inclination of the confocal probe axis with respect to the horizontal orientation by an angle of φ.

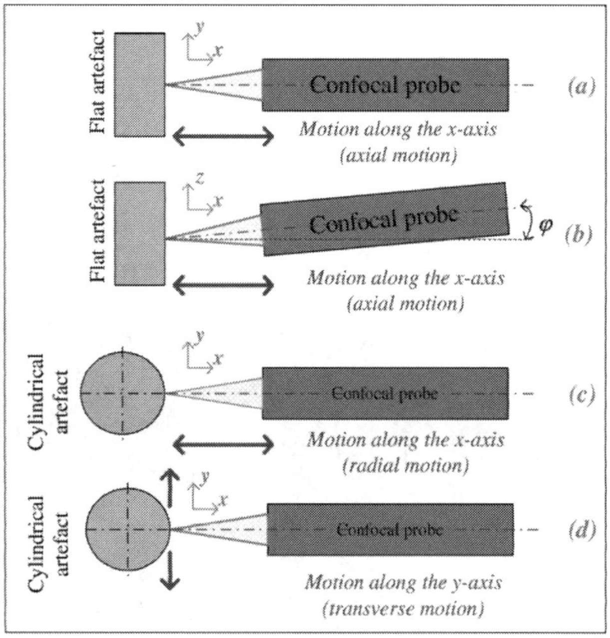

Figure 2. Description of the different tests

The interaction probe-material, roughness, speed and geometry of the artefact impact the behaviour of the probe, and are investigated here. To complete all these tests, four experiment configurations are devised as described in Figure 2(a, b, c and d). Note that when using flat artefacts, the motions along the x and y-axes are named respectively axial and transverse motions. When using cylindrical artefacts, the motions along the x, y and z-axes are named respectively radial, transverse and axial motions.

4. Description of the experimental test bench

The design of the test bench ensures two independent motions along the x and y-axes and two rotations around the z and y-axes. Along the x-axis two independent mechanical motions can be generated by using two independent and superposed mechanical stages. The first one produces a micrometric motion step over 63 mm range. The second stage incorporates a piezoelectric actuator, which is used with an electronic feedback, and a mechanical structure based on flexible blades to limit any hysteresis phenomenon. It generates nanometric motion steps of 5 nm over 90 μm entire range. A precision mechanical linear stage equipped with a stepper motor ensures the motion T_y over 50 mm range with a minimum incremental motion of 0.1 μm.

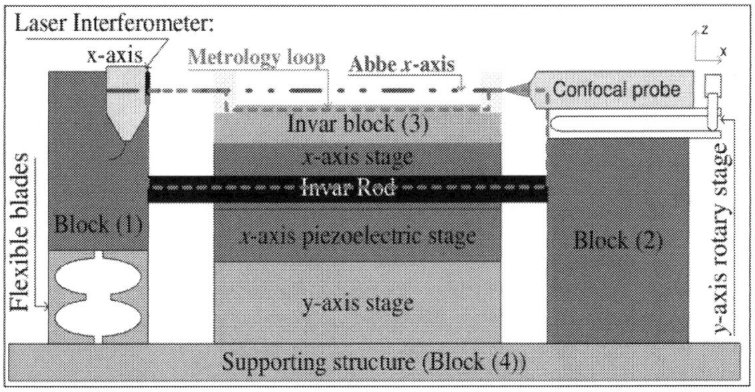

Figure 3. Architecture of the test bench

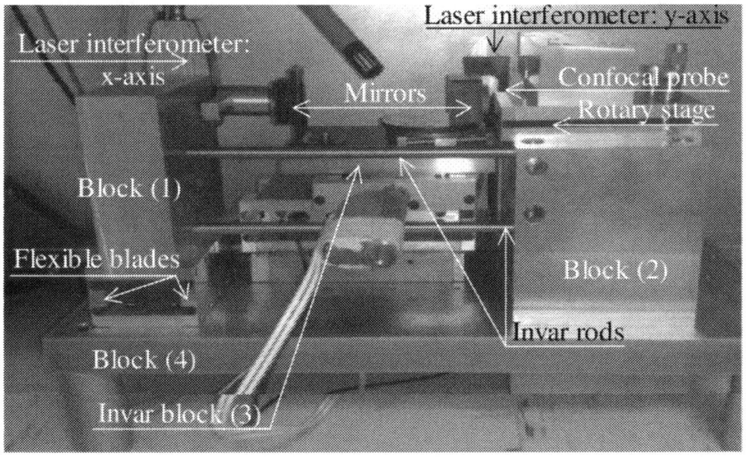

Figure 4. Photograph of the experimental test bench

A manual rotary stage based on flexible blades guarantees the rotation R_y of the confocal probe over a working range of ±3° and a second manual rotary stage ensures the rotation R_z of the target over a working range of 180° (Figure 3 and Figure 4).

The design of the test bench applies the Abbe principles along both the x- and y-axes. The Abbe x-axis is mingled with the x-axis laser beam in correspondence with the confocal probe's axis when perfectly oriented horizontally. The Abbe y-axis is collinear to the y-axis laser interferometer beam.

Two laser interferometers with sub-nanometric resolution are used to measure the generated displacements T_x and T_y of two Zerodur plane mirrors. These mirrors have an overall flatness of $\lambda/20$. The perpendicularity of the mirrors surfaces to their respective laser beam axes is insured by a nanometric sensitive contact probe. The parallelism of the confocal chromatic probe and the x-axis laser, as well as the perpendicularity between the x-axis and y-axis laser interferometers are carried out carefully by CMM measurements.

The architecture of the test bench perfectly applies the dissociated metrological structure principle [6]. Block (1) supporting the x-axis laser interferometer is decoupled from block (4), made of aluminium, by using flexible blades. Two Invar rods have been mounted to rigidly link both blocks (1) and (2) and shorten the metrology loop [6]. Hence, any thermal expansion and mechanical deformation of block (4) never influences the metrology loop. Block (3) is also made of Invar, which improves the stability of the metrological structure.

The experiment is performed inside the LNE clean room, where both temperature and humidity are controlled to 20±0.3 °C and 50±5 %HR, respectively. The experiment is also enclosed in an aluminium shell and installed on the anti-vibration structure.

All the motions and acquired data are controlled by a developed LabView program. A synchronous record of the laser interferometers, confocal probe, temperature, hygrometry and pressure data are carried-out during the test. The environment parameters are used to introduce correction for the refractive index of air (updated Edlen formula [8] and dead-path [9]).

5. Experimental results

5.1. Metrological Analysis

Two kinds of errors that influence the measurement can be distinguished: systematic errors that can be characterized and compensated, and random errors that are merely random (incorrigible errors). The qualified systematic errors correspond to the thermal expansion of the test bench, thermal variation of the refractive index of air, cosine errors, Abbe errors and resolutions of the laser interferometer and confocal chromatic probe. The analysis of these errors reveals that the resolution of the confocal probe presents the most important contribution of 12 nm. However, the test bench presents a nanometric level of accuracy (<5 nm), when the working range doesn't exceed 90 μm.

5.2. Stability of the 20 μm confocal chromatic probe

To accurately characterize the probe, the stability of the laser interferometer and the stability of the confocal probe are firstly studied over half an hour and reveal ±2 and ±20 nm of variation, respectively (Figure 5).

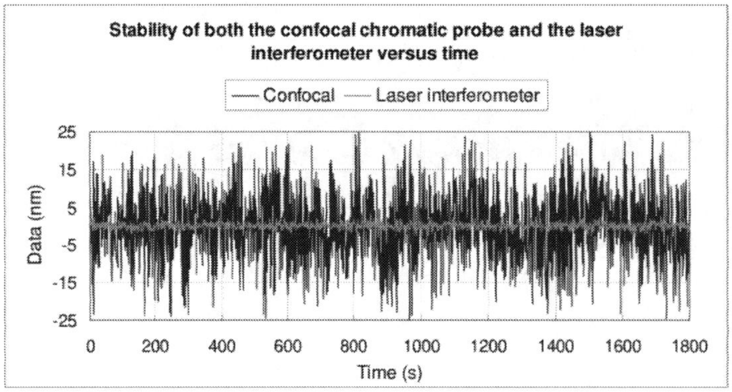

Figure 5. Stability of the confocal probe and x-axis laser interferometer

5.3. Effect of the material and colour on the linearity of the probe

The linearity tests are performed with different kinds of materials (aluminium, steel, gold, copper, silicon and ceramic) with respect to Figure 2(a). They have different absorption and reflection values according to the wavelengths in the white light source. The reflectance values of the selected materials are respectively equal to: 93%, 53%, 92%, 60%, 30% and 95%. From Figure 6, the gold-deposit mirror has large linear residuals, values near the start seem to be "compressed" and the useful range for the gold-deposit mirror is not so large as compared to other materials. Ceramic materials show low linear residuals (less than 40 nm) but are limited to the smallest travel range compared to all tested materials. The other four materials, silicon, copper, steel and aluminium behave approximately the same and exhibit similar linear residuals (between –150 and 200 nm) with a large working range.

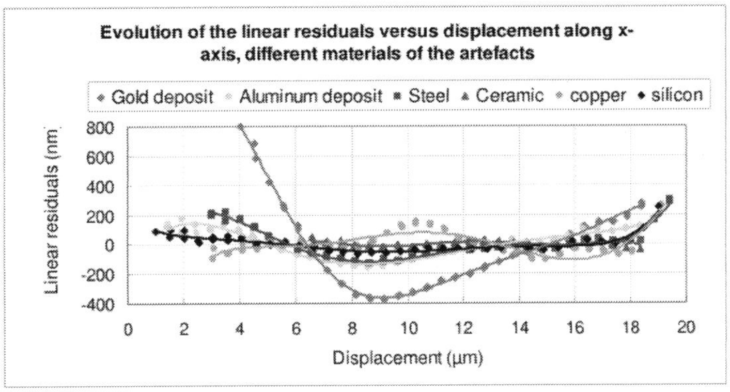

Figure 6. Linear residuals versus displacement along x-axis (material)

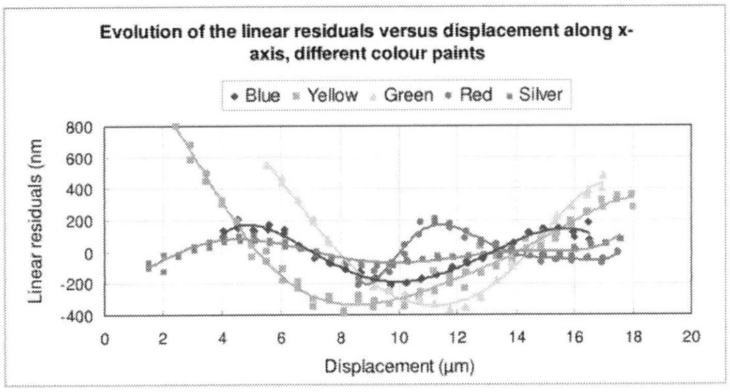

Figure 7. Linear residuals versus displacement along x-axis (colour)

Since the observed behaviours are different in Figure 6, the impact of colour is then investigated. Five colour paints for the same steel material are chosen for these tests: yellow, silver, red, blue and green. The recorded linear residuals are illustrated in Figure 7 and show dissimilar behaviours. For red, all wavelengths are absorbed except the one corresponding to the red colour, that's why there is no signal for the red colour sample from 0 to 9 µm. It means that the 380-500 nm wavelengths, which correspond to the partial range until 9 µm, match with blue and violet colours. For yellow and silver, they behave the same as the gold-deposit and aluminium-deposit mirrors, which proves that a certain amount of wavelengths is absorbed by the tested artefacts. These results confirm that the confocal probe's behaviour depends on the colour rather than on the material.

5.4. Impact of the roughness of the aluminium artefact

The linearity of the confocal chromatic probe is investigated with three selected rough flat artefacts made of aluminium and measured by the LNE ultra-high precision profilometer for roughness measurements. Table 1 gives the average of 16 measurements of the Rz parameter (ISO 4287:1997) and the standard deviation (σ).

Table 1. Roughness of the tested aluminium materials

	Roughness parameter Rz (μm)	Standard deviation σ (μm)
Aluminium deposit	0.020	0.0004
Aluminium R1	2.50	0.15
Aluminium R2	4.40	0.70
Aluminium R3	6.50	0.50

From Figure 8, it is obvious that the effective working range depends on the roughness of the artefact. Then, adding roughness induces a less effective working range. As a consequence, the linear residuals vary according to the change of the effective working range change. This observation is also related to the power density spectrum. For instance, at the beginning of the travel range, the power density could not reach its sufficient level which lead the power spectrometer to fail to detect the signal.

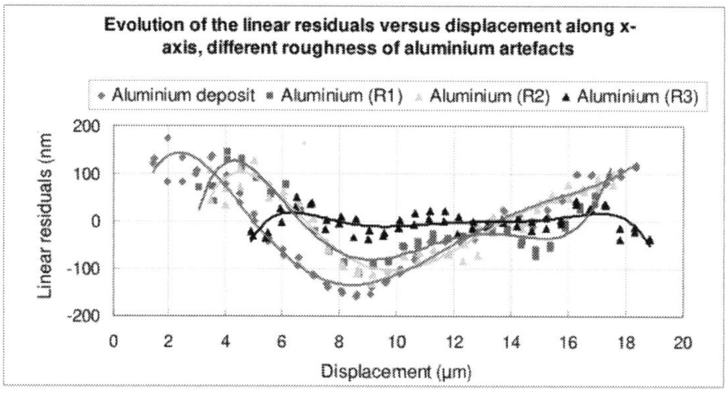

Figure 8. Linear residuals versus displacement along x-axis (roughness).

5.5. Influence of the inclination of the 20 μm confocal probe

The influence of the inclination is investigated here in order to be able to estimate the correction to be taken into account when the probe is focussed on a part with one slope and/or v-groove or/and cylindrical and/or aspherical shapes.

The impact of the angle φ between the probe and the flat artefact is investigated within the range of ±16°, as shown in Figure 2(b). For each test, the parameters of linearization are identified and the linear residual is deduced. Since the results from 0° to –16° are symmetric to the ones between 0° and 16°, only results from 0° to 16° are presented in Figure 9.

At 0°, the linear residuals are between ±150 nm. With the increase in angle, the linear residuals decrease to less than ± 50 nm at 6°, then increase again. Up until 16°, the linear residuals vary in-between -600 and 130 nm. This observation is related to the power density of each wave reflected back into the spectrometer (diffraction grating).

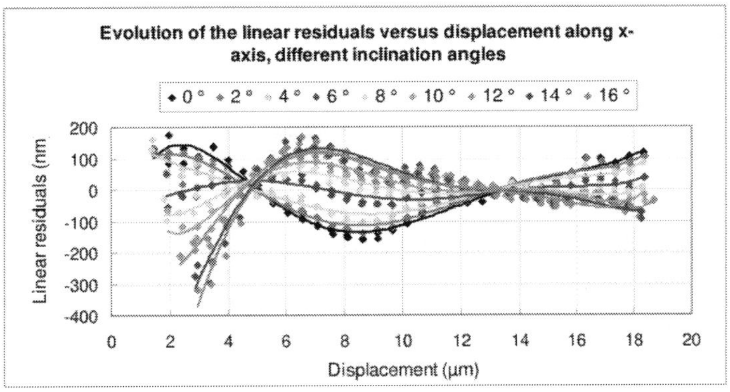

Figure 9. Linear residuals versus displacement along x-axis (inclination).

5.6. Influence of the artefact geometry and transverse offset

The linearity tests are completed when the confocal chromatic probe is well aligned on the generator of the cylindrical artefacts with diameters of 50 and 75 mm as presented in the Figure 2(c).

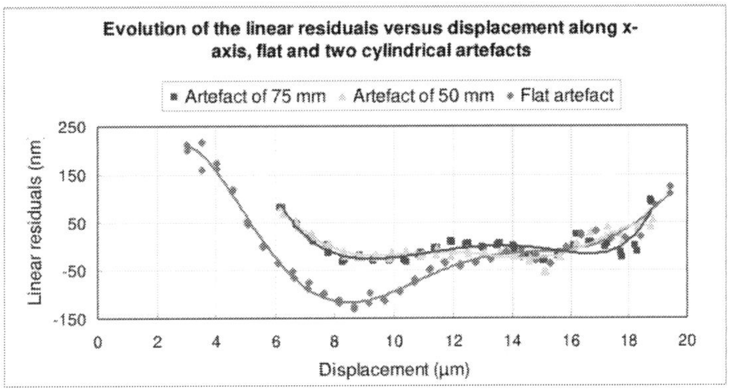

Figure 10. Linear residuals of 50 mm cylinder, 75 mm cylinder and flat artefact versus displacement along x-axis.

From Figure 10, both the working range and linear residuals of the confocal chromatic probe are higher when it is focused on the steel flat artefact. However, if we consider the same measuring range for all tests, the linear residuals perform perfectly the same. It means that the linear residuals variations are due to the effective working range change.

As the confocal probe is integrated into the high-precision profilometer which is equipped with mechanical guiding systems [6], the guided movements are never achieved perfectly and usually present repeatable and un-repeatable transverse, axial and radial error motions. The amplitude of these error motions varies according to the adopted technology for the linear mechanical guiding systems. Therefore, when the artefact presents cylindrical or/and aspherical or/and freeform shapes, the probe has to be qualified before any measurement in order to estimate the effect of errors in relation to the combination of both motion and shape.

Tests are completed by moving the confocal probe, while it is focussed on cylindrical artefacts, along the y-axis and over a range of ± 200 µm. The position corresponding to y = 0 µm represents the probe when it is well focused on the generator of the cylindrical artefact. The experimental results are compared to the theoretical model detailed in [6], obtained when the probe undergoes a transverse motion and is located at position (B) instead of position (A) (Figure 11).

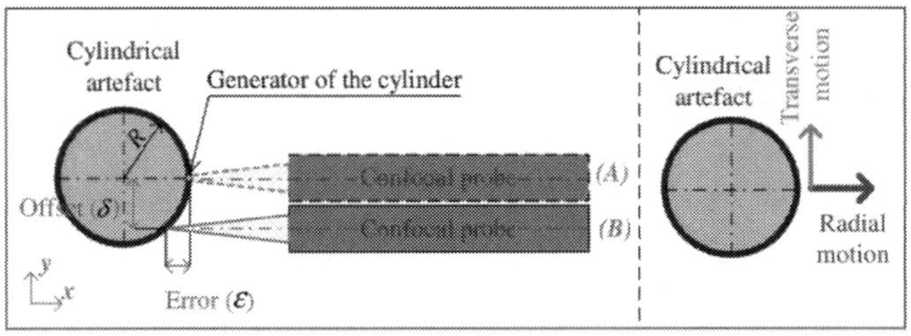

Figure 11. Description of the transverse offset test

To complete the transverse offset tests, the value of the gap is fixed and only the transverse motion is generated as presented in Figure 2(d). The tests are achieved with three values of the gap (distance between the probe and the artefact): 7, 12 and 17 µm, and two cylinder diameters: 50 and 75 mm. The experimental results are compared to the theoretical results obtained with a cylinder diameter of 50 mm and shown in black in Figure 12. Results are presented according to y-displacement measured by the y-axis laser interferometer, perpendicular to the confocal probe's axis. A good agreement between the model and the experiment is observed despite the noise, which is slightly high.

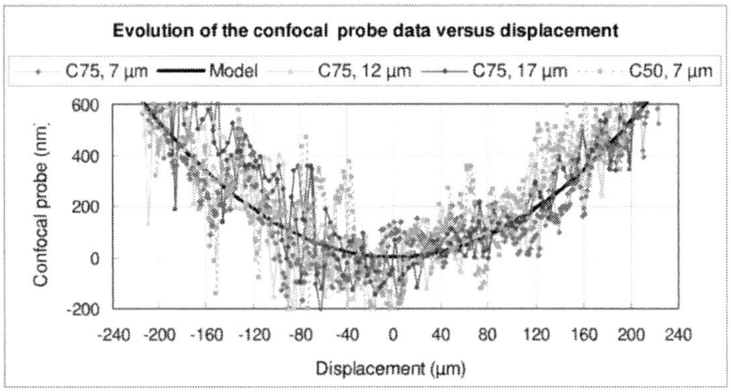

Figure 12. Influence of the artefact geometry on the confocal probe measurement (C75, 7 µm→cylinder diameter of 75 mm, gap of 7 µm) versus displacement along y-axis

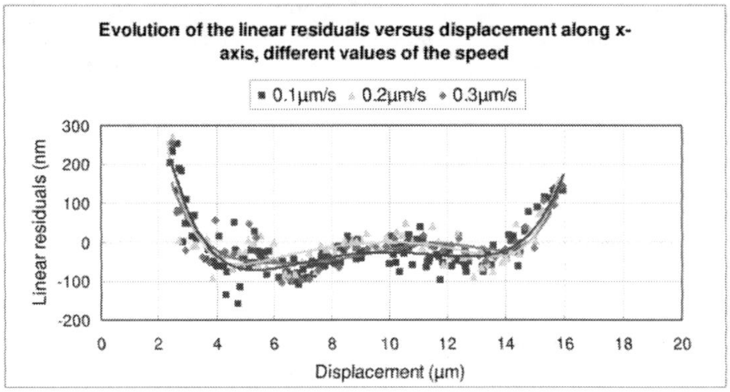

Figure 13. Evolution of the linear residuals when the confocal probe is used in dynamic mode: speed values of 0.1, 0.2 and 0.3 µm/s

On CMM, confocal probes can be used in quasi-static or dynamic modes, that is why the impact of the motion speed of the artefact on the behaviour of the probe is explored. Three values of the speed were selected between 0.1 and 0.3 µm/s. When the motion speed increases, the number of acquisitions deceases. It is the reason why it is impossible to reach high speed values.
The obtained linear residuals are similar (±200 µm) and in the same order as the previous linear residuals obtained when the probe is employed in the quasi-static mode (Figure 6). However, the travel range is reduced and is equal to 14 µm (Figure 13) instead of 17 µm as before.

5.7. Confocal chromatic probe of 350 µm range

The confocal probe of 350 µm range is characterized under similar conditions. Only results concerning linear residuals evolution when changing the part materials is presented her. Figure 14 reveals that the linear residuals vary between ± 600 nm. The behaviour of the confocal probe of 350 µm is summarized in Table.2.

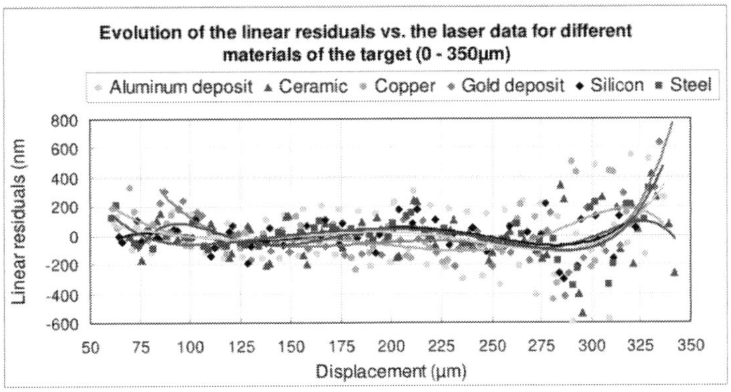

Figure 14. Linear residuals versus displacement along x-axis (material)

Table 2. Comparison of the confocal probe of 20 and 350 µm range

Confocal probe of:	350 µm	20 µm
Stability	100 nm	20 nm
Linear residuals	± 400 nm	± 150 nm
Effect of materials	Not sensitive	Not sensitive
Effect of colours	Not sensitive	sensitive
Effect of roughness	Not sensitive	Not sensitive
Effect of inclinations	Not sensitive	sensitive
Effect of transverse offsets	Not sensitive	Not sensitive
Effect of speeds	sensitive	Not sensitive

The confocal probe of 350 µm is in-situ calibrated again on the ultra-high precision profilometer with different values of the speed. 2048 linear models and parameters of linearization are calculated separately over 2048 partial ranges to cover the whole travel range of 350 µm. The linear residuals are obtained for the entire travel range and results are presented in Figure 15. The linear residuals are less than 14 nm at the beginning of the travel range (0 µm) and decrease to 4 nm at the travel range's end (350 µm).

By using this technique it is possible to reduce the values of the linear residuals to few nanometres which is very beneficial when performing measurements at the nanometric level of uncertainty. However, this solution needs to be able to process a large number of linear models and parameters of linearization in real time.

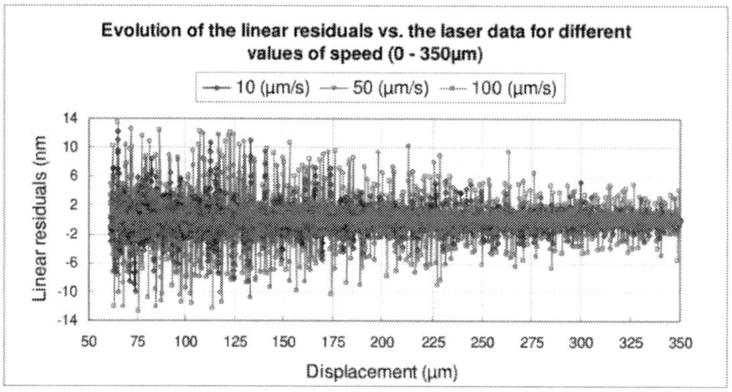

Figure 15. Linear residuals versus displacement along x-axis (speed)

The variation of the speed from 10 to 100 µm/s does not influence the evaluation of the linear residuals, which means that it is possible to perform quick measurements with similar level of accuracy. When using the confocal sensor of 350 µm over a partial travel range of 150 µm the uncertainty is evaluated to 60 nm.

6. Conclusion

In this paper a novel approach to characterize confocal probes with nano-scale level of accuracy was proposed. A new developed test bench has been used to perform different experiments that are related to different error sources impact parameters: material, roughness, colour, inclination angles, target geometry and speed. The basic design principles are respected: the metrology loop is optimized, the Abbe principle is respected and specific materials are carefully selected. The results reveal that the confocal probe with a travel range of 20 µm is more sensitive to colours and less sensitive to materials. The impact of the geometry on the behaviour of the confocal probes is also investigated by using two cylindrical artefacts with the diameters of 50 and 75 mm, respectively. So, when the diameter of the artefact decreases, the amplitude of noise increases, and inversely. The impact of the motion speed is also investigated and can also influence the linear residuals and the working range of the confocal probes.

7. Acknowledgements: This work is part of EMRP Joint Research Project-IND10 “Optical and tactile metrology for absolute form characterization” project. The EMRP is jointly funded by the EMRP participating countries within EURAMET and the European Union.

8. References

[1] Hocken R J, Chakraborty N and Brown C 2005 Optical metrology of surfaces *CIRP Ann.* **54** 705–719

[2] Weckenmann A, Jiang X, Sommer K, Neuschaefer-Rube U, Seewig J, Shaw L and Estler T 2009 Multisensor data fusion in dimensional metrology, *CIRP Ann. Manuf. Technol.***58** 701–721.

[3] Cohen-Sabban J, Gaillard-Groleas J and Crepin P J 2001, Quasi confocal extended field surface sensing, *Proc. SPIE 4449* 178-183.

[4] Boltryk P J, Hill M, Mcbride J W and Nasce A 2008 A comparison of precision optical

displacement sensors for the 3D measurement of complex surface profiles *Sens. Actuators A* **142** 2–11.

[5] Boltryk PJ, Hill M and Mcbride J W 2009 Comparing laser and polychromatic confocal optical displacement sensors for the 3D measurement of cylindrical artefacts containing microscopic grooved structures *Wear* **266** 498-501.

[6] Vissiere A, Nouira H, Damak M, Gibaru O and David J M 2012 Concept and architecture of a new apparatus for cylindrical form measurement with a nanometric level of accuracy *Meas. Sci. Technol.* **23** 9pp.

[7] Abbe E 1890 Meßapparate für Physiker *Zeitschrift Fur Instr.* **10** 446 – 448

[8] Birch KP and Downs M J 1993 An updated Edlen equation for the refractive index of air *Metrologia* **30** 155-162.

[9] Stone J, Phillips S and Mandolfo G 1996 Corrections for Wavelength Variations in Precision Interferometric Displacement Measurements *J. Res. Natl. Inst. Stand. Technol.* **101** 671-674.

Journal of Physics: Conference Series **483** (2014) 012016 doi:10.1088/1742-6596/483/1/012016

A Robust Gaussian Filter Corresponding to the Transmisson Characterisic of the Gaussian Filter

Yuki Kondo, Munetoshi Numada, Hiroyasu Koshimizu

Chukyo University, 101 Tokodachi, Kaizu-cho, Toyota-shi, 470-0393, Japan

E-mail : h11211m@st.chukyo-u.ac.jp

Abstract. A surface roughness profile of an object can be measured by extracting a mean line of the long wavelength component from the primary profile, and by subtracting it from the primary profile. This mean line is usually computed by convolving the traditional Gaussian filter (GF) and the primary profile. However, if an outlier exists in the primary profile, the output of a Gaussian filter will be greatly affected by the outlier. To solve the outlier problem, several schemes of robust Gaussian filter have been proposed. However there are several fatal problems that a mean line determined with respect to the measurement data containing no outliers does not agree with the mean line of the Gaussian filter output. To solve these problems, this paper proposes a new robust Gaussian filter based on a fast M-estimation method (FMGF) and the performance of the new robust Gaussian filter was experimentally clarified. As a result, if an outlier exist, the proposed method behaves a robust performance. If no outlier exists, the output wave pattern, RMSE and transmission characteristic accorded mutually with Gaussian filter.

1. Introduction

A surface roughness profile of an object to be measured is obtained by extracting a mean line for the long wavelength component from a primary profile by means of a low-pass filter of cut-off value λc, and subtracting it from the primary profile. This mean line is usually determined by convolving a Gaussian filter with the primary profile [1] [2]. However, if an outlier exists in the primary profile, the output of a Gaussian filter will be fatally affected from it [3].

To solve the outlier problem, S.Brinkmann et al. proposed a robust Gaussian filter(RGF) [4]. However this RGF has a serious problem that its output becomes quite different from the GF output if no outliers exists in the measurement data. Therefore the filter is still not popularized in actual sites.

To solve these problems, the authors proposed a new robust Gaussian filter based on a fast M-estimation method (FMGF). If no outlier exists in measurement data, the FMGF output wave form accorded mutually with GF output [5]. The deviation between GF output and FMGF output is minimized when the data without outlier is applied. However a distribution characteristic of transmission ratio is not yet known completely. If this small deviation is caused by the difference in transmission rate of a specific wavelength, then this transmission characteristic of FMGF has also the deflection. If FMGF has such a deflection, FMGF is not used in the actual sites. And, it is not yet took into consideration that FMGF cannot realize 50% of transmission ratio at cut-off value λc. Therefore, the transmission characteristic of the FMGF must be verified.

Generally, the transmission characteristic of the low-path filter is calculated by implementing Fourier transform of the weighting function of the filter. However, in such a case of RGF that it is difficult to calculate the transmission characteristic because, in addition to a weighting function of RGF which cannot be directly expressed in a mathematical formula, the coordinate axis of the estimation function is the unuque clue because this coordinate axis is perpendicular to the coordinate axis of the weighting function. Because FMGF weighting function is not directly given as a mathematical formula either, the above circumstances are the same. This problem may be solved by so-called Fast Convolution that Fourier transform of the convolution output of these two functions is equal to the product of the Fourier transform of each function. A mean line is obtained as the output of the convolution between the low-pass filter ad the h measurement data. However, a mean line is equal to the product between the Fourier transform of measurement data and the Fourier transform of the low-pass filter. From this fact, the Fourier transform of the low-pass filter can be provided by dividing the Fourier transform of measurement data from the Fourier transform of the mean line. This will be the transmission characteristic of the low-pass filter. However,as DFT(Discrete Fourier Transform) must be used as the substitute of Fourier Transform, the related problem [5)] inherent to DFT called ringing or aliasing must be coped with.

Therefore, this paper established a computation method and verify its transmission characteristic of FMGF equal to transmission characteristic of GF.

2. Computation for transmission characteristic using discrete Fourier transform

2.1 Basic computation model

Here, N pieces of the measurement data is defined as $z_i = f(x_i)$ just in the same way of the weighting function of discretized low-pass filter as $l(x_i)$, and of a mean line of output by convolution of weighting function with $f(x_i)$, as $h(x_i)$, respectively.

$$h(x_i) = \sum f(x_k) l(x_i - x_k) = (f \otimes l)(x_i) \tag{1}$$

$\tilde{F}(u_k), \tilde{L}(u_k)$, and $\tilde{H}(u_k)$ are defined as the discrete Fourier transform of $f(x_i)$, $l(x_i)$, and $h(x_i)$, respectively. Assuming that $u_k = k/N$ leads to the following equation:

$$\tilde{H}(u_k) = \tilde{F}(u_k)\tilde{L}(u_k) \tag{2}$$

Using this equation let us go to the following equation for the low-pass filter the transmission function ($\tilde{L}(u_k)$).

$$\tilde{L}(u_k) = \frac{\tilde{H}(u_k)}{\tilde{F}(u_k)} \tag{3}$$

2.2 How to give measurement data

The measurement data to calculate transmission characteristic is enforced by the random data of the small amplitude to base line which is designed by considering following condituions carefully : An original function does not become discontinuous, the discrete Fourier transform has a periodicity, and the both ends of the data coincide and are connected smoothly. In order to do this, base curve was designed to be becomes point symmetric as shown in Eq.(4) below, and a 2nd-order B-spline function was utilized and 1st-order differential coefficient was introduced for coinciding the values at both ends of data. Here, L is an evaluation length and C is a constant of amplitude.

$$z = \begin{cases} -C\left(x - \frac{1}{4}L\right)^2 + \frac{CL^2}{16} & (0 \le x \le \frac{L}{2}) \\ C\left(x - \frac{3}{4}L\right)^2 - \frac{CL^2}{16} & (\frac{L}{2} < x \le L) \end{cases} \tag{4}$$

3. Experiments

Experiment is conducted to determine a transmission characteristic of FMGF. Figure 1 shows the primary profile $f(x_i)$, of which evaluation length L=4mm in the x-direction. This primary profile has a base line, which is enforced by the random data where max. 1μm to a base curve becomes as a periodic function of 25μm amplitude as shown in Eq.(4). The number N of data is 200, and the GF width and the cut-off value are both λc= 0.8mm. The sampling interval in the z-direction of FMGF was set at Δz= 4μm, and the parameters for dividing the basic width were set set at m= 7, 9 and 11, respectively.

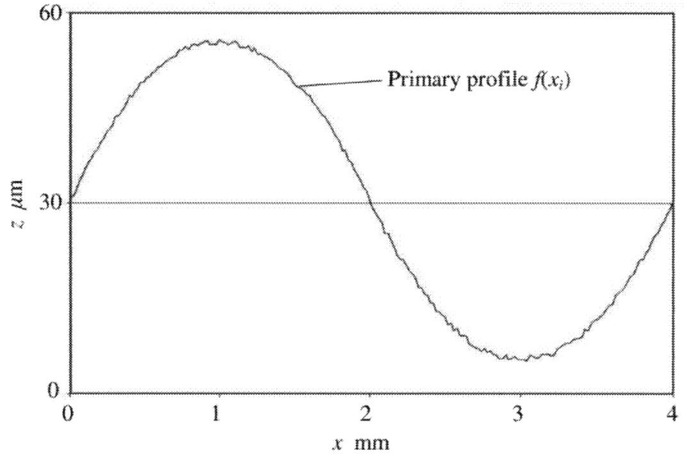

Figure 1 Test data (Primary profile)

3.1 Computation for determination of transmission characteristic

Figure 2 shows transmission characteristic of FMGF, which is computed by Eq.(3) from primary profile $f(x_i)$ and FMGF output $l(x_i)$ and the mean line $h(x_i)$. The computation of the discrete Fourier transform of the data of N=200 was executed by Chirp-Z transform [6]. The horizontal axis is a logarithm axis showing the ratio of space frequency to cut-off value λc, and the vertical axis is the transmission ratio.

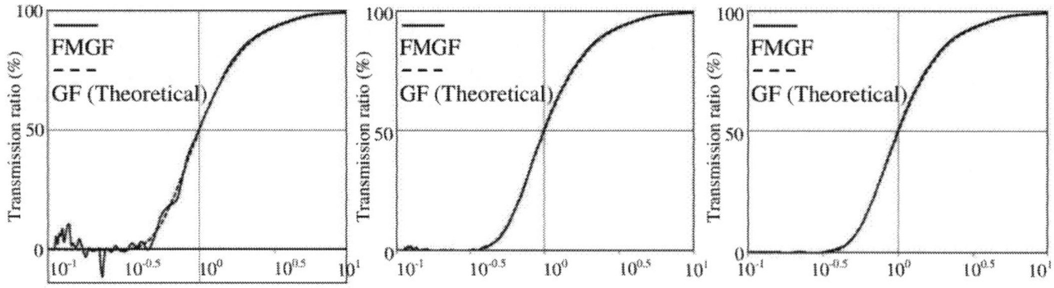

Figure 2 Transmission characteristic of the FMGF

In the case of m=7, Figure. 2(a) shows that similar to transmission characteristic of theoretical GF generally. However characteristic of the short wavelength area is different from theoretical GF. However Table 1 shows a transmission ratio in cut-off value λc is 0.5044 and the error with transmission characteristic of filter GF of width λc is less than 1%, and almost agree in the band of the long wavelength than λc.

In the case of m=9, Figure 2(b) shows that almost all characteristcs becomes coincident except that some characteristics of the low wavelength area are different from GF. The transmission ratio at the cut-off value λc is 0.5090 where the error with the transmission ratio of filter GF of width λc is less than 0.1% (Table 1).

In the case of m=11, Figure 2(c) shows that a transmission ratio becomes almost perfectly coincident with theoretical GF in all wavelengths area. The transmission ratio at cut-off level λc is 0.5085 (Table 1) twhere the transmission ratio of filter GF of width λc.

The transmission ratio differs slightly from the theoretical GF, however, it coincides with the theoretical GF at the cut-off wave length level λc. Figure 3 shows the deviation of the transmission ratio among FMGF and theoretical GF and the theoretical GF at the cut-off wave length level λc. The deviation with the transmission ratio of FMGF and theoretical GF becomes up to 1%. However, the deviation with the transmission ratio of FMGF and transmission ratio of the theoretical GF at the cut-off wave length level λc is very few less than 10^{-3}%. It was clarified that such an agreement occurs in the first calculation step of FMGF to apply the the theoretical GF at the cut-off wave length level λc. If the width of GF is wider than λc then the consistency between the transmission characteristic of FMGF and transmission characteristic of theoretical GF become highly precise.

Table 1 Transmission ratio at cut-off wave length λc

Filter	Fast M-estimation type			GF
Remarks	*m*= 7	*m*= 9	*m*= 11	Filter size λc
Transmission ratio	0.5044	0.5090	0.5085	0.5085

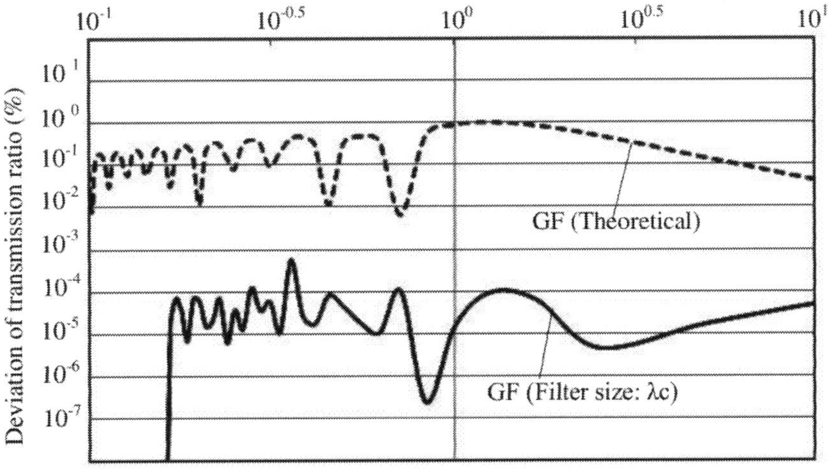

Figure 3 Deviation of transmission characteristic between the FMGF (*m*= 11) and Gaussian Filter

4. Conclusion

The method for determining the transmission characteristic of FMGF by the computation using the discrete Fourier transform has established in this paper. Therefore, FMGF became practical to the actual site, and it is verified that the filter actually behaves in a robust way against outliers [5], and that its output value became in good agreement with that of a GF in terms of measurement data other than outliers.

We would like express many thanks to that this work was supported by JSPS KAKENHI Grant Number 23560271.

References

[1] Yanagi K, Hara S 2003 Technical committee for standardizing the software to characterize surface topographic data — In concert with the geometrical product specifications: Surface texture in ISO *Precision Engineering* **69**/08 1057-60 [in Japanese]

[2] Hara S, Sasajima K, Uchijima T 2005 Handling of the Outliers in 3-D Surface Topography Data *proc. Of JSPE spring conference* 109 [in Japanese]

[3] Brinkmann S., Bodschwinna H., Lemke H.-W. 2001 Assessing roughness in three-dimensions using Gaussian regression filtering *International J. of Machine Tools and Manufacture* **41**/13 2153-61

[4] Numada M, Fujiwara T, hunahashi T, Koshimizu H 2010 A Proposal of Robust Gaussian Filter by Using Fast M-Estimation Method *Precision Engineering* **76**/6 684-8 [in Japanese]

[5] Nakamura S 2003 *Fourier analysis* (Tokyo: Asakura Publishing) pp29-34 [in Japanese]

[6] Numada M, Nomura T, Kamiya K, Tashiro H, Koshimizu H 2006 Filter with variable transmission characteristics for determination of three-dimensional roughness *Precision Engineering* **30** 431

Journal of Physics: Conference Series **483** (2014) 012017 doi:10.1088/1742-6596/483/1/012017

Surface texture generation during cylindrical milling in the aspect of cutting force variations

S Wojciechowski[1], P Twardowski, M Pelic
Poznan University of Technology, Faculty of Mechanical Engineering, Piotrowo 3, 60-965 Poznan, Poland,

E-mail: sjwojciechowski@o2.pl

Abstract. The work presented here concentrates on surface texture analysis, after cylindrical milling of hardened steel. Cutting force variations occurring in the machining process have direct influence on the cutter displacements and thus on the generated surface texture. Therefore, in these experiments, the influence of active number of teeth (z_c) on the cutting force variations was investigated. Cutting forces and cutter displacements were measured during machining process (online) using, namely piezoelectric force dynamometer and 3D laser vibrometer. Surface roughness parameters were measured using stylus surface profiler. The surface roughness model including cutting parameters (f_z, D) and cutting force variations was also developed. The research revealed that in cylindrical milling process, cutting force variations have immediate influence on surface texture generation.

1. Introduction

The following work is an extension of the previous research [1], related to the surface roughness modelling after finish milling of hardened 55NiCrMoV6 steel. Milling of hardened steels is very often applied to production of drop forging dies and casting moulds [2], because it enables the reduction of technological process through the elimination of some machining operations conducted in a tempered state [3]. Nevertheless, the primary objective of the finish milling process is related to high surface texture quality, which is dependent on the appropriate selection of cutting conditions and tool's geometry.

Previous researches related to the hardened steels milling [4, 5] reveal that basic kinematical – geometrical model is insufficient to the accurate surface roughness estimation. The differences between the measured and calculated surface roughness height values are from 15 to 40 times. These discrepancies are resulting from many various phenomena e.g. plastic and elastic deformations of work material, cutting edge notch, variation of the temperature in the cutting zone, some random factors [6]. However during finish cylindrical milling they are mainly affected by the tool vibrations related to cutter's run out. The majority of works related to surface roughness modelling in milling focus only on static value of cutter's run out [7, 8], which results only from the offset between the position of the tool rotation axis and the spindle rotation axis. Nevertheless, during the machining process, cutter's run out is also affected by some dynamical phenomena related to the instantaneous cutting force values. These cutting forces induce cutter's deflections. Thus, dynamical run out instead of static one

[1] Corresponding author.

affects machined surface roughness. In order to measure cutter's vibrations, during the milling process (online), one can apply laser vibrometer [9] or capacitive gap sensor [10].

The surface roughness model including cutter displacements developed by the authors of this paper significantly increases the accuracy of the surface roughness parameters estimation, in comparison to theoretic basic model [11]. However this model was only validated in the range of variable feed per tooth values. In this paper, developed model is applied to the surface roughness parameters estimation in the range of variable cutting depth (a_p, a_e), and consequently active number of teeth values. The active number of teeth influences cutting force variations and thus cutter's displacements which have direct effect on the generated surface texture. The calculated cutter's displacements and surface roughness parameters values are compared to those measured by 3D laser vibrometer and stylus surface profiler.

2. Surface roughness model including cutter's displacements in cylindrical milling

Figure 1 depicts surface roughness models including cutter's displacements. The model depicted in Figure 1 is related to the case when the tool rotates around the spindle axis with an eccentricity and the maximum cutter's displacement y_{emax} per one tool revolution, related to the run out is significantly lower than the tool radius R, $y_{emax} << R$, as well as the radial depth of cut a_e, $y_{emax} << a_e$. For simplification it was assumed that teeth location errors were neglected, which means that partial displacements for consecutive teeth are uniform. It is necessary to emphasize that this assumption can be valid only for monolithic tools. In the case of inserted tools, the influence of insert setting error must be included. From the Figure 1 it is resulting, that in the cylindrical milling process, two different mechanisms of surface irregularities generation can occur. Their appearance (in case when number of teeth $z = 6$) depends on the following condition:

$$f_z < \frac{\sqrt{y_{emax}(3R - y_{emax})}}{3} \tag{1}$$

where: R – tool radius.

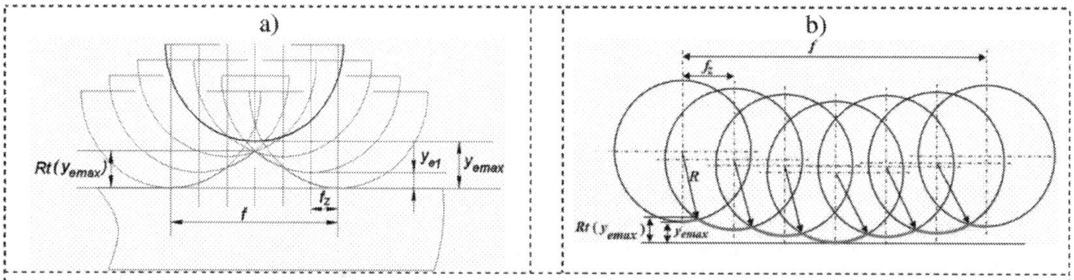

Figure 1. Projection of cutter into work piece including cutter displacement y_{emax} induced by run out: a) when the condition (1) is fulfilled, b) when the condition (1) is not fulfilled [11].

When the condition (1) is fulfilled (i.e. for small values of feed per tooth and/or large values of cutter's displacement) then surface irregularities are generated by the mechanism presented in figure 1a. In this mechanism surface irregularities formed during cutting process are removed by subsequent positions of teeth, and finally theoretic surface roughness height is lower than cutter's displacement y_{emax}, which from the point of view of surface texture is highly eligible. According to figure 1a, surface roughness height for the $z = 6$ can be calculated from the following equation:

$$Rt(y_{emax}) = \frac{y_{emax}}{3} + R - \sqrt{R^2 - 4f_z^2} \tag{2}$$

When the condition (1) is not fulfilled (i.e. for larger values of feed per tooth and/or smaller values of cutter's displacement) then surface irregularities are generated by the mechanism presented in figure 1b. In this mechanism surface irregularities are at least equal to cutter's displacement y_{emax}. According to figure 1a, surface roughness height can be calculated from the following equation:

$$Rt(y_{emax}) = \frac{2\left(4y_{emax}^{\ 2}+f_z^2\cdot z^2\right)\cdot\left(y_{emax}\cdot z\right)-f_z\cdot z\cdot\sqrt{4y_{emax}^{\ 2}+f_z^2\cdot z^2}\cdot\sqrt{z^2\left(4R^2-f_z^2\right)-4y_{emax}^{\ 2}}}{2z\cdot\left(4y_{emax}^{\ 2}+f_z^2\cdot z^2\right)}-$$
$$-\frac{2\left(4y_{emax}^{\ 3}-4R\cdot y_{emax}^{\ 2}\cdot z+f_z^2\cdot y_{emax}\cdot z^2-f_z^2\cdot R\cdot z^3\right)}{2z\cdot\left(4y_{emax}^{\ 2}+f_z^2\cdot z^2\right)}, \quad (3)$$

The value of cutter's displacement y_{emax} is resulting both, from cutter's static run out and dynamical phenomena occurring in machining process. Cutter's static run out e_r can be due to tool itself (wear, asymmetry, insert setting, dynamic imbalance and thermal deformation) but it is mainly due to the offset between the position of the tool rotation axis and the spindle rotation axis. The consequence is a tool rotation around the spindle axis with an eccentricity, which induces cutter displacements. Apart from the static run out, cutter's displacement is also affected by some dynamical phenomena related to the cutting and centrifugal force interactions. During the machining process, cutting tool is deflected by cutting forces. This deflection affects the dynamical cutter's run out (which directly affects maximal cutter's displacement per one tool revolution y_{emax}). Moreover, at higher rotational speeds y_{emax} value can be also affected by centrifugal forces.

In order to determine cutter's instantaneous displacements $y\ (t)$ related to cutter's static run and deflections induced by forces one should solve the following equation:

$$y(t) = e_r\cdot\cos\left(\frac{\pi\cdot n\cdot t}{30}\right)-y_d(t) \quad (4)$$

In equation (4) $y_d(t)$ denotes cutter's deflections induced by cutting forces, which can be solved on the basis of differential motion equation:

$$m\cdot\ddot{y}_d(t)+c\cdot\dot{y}_d(t)+k\cdot y_d(t)=F_{fNe}(t) \quad (5)$$

The method of instantaneous feed normal force $F_{fNe}(t)$ determination is discussed in details in the previous work [11]. The example of cutter's instantaneous displacement $y\ (t)$ chart obtained on the basis of equation (4) is presented in Figure 2. Displacement model parameters were selected from [11]. From figure 2b it can be seen, that during milling with static run out $e_r > 0$ alterations of instantaneous maximal displacements produce the envelope, which has a period equal to tool revolution time T. The height of this envelope is equal to cutter's displacement y_{emax}. In case, when static run out $e_r = 0$, no envelope is found and instantaneous maximal displacements per each tooth are uniform (Figure 2a).

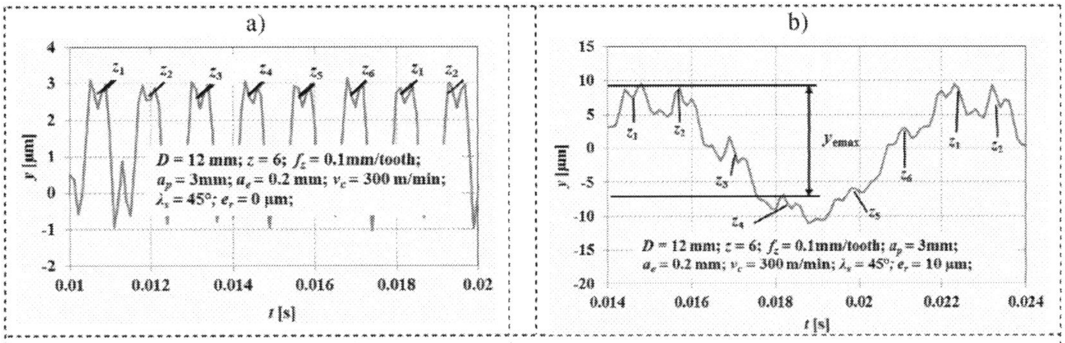

Figure 2. Theoretic cutter's displacement $y\ (t)$ chart: a) $e_r = 0$, b) $e_r > 0$.

Journal of Physics: Conference Series **483** (2014) 012017 doi:10.1088/1742-6596/483/1/012017

3. Research range and method

The investigations have been carried out on hardened hot-work tool steel plate (55NiCrMoV6, hardness approx. 55 HRC). The monolithic torus mill (z = 6 number of teeth, D = 12 mm diameter, r_ε = 1 mm corner radius, λ_s = 45° tool major cutting edge inclination angle) was selected as a milling cutter. The cutting edges were made from fine-grained sintered carbide (mean grain size approx. 0.4 µm) within TiAlCN anti-wear resistance coating. Experiments were conducted in free, down milling conditions, on 5-axes CNC milling workstation (DECKEL MAHO Co., model DMU 60monoBLOCK). Cutting parameters applied in the research are presented in the Table 1.

Table 1. Cutting parameters applied in the finish milling.

a_e [mm]	a_p [mm]	z_c	a_p/p_o	f_z[mm/tooth]	n [rev/min]	v_c [m/min]	v_f [mm/min]
0.1	2	0.5	0.32	0.1	5305	200	3183
0.2	6.28	1.25	1	0.1	5305	200	3183
0.5	7	1.5	1.11	0.1	5305	200	3183

In the carried out research, active number of teeth z_c (related to the value of axial and radial depth of cut: a_p, a_e) was variable. This parameter significantly influences cutting force variations in function of time. Figure 3 depicts feed normal force $F_{fNe}(t)$ time courses obtained for different z_c values, calculated on the basis of model presented in [11].

In case, when active number of teeth z_c = 0.5, cutting forces per each tooth pulsate from 0 to the maximal value (Figure 3a). When active number of teeth z_c = 1.25 and a load constancy condition (quotient of axial depth of cut to axial pitch a_p/p_o = 1) is fulfilled, cutting force is almost constant in time. This phenomenon results from the projection of cutter with the helical cutting edges into the work piece. For the active number of teeth z_c = 1.5, cutting force is periodically variable from the minimal to the maximal values. In this case minimal force is different than zero, because in the whole cutting time at least 1 tooth cuts the material. Furthermore, the growth of z_c value induces the increase of maximal cutting force value, which is attributed to the growth of axial and radial depth of cut values. It is worth indicating that, when cutter's static run out is e_r>0 (Figure 3b), instantaneous maximal forces per each tooth are not uniform. The alterations of these forces produce the envelope, which has a period equal to tool revolution time T. Above mentioned force variations affect cutter's displacements and thus surface texture.

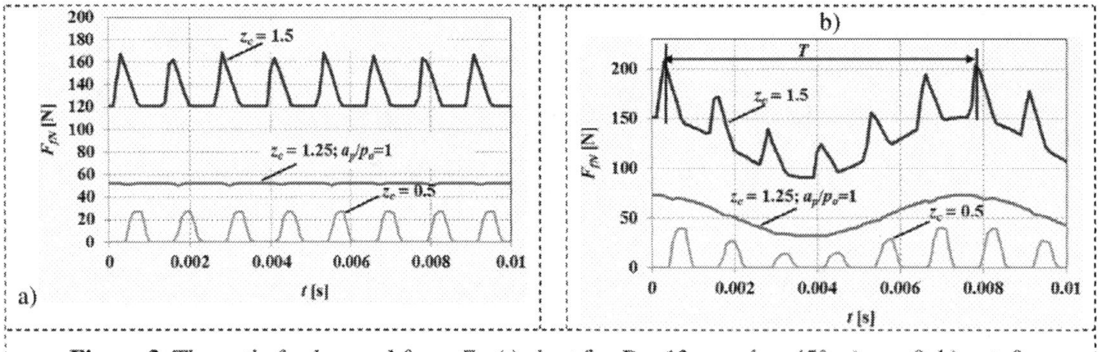

Figure 3. Theoretic feed normal force F_{fN} (t) chart for D = 12 mm, λ_s = 45°: a) e_r = 0, b) e_r > 0.

In the carried out research surface roughness topographies, cutting force components and tool displacements were measured for various z_c values. Surface roughness measurements were made using T8000 stationary surface profile meter (Hommelwerke) in the direction parallel to feed motion vector v_f. After each trial, measuring probe was moved in the direction perpendicular to the feed motion. The measurements were repeated 50 times. As a result of these measurements 3D surface topography charts were received. The value of *Rz* parameter was calculated using Turbo DATAWIN software for the each measurement trial. These *Rz* parameter values were averaged for the 50 carried out trials. Parallel to surface roughness measurements, cutting force components were measured (in machine tool coordinates – Figure 4), in feed force F_f and feed normal force F_{fN} direction.

Figure 4. Milling process experimental stand and block scheme of cutting force experimental apparatus.

In order to solve differential motion equation (5), modal parameters (m, c, k) were determined using impact test, and thus the following parameters were received: m = 0.079 Ns^2/m, c = 40.8 Ns/m, k = 19492469 N/m. Specific cutting pressures k_c, k_{cN} for the cutting force model were obtained empirically from calibration tests as: k_c = 5440 MPa and k_{cN} = 2634 MPa. Differential motion equation (5) was solved in **Simulink** software equipped with **ode45** solver.

In order to measure cutter displacements during milling process, scanning laser vibrometer Polytec PSV-400 was applied. The method of measurement is described in details in [1]. The diagram depicting displacement measurement is shown in figure 5. Cutter displacement measurements were conducted in the joining part of the tool (at the distance l_m from the toolholder), because it is impossible to conduct the measurement on the working part of tool during milling. Cutter's static run out (expressed as the offset between the position of the tool rotation axis and the spindle rotation axis) was measured offline using dial indicator, and its value was e_r =9 μm.

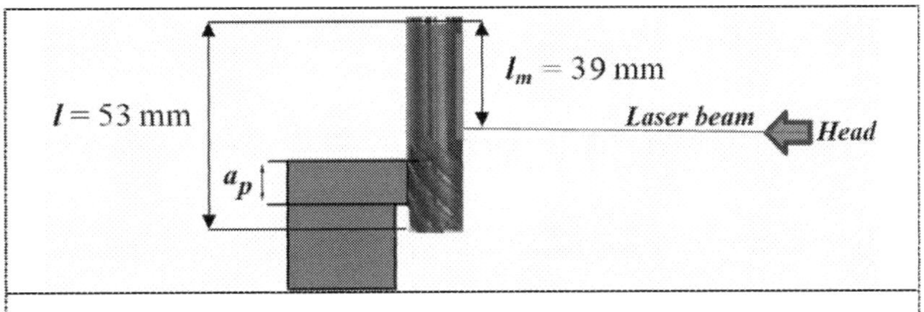

Figure 5. Tool displacement measurement diagram.

4. Research results and analysis

Figure 6 depicts measured and calculated time courses of the feed normal force component F_{fN} =f(t) for the active number of teeth z_c = 0.5. It was found that maximal instantaneous force F_{fN} values per consecutive teeth are not uniform. Alterations of these instantaneous maximal forces produce the envelope, which has a period equal to tool revolution time. This is caused by the cutter's radial run out

Journal of Physics: Conference Series **483** (2014) 012017 doi:10.1088/1742-6596/483/1/012017

phenomenon, related directly to the tool revolution period (see also Figure 3b). Similar dependencies were observed for all investigated active number of teeth. It can be also seen that instantaneous feed normal force values calculated on the basis of the developed model stay in good agreement with the measured ones.

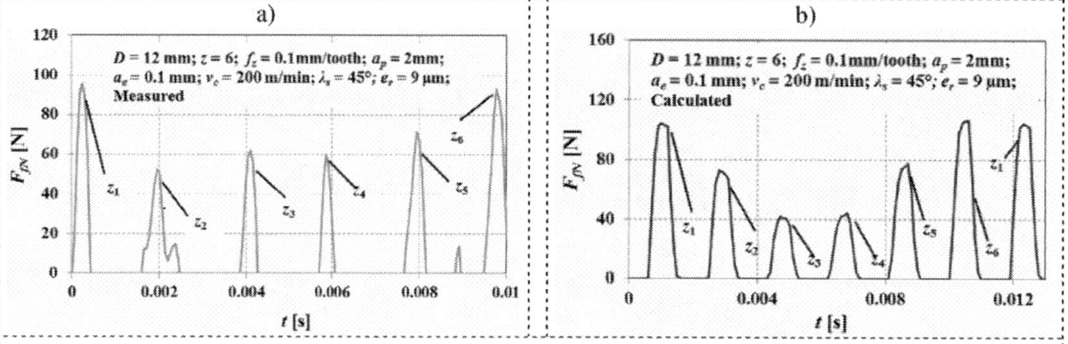

Figure 6. Feed normal force F_{fN} (t) chart for $z_c = 0.5$: a) measured, b) calculated.

Figure 7 depicts the time course of simulated (based on equation 4), and measured cutter's displacements y (t) chart for the active number of teeth $z_c = 1.25$ and $z_c = 1.5$.

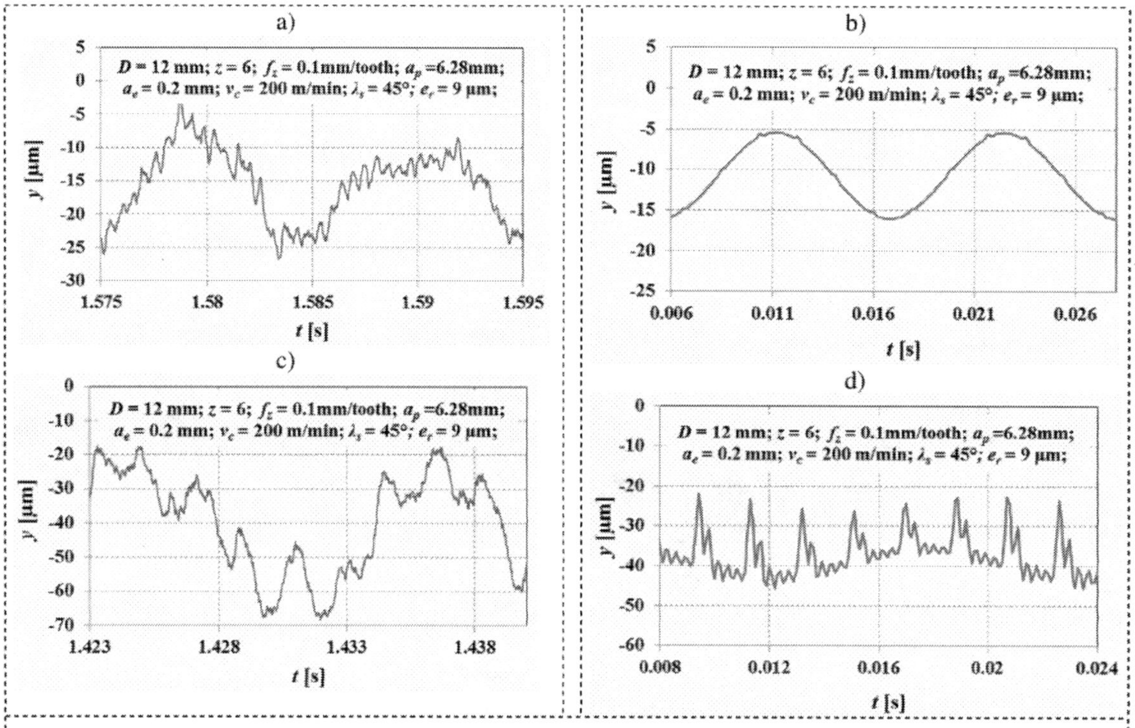

Figure 7. The time course of cutter's displacement: a) measured for $z_c = 1.25$, b) calculated for $z_c = 1.25$, c) measured for $z_c = 1.5$, d) calculated for $z_c = 1.5$.

In case, when active number of teeth z_c = 1.25 and a load constancy condition is fulfilled, the time course of calculated and measured displacements is similar to sine function, without any peaks related to the projection of teeth into the work piece. This phenomenon is attributed to the shape of the feed normal force time course (see Figure 3b), which forces cutter's displacements. Therefore, for the z_c = 1.25 cutter's displacements are induced mainly by the run out phenomenon. In case, when z_c = 1.5, cutter's displacements have different shape than those for z_c = 1.25. The displacement chart consists also of peaks related to the projection of teeth into the work piece, which stays in agreement with the shape of the feed normal force time course for the z_c = 1.5 (see figure 3b). It can be also seen that maximal instantaneous displacement values per consecutive teeth are not uniform. Alterations of these instantaneous maximal displacements produce the envelope, which results from the run out phenomenon. It is worth indicating, that displacement amplitudes have higher values for the z_c = 1.5. The increase in active number of teeth z_c induces higher values of cutting forces and thus higher cutter's displacement amplitudes. The cutter's displacement amplitudes have direct influence on the surface texture (e.g. surface location error – SLE).

From the figure 7 it is resulting, that simulated time course of cutter's displacements is similar to measured one, both in quantitative and qualitative aspect. Nevertheless, some discrepancies between these courses are found. They are probably resulting from the method of measurement using laser vibrometer, as well as some phenomena not included in model, i.a. the damping effect of the rigid work piece.

In all investigated cases, the condition (1) was fulfilled, and therefore surface texture was generated according to mechanism presented in Figure 1a. From the comparison of theoretical surface texture generation mechanism (Figure 1a) and measured surface topography (Figure 8a) it is resulting, that developed model including cutter's displacements stays in good agreement with the experiment. The distance between the surface irregularity peaks is equal to feed per revolution f, instead of f_z value resulting from kinematical – geometrical model. Figure 8b depicts the comparison of calculated $Rt(y_{emax})$ and measured Rz parameter values for various active number of teeth z_c. It can be found, that surface roughness parameters Rz calculated on the basis of equation (2) are similar to those measured using surface profiler, independently of active number of teeth z_c. It is worth indicating, that surface roughness value calculated on the basis of kinematical – geometrical model is barely Rzt = 0.21 µm. It proves that during finish milling this model is insufficient to the accurate surface roughness estimation. From the Figure 8b it is also resulting, that active number of teeth z_c have insignificant influence on the surface roughness parameter Rz values. The increase in active number of teeth induces the growth of cutting forces and tool's displacement (vibrations) amplitudes, which in turn affects surface location errors (SLE) and surface waviness. Nevertheless, investigations revealed, that surface roughness height Rz is affected mainly by the cutter's displacements y_e resulting from the occurrence of radial run out.

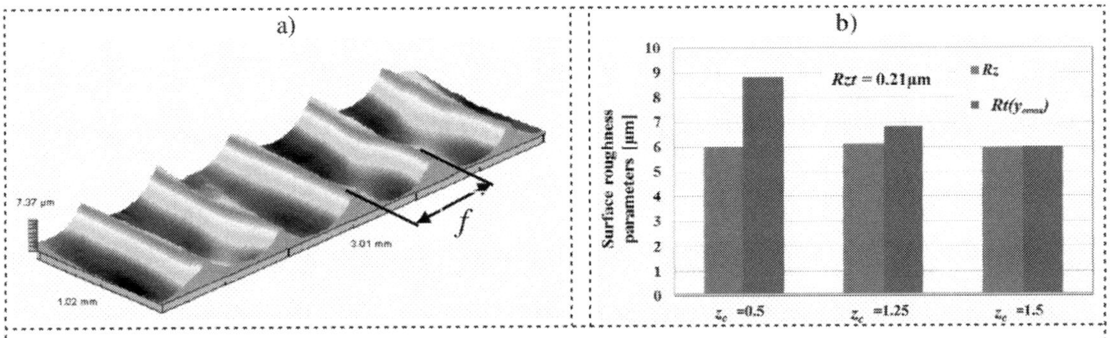

Figure 8. Surface roughness after milling: a) the view of surface topography for z_c = 1.25, b) the comparison of calculated $Rt(y_{emax})$ and measured Rz parameters.

5. Conclusions

The research revealed that in finish cylindrical milling process, cutting force variations have immediate influence on surface texture generation. The increase in depths of cut, and thus active number of teeth values induces the growth of cutting forces and tool's displacement (vibrations) amplitudes, which can affect surface location errors (SLE) and surface waviness. However, surface roughness height is affected mainly by the cutter's displacements resulting from the occurrence of radial run out.

The surface roughness values *Rz* estimated on the basis of the developed model including cutter's displacements stay in good agreement with measured ones, what confirms the validity of applied model. The surface roughness value calculated on the basis of kinematical – geometrical model is approximately 28 times lower than the measured value, which proves that during finish milling this model is insufficient to the accurate surface roughness estimation.

The developed cutter's displacements model can be also applied to the estimation of surface location errors (SLE) and surface waviness.

6. References

[1] Wojciechowski S, 2011 Machined surface roughness including cutter displacements in milling of hardened steel, *Metrology and Measurement Systems* Vol. **18**, **3**, 429 – 440

[2] Urbanski J P, Koshy P, Dewes R C, Aspinwall D K, 2000 High speed machining of moulds and dies for net shape manufacture, *Materials and design* **21**, 2000, s. 395 – 402

[3] Kawalec M, 1990 Cutting of hardened steels and cast iron with tools with defined geometry. *Dissertation* no 234, Poznan University of Technology, Poznan

[4] Twardowski P, Wojciechowski S, Wieczorowski M, Mathia T G 2011 Selected Aspects of High Speed Milling Process Dynamics Affecting Machined Surface Roughness of Hardened Steel, *Scanning* VOL. **33**, 386–395

[5] Wojciechowski S, Twardowski P 2011 Machined surface roughness in the aspect of milling process dynamics, *13th International Conference on Metrology and Properties of Engineering Surfaces*. 12 – 15 April 2011 Twickenham Stadium, Great Britain, p. 87 – 91

[6] Benardos P G, Vosniakos G C 2003 Predicting surface roughness in machining: a review. *International Journal of Machine Tools & Manufacture*, **43**, p. 833-844

[7] Schmitz T L, Couey J, Marsh E, Mauntler N, Hughes D 2007 Runout effects in milling: Surface finish, surface location error, and stability. *International Journal of Machine Tools & Manufacture* **47**, 841–851

[8] Franco P, Estrems M, Faura F, 2004 Influence of radial and axial runouts on surface roughness in face milling with round insert cutting tools. *International Journal of Machine Tools & Manufacture* **44**, p. 1555 – 1565

[9] Tatar K, Gren P, 2008 Measurement of milling tool vibrations during cutting using laser vibrometry. *International Journal of Machine Tools & Manufacture* **48**, p. 380 – 387

[10] Miyaguchi T, Masuda M, Takeoka E, Iwabe H, 2001 Effect of tool stiffness upon tool wear in high spindle speed milling using small ball end mill. *Journal of the International Societies for Precision Engineering and Nanotechnology*, **25**, 145–154

[11] Wojciechowski S, Pelic M, Twardowski P 2012 Cutter displacements as a main factor in surface texture generation during cylindrical milling of hardened steel. *ICSM 2012 International Conference on Surface Metrology, Annecy – Mont Blanc*, France, March 21-23 2012

Surface roughness analysis after machining of direct laser deposited tungsten carbide

S Wojciechowski[1], P Twardowski, T Chwalczuk
Poznan University of Technology, Faculty of Mechanical Engineering,
Piotrowo 3, 60-965 Poznan, Poland,

E-mail: sjwojciechowski@o2.pl

Abstract. In this paper, an experimental surface roughness analysis in machining of tungsten carbide is presented. The tungsten carbide was received using direct laser deposition technology (DLD). Experiments carried out included milling of tungsten carbide samples using monolithic torus cubic boron nitride (CBN) tool and grinding with the diamond cup wheel. The effect of machining method on the generated surface topography was analysed. The 3D surface topographies were measured using optical surface profiler. The research revealed, that surface roughness generated after the machining of tungsten carbide is affected by feed per tooth (f_z) value related to kinematic-geometric projection only in a minor extent. The main factor affecting machined surface roughness is the occurrence of micro grooves and protuberances on the machined surface, as well as other phenomena connected, inter alia, with the mechanism for material removal.

1. Introduction

Tungsten carbide is a material applied mainly in the cutting tools or moulds and dies industries. This field of application is attributed to tungsten carbide's excellent physicochemical properties such as, superior strength, high hardness, high fracture toughness, and high abrasion wear-resistance. However, these unique properties can cause substantial difficulties during machining process, which can result in low machinability. Therefore, machining of tungsten carbide requires the knowledge about the physical effects of the process, as well as appropriate selection of machining method and cutting conditions, enabling desired technological effects. The primary objective of post-process machining of tungsten carbide is to achieve satisfactory geometric and physical properties of its surface texture.

The most popular finishing method of tungsten carbides manufactured by powder metallurgy technology is grinding with the diamond and CBN (cubic boron nitride) wheels. However, in order to produce optical components made of cemented carbide (e.g. spherical mirrors) the profile quality requires a low surface roughness, stringent form accuracy on the submicron scale, as well as a low amount of surface damage [1]. Traditional grinding with the diamond wheels can cause machining-induced cracks and damages to the material. To remove these cracks and damage the ultraprecision grinding can be applied [2, 3]. This kind of process is conducted on the ultraprecision CNC grinding machines, with three – axes movements, and micro-system to deterministically generate, fine, and pre-polish surface. Apart of grinding, recently are seen tendencies to cutting (mainly turning and milling) brittle materials such as, tungsten carbide and reaction-bonded silicon carbide (RB-SiC) by a

[1] Corresponding author.

superhard CBN (cubic boron nitride) and PCD (polycrystalline diamond) cutters in cutting conditions assuring ductile cutting [4, 5]. This technique of cutting can be achieved when uncut chip thicknesses are extremely low and a quotient of the tool cutting edge inclination angle to uncut chip thickness is greater than unity ($r_n/h>1$). In milling process of tungsten carbide by CBN tools, the transition from ductile to brittle cutting occurs at critical depth of cut equal to approx. 5 μm. Therefore, machining with very low cutting conditions is feasible only on ultraprecision machine tools with high rigidity, what is substantial limitation of this technique.

The most popular method for producing tungsten carbide components is by powder metallurgy technology. Nonetheless, for individual, small quantity production or product prototyping this method is too costly and time consuming. The alternative to powder metallurgy, which can be applied to manufacturing of tungsten carbide parts, is Direct Laser Deposition (DLD) technology. This kind of process can be used to quickly produce metallic powder prototypes by a layer manufacturing method [6, 7]. The primary objective of DLD technology is the regeneration of machine parts or machine parts manufacturing with the improved surface layer properties, e.g. higher corrosion, erosion and abrasion resistance. Direct Laser Deposition is an extension of the laser cladding process, which enables three dimensional fully-dense prototype building by cladding consecutive layers on top of one another [8]. Unfortunately parts produced by DLD technology (i.a. tungsten carbide) has an unsatisfactory geometric accuracy as well as surface roughness and requires some post-process machining to finish them to required tolerances [9].

From the previous own – research [10, 11] it is resulting, that machining of direct laser deposited tungsten carbide is feasible and significantly increases surface quality (see Fig. 1).

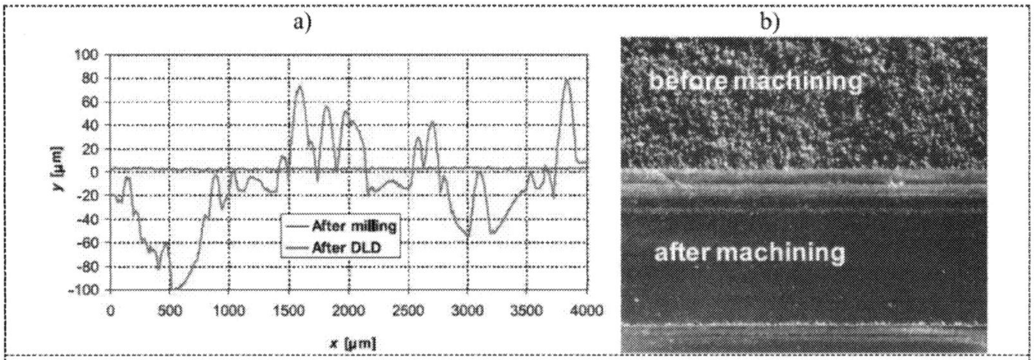

Figure 1. Surface texture of direct laser deposited tungsten carbide: a) 2D surface profile, b) image of surface [12].

However the problem of surface texture generation during machining of direct laser deposited tungsten carbide is still insufficiently examined. Therefore, in this paper the analysis of surface topographies after machining of tungsten carbide is presented. The investigations have also focused on the comparison of surface topographies generated after finish milling and grinding.

2. Research range and method

2.1. Finish milling process

Tungsten carbide received employing direct laser deposition technology DLD (fig. 2) was used as the machined workpiece in this study. The sample was machined during symmetrical face milling process (radial depth of cut – $a_e = D$) with the application of 2-toothed cubic boron nitride monolithic torus end mil (fig. 3). Tool diameter was D = 12 mm, corner radius was r_ε = 0.5 mm, cutting edge inclination angle was $\lambda_s = 0^0$ and the radius of tool arc cutting edge was r_n = 4μm. The cutting tests were conducted in face milling conditions on 5-axis DECKEL MAHO model DMU 60monoBLOCK milling centre.

Figure 2. The view of the machined sample.

Figure 3. CBN torus end mill applied in the research.

Surface texture of the tungsten carbide manufactured with the application of the DLD technology is very often characterized by the occurrence of asperities which height usually exceeds 100 μm (see – fig. 1). In order to finish this kind of surface, series of face milling tests with low axial depths of cut a_p and feeds f_z were carried out. During the first pass, cutter was in contact with the highest asperity of the machined sample (see fig. 4). Subsequently, series of passes (number of passes – $i = 10$) with the constant machining allowance in the tool's axial direction were conducted. Cutting parameters applied in the milling experiments are presented in table 1.

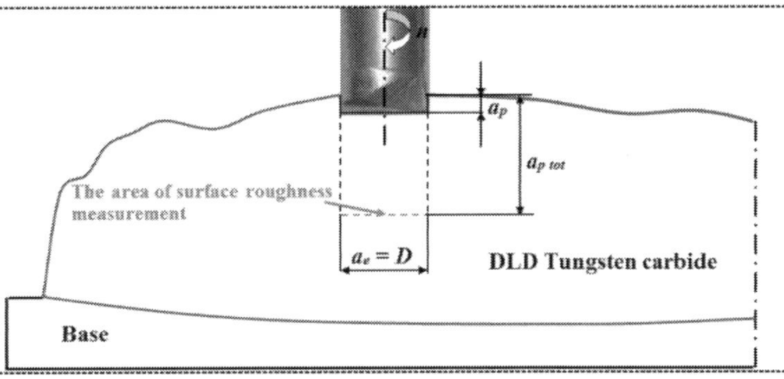

Figure 4. The scheme of face milling process.

Table 1. Cutting conditions applied in the finish milling.

a_e [mm]	a_p [mm]	i	$a_{p\ tot}$ [mm]	f_z[mm/tooth]	n [rev/min]	v_c [m/min]	v_f [mm/min]
12	0.02	10	0.2	0.01	1804	68	36

2.2. Grinding process

Tungsten carbide samples were also grounded during plane grinding process without coolant. The FUM SPC 20B grinding machine was applied. A metal – bond diamond cup wheel with a diameter of 200 mm was used in grinding. The grinding set – up is shown in figure 5. The selected grinding process included two phases, i.e. stock removal and spark out. During the stock removal part axial depth of grinding a_p was selected as 2 μm. Subsequently, sample was sparked out. The number of spark out passes was equalled to 30. These two phases of grinding process were repeated for 100 times in order to finish sample's surface and to obtain total grinding allowance $a_{p\ tot}$ equalled to 200 μm. The detailed grinding conditions are presented in table 2.

Journal of Physics: Conference Series **483** (2014) 012018 doi:10.1088/1742-6596/483/1/012018

Figure 5. The set – up of plane grinding process.

Table 2. Cutting conditions applied in the grinding.

a_p [μm]	$a_{p\ tot}$[μm]	n [rev/min]	v_c [m/s]	v_f [mm/min]
2	200	2400	25	5000

2.3. Surface topography measurement

The 3D surface topographies were measured using optical surface profiler Veeco NT 1100 (see fig. 6). The area of surface roughness measurement is presented in figure 4. The measurements were made with 5x and 20x magnification. During the measurement with 5x magnification the scanning area was equalled to 0.9 mm x 1.2 mm and the distance of vertical points was 1.65 μm. The measurements with 20x magnification were carried out with the scanning area equalled to 0.23 mm x 0.3 mm and the distance of vertical points 0.41 μm.

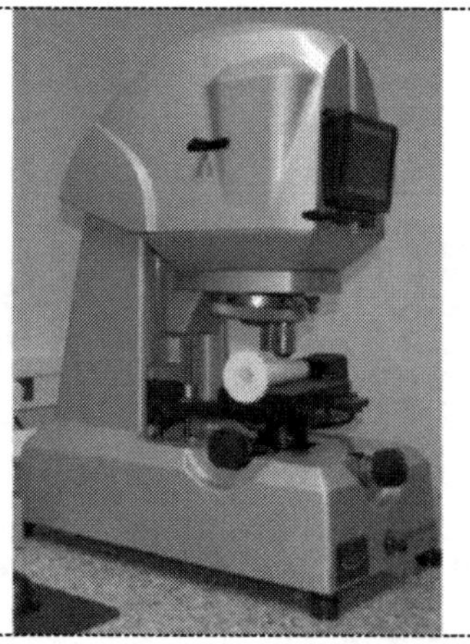

Figure 6. The optical surface profiler Veeco NT 1100.

As a result of 3D measurements, surface topography charts were received. On the basis of surface topography charts the 2D and 3D surface roughness parameters (*Ra*, *Rz*, *Sa*, *Sz*) were calculated and the Power Density Spectra (PDS) were evaluated using Veeco Vision 32 software.

3. Research results and analysis

Figures 7 and 8 depict namely surface topography and surface roughness parameters of the tungsten carbide sample after manufacturing with direct laser deposition technology (DLD).

Figure 7. The surface topography of tungsten carbide after manufacturing with DLD technology.

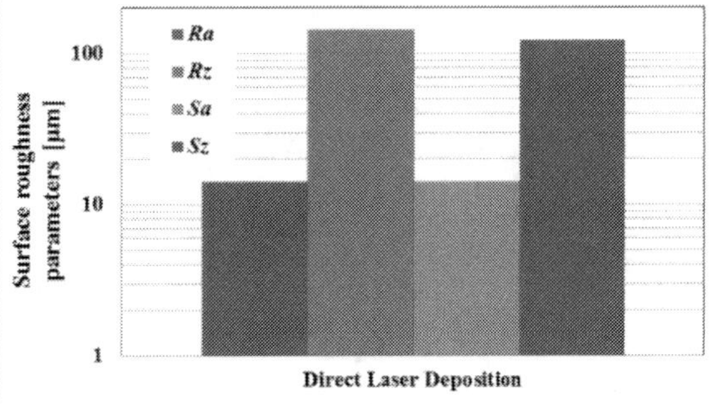

Figure 8. The surface roughness parameters of tungsten carbide after manufacturing with DLD technology.

It can be seen, that tungsten carbide's surface (after DLD process) is rough and characterized by the occurrence of irregularities which height usually exceeds 100 µm (see fig. 8). This kind of surface is typical for the materials manufactured by the layer manufacturing methods and requires some post-process machining. Nevertheless these protruding irregularities are remarkably hampering machining by the generation of the sudden cutting force oscillations [12], which as a result can cause cutting edge chipping during milling or fracture of grinding wheels.

The figures 9 and 10 present surface topographies of tungsten carbide samples after grinding and milling. Surface topography after grinding reveals the occurrence of irregularities orientation perpendicular to the feed motion vector v_f. This phenomenon is probably resulting from the projection of the grinding wheel grains into the workpiece. In case of milled sample, the orientation of surface irregularities stays in agreement with the kinematic – geometric projection of cutter into the workpiece. The distance between the irregularities is approximately equal to feed per tooth f_z value, what is especially seen on the sample measured with the 20-fold magnification. It was also observed, that independently of machining method some micro grooves and protuberances on the machined surfaces can be found (fig. 11). Their height very often exceeds 5 µm and thus have direct influence on the machined surface topography. These irregularities are probably resulting from the material discontinuities formed during direct laser deposition process. However this problem requires further studies.

The figure 12 depicts the comparison of surface roughness parameters generated after grinding and milling of tungsten carbide samples. It can be seen that all investigated parameters (*Ra*, *Rz*, *Sa*, *Sz*) have lower values after finish milling using CBN torus mill in comparison to those obtained after grinding. It means that finish milling of hard tungsten carbide parts can be applied as an alternative to grinding process.

It was also observed, that the value of surface roughness height after milling (described by the *Rz* and *Sz* parameters) exceeds 4 µm, whereas its theoretic value *Rzt*, resulting from kinematic model is scarcely equal to 0.002 µm. This means that surface roughness generation during milling is dependent on the kinematic – geometric projection only in the minor extent. One of the factors, substantially affecting surface roughness height is the presence of previously discussed micro grooves and protuberances on the machined surface. This observation is also confirmed by the power density spectra (Fig. 13, 14) which represent wavelengths of surface irregularities generated during machining. Surface topographies don't consist of wavelengths related to the feed per tooth (f_z = 0.01

Journal of Physics: Conference Series **483** (2014) 012018 doi:10.1088/1742-6596/483/1/012018

mm) or feed per revolution (f = 0.02 mm) values, which occurrence is related to the kinematic-geometric projection of cutter into the workpiece.

Magnification 5x | Magnification 20x

Figure 9. The surface topography of tungsten carbide after grinding.

Magnification 5x | Magnification 20x

Figure 10. The surface topography of tungsten carbide after finish milling.

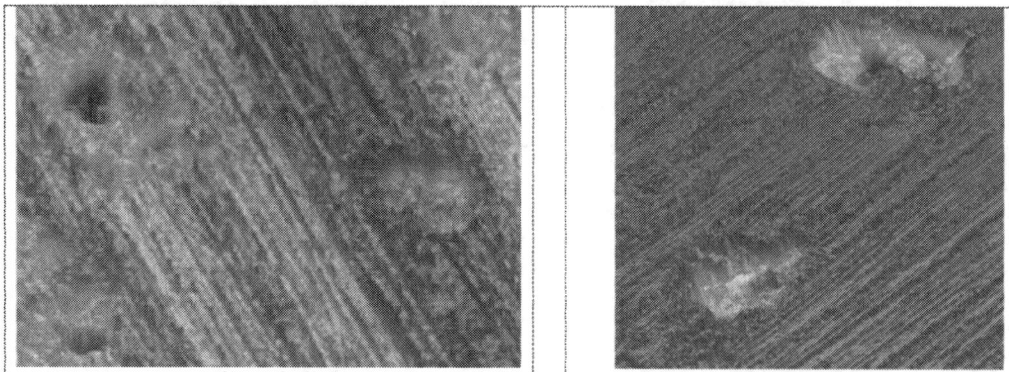

After grinding | After milling

Figure 11. The images of micro grooves observed on the machined surface.

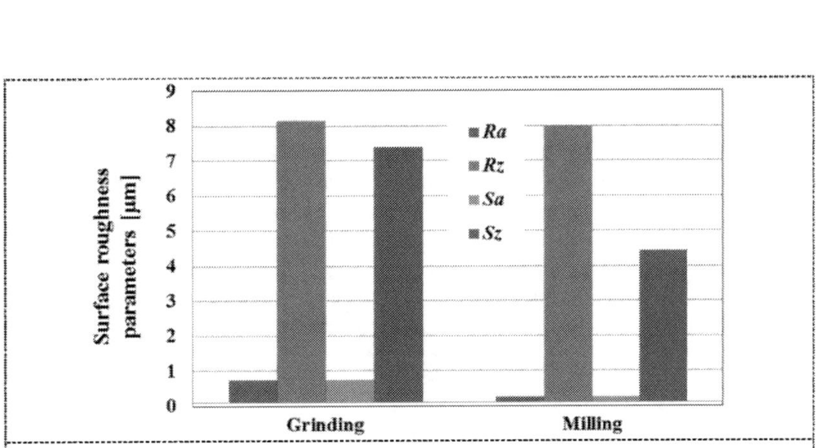

Figure 12. The comparison of surface roughness parameters generated after grinding and milling of tungsten carbide samples.

Instead of these constituents, surface topography consists of wavelengths within the range of 0.05 mm to 0.15 mm. This means that surface topography is affected by the micro grooves and protuberances on the machined surface, as well as the other phenomena (e.g. plastic and elastic deformations of work material, cutting edge notch, variation of the temperature in the cutting zone and some random factors). Surface topography after grinding (Fig. 14) consists also of many wavelengths, which means that it has a random character.

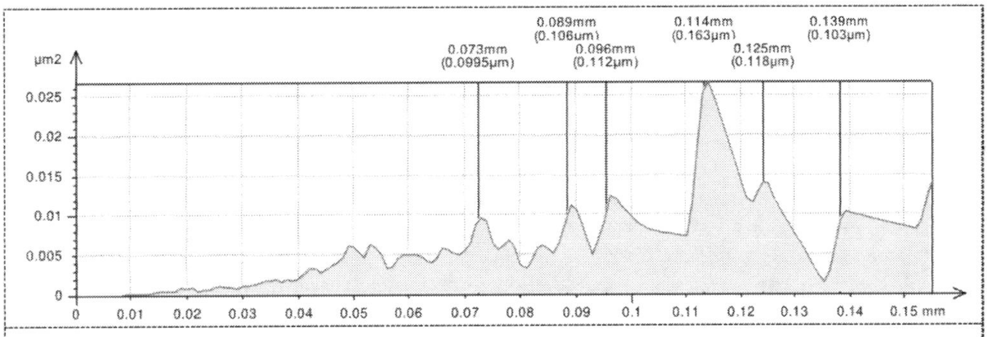

Figure 13. Power density spectrum chart of surface topography after milling.

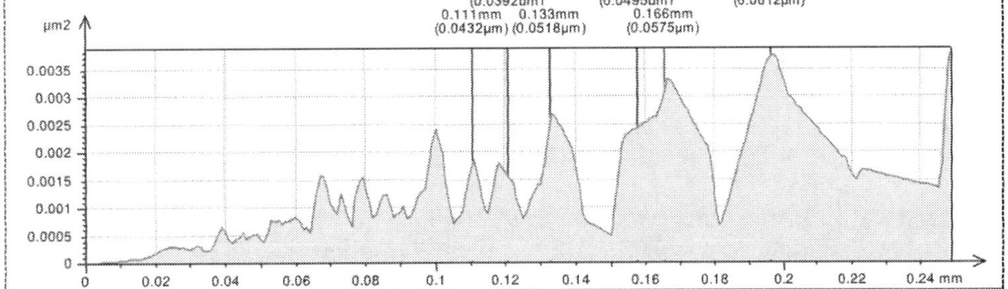

Figure 14. Power density spectrum chart of surface topography after grinding.

4. Conclusions

The investigation revealed, that surface topographies after milling and grinding have different irregularities orientation. These distinctions are resulting from the different milling and grinding kinematics. In case of grinding, irregularities are oriented perpendicularly to the feed motion vector v_f, whereas after milling they stay in agreement with the kinematic – geometric model (the distance between the irregularities is approximately equal to feed per tooth f_z value). Nevertheless, power density spectrum analysis reveals, that surface roughness generation during milling is dependent on the kinematic – geometric projection only in the minor extent.

It was also observed that independently of machining method some micro grooves and protuberances on the machined surfaces can be found. Their height very often exceeds 5 μm and thus have direct influence on the machined surface topography.

It can be seen that all investigated parameters (*Ra, Rz, Sa, Sz*) have lower values after finish milling using CBN torus mill in comparison to those obtained after grinding. It means that finish milling of hard tungsten carbide parts can be applied as an alternative to grinding process. Nevertheless it should be carried out on rigid machine tools and with very low depths of cut and feeds, ensuring moderate tool wear.

5. References

[1] Yin L, Spowage A C, Ramesh K, Huang H, Pickering J P, Vancoille E Y J 2004 Influence of microstructure on ultraprecision grinding of cemented carbides *International Journal of Machine Tools & Manufacture* **44**: 533–543

[2] Stephenson D J, Vaselovac D, Manley S, Corbett J 2001 Ultra-precision grinding of hard steels *Precision Engineering* **25 (4)** 336–345

[3] Yin L, Vancoille E Y J, Lee L C, Huang H, Ramesh K, Liu X D 2004 High-quality grinding of polycrystalline silicon carbide spherical surfaces *Wear* **256**: 197–207

[4] Liu K, Li X P 2001 Ductile cutting of tungsten carbide, *Journal of Materials Processing Technology*, **113**: 348-354

[5] Liu K, Li X P 2001 Modelling of ductile cutting of tungsten carbide *Trans. NAMRI/SME*, **XXIX** 251-258

[6] Banerjee R, Collins P C, Genc A 2003 Direct laser deposition of in situ Ti-6Al-4V-TiB composites *Materials Science and Engineering*, **A358**, 343–349

[7] Fearon E, Watkins K G, 2004 Optimisation of layer height control in direct laser deposition *23nd International Congress on Applications of Lasers & Electro–Optics (ICALEO 2004), San Francisco, California, Paper No. 1708, Laser Institute of America*, Publication No 597, Vol. **97**

[8] Murphy M, Lee C, Steen W M 1993 Studies in rapid prototyping by laser surface cladding. *Proceedings of ICALEO*, 882–891.

[9] Choi J, Sundaram R 2002 A process planning for 5 – axis laser-aided DMD process *Proceedings of International Conference on Metal Powder Deposition for Rapid Manufacturing*, San Antonio, 112–120.

[10] Twardowski P 2011 Surface roughness analysis in milling of tungsten carbide with CBN cutters *Metrology and Measurement Systems*, **18(1)**, 105-114

[11] Twardowski P, Wojciechowski S 2010 Surface roughness analysis in milling of tungsten carbide with CBN cutters *2nd International Conference, Process Machine Interactions*, 10 – 11 June 2010, The University of British Columbia, Vancouver, Canada

[12] Twardowski P, Wojciechowski S, 2012 Machining Characteristics of Direct Laser Deposited Tungsten Carbide *In: Tungsten Carbide - Processing and Applications*, red. Kui Liu. InTech ISBN 978-953-51-0902-0, p. 103 – 120.

Journal of Physics: Conference Series **483** (2014) 012019 doi:10.1088/1742-6596/483/1/012019

Surface roughness analysis after laser assisted machining of hard to cut materials

D Przestacki[1], **M Jankowiak**
Poznan University of Technology, Faculty of Mechanical Engineering,
Piotrowo 3, 60-965 Poznan, Poland,

E-mail: damian.przestacki@put.poznan.pl

Abstract. Metal matrix composites and Si_3N_4 ceramics are very attractive materials for various industry applications due to extremely high hardness and abrasive wear resistance. However because of these features they are problematic for the conventional turning process. The machining on a classic lathe still requires special polycrystalline diamond (PCD) or cubic boron nitride (CBN) cutting inserts which are very expensive. In the paper an experimental surface roughness analysis of laser assisted machining (LAM) for two tapes of hard-to-cut materials was presented. In LAM, the surface of work piece is heated directly by a laser beam in order to facilitate, the decohesion of material. Surface analysis concentrates on the influence of laser assisted machining on the surface quality of the silicon nitride ceramic Si_3N_4 and metal matrix composite (MMC). The effect of the laser assisted machining was compared to the conventional machining. The machining parameters influence on surface roughness parameters was also investigated. The 3D surface topographies were measured using optical surface profiler. The analysis of power spectrum density (PSD) roughness profile were analyzed.

1. Introduction

Among the many advanced engineering materials, SiC reinforced aluminium-based composites and Si_3N_4 ceramics are one of the most important lightweight metals. Mechanical and physical properties of these materials such as high specific strength, very good corrosive resistance and wear resistance indicate that these materials are becoming increasingly attractive for engineering applications [1, 2, 3]. Silicon nitride, as well as metal matrix composites were also successfully applied to sliding elements of cars. However, in general these materials are extremely difficult to cut due to high hardness of ceramics.

One of the possibilities to improve the machining properties of difficult to machine materials is to employ the thermal softening ability of a heat source to heat the material during cutting. The heat source is focused in front of a cutting tool to soften the work piece material, just ahead of the cutting tool, thereby lowering the forces required to cut the material.

The technology of hot machining is not new: the first patent on hot machining was issued before the introduction of high speed steel. Previous generations of this technology employed low-grade heat sources such as electrical resistance, induction and plasma arcs. With the advent of several advanced difficult to cut materials, and with the availability of heat resistant tool materials and cost-effective

[1] Corresponding author.

lasers, there has been a renewed interest in this technology. Most of the hybrid machining processes has been developed by combining conventional machining processes with laser beam machining. Successful application of LAM relates to the following advantages: increase in material removal rates, reduction in the occurrence of chatter and catastrophic tool failure, decrease in cutting forces and tool wear, and the capability to cut brittle work materials without extensive cracking. Therefore, laser hybrid machining processes are becoming more popular in industries in recent time. Many attempts have been made to combine LAM with other machining processes. Some of them have been found very effective. The idea of LAM is that laser source of energy is used for softening the work piece material when it is combined with conventional machining processes such as turning, milling, shaping and grinding. In laser assisted turning, the laser heat source is focused on the un-machined section of the work piece directly in front of the cutting tool. The addition of heat softens the surface layer of difficult to cut materials, so that ductile deformation takes place rather than brittle deformation during cutting. This process substantially reduces the tool wear and cost of machining by reducing man and machine hours per part. Chang and Kuo [4] additionally found a reduction of 20–22% in cutting force with a better surface quality during laser-assisted planning of alumina ceramics.
The work described in this paper is the extension of the previous works made by the authors [5, 6, 7] to further understand the process of laser assisted turning.

2. Research range and method

The work piece material consisted of a AlSi9Mg aluminum alloy matrix (composition: 9,2% silicon, 0.6% magnesium, 0,1% iron) to which 20% of silicon carbide with a particle size of between 8 to 15 μm had been added (Fig.1). The investigated material was fabricated by molten metal mixing and direct chill casting by DURALCAN Inc. Material was obtained in the as-cast condition as well as after hot extrusion. The work pieces (cylindrical shape of 10 mm length and 50 mm in diameter) were painted by an absorptive coating each time to increase laser absorption.

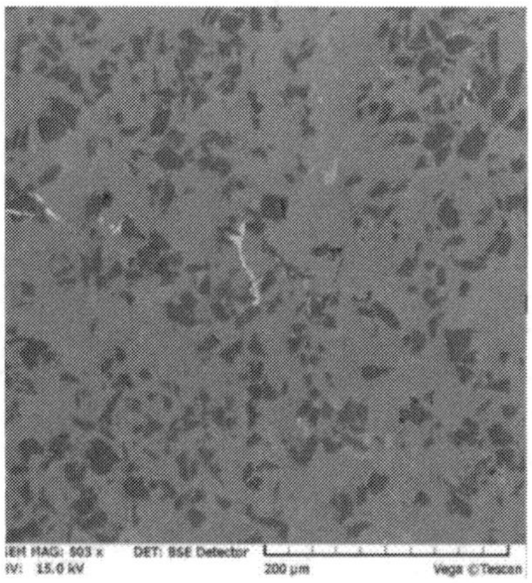

Figure 1. Microscope image of the Al/SiCp material microstructure. The SiC particles (have darker colour on the pictures) are distributed through the matrix.

Second research material was hard ceramics Si_3N_4 (2125 HV). The manufacturer of ceramics was Norton Dias (Czech Republic). Experiments were carried out by turning on the TUM 35D1 lathe with infinitely variable adjustment of rotational speed.

Conventional and laser assisted turning tests were carried out using three different wedges:

1. SNMG 120408 MS from fine-grained tungsten carbide KC5510, coated with TiAlN (PVD method), manufactured by Kennametal (USA)
2. SNMG 120408 MS fine-grained tungsten carbide KC5525 with enlarge of cobalt contents, coated with TiAlN (PVD method)

3. SNGN 120708T 02020 from oxide ceramics (Al_2O_3 + ZrO_2) AC5 of Kennametal manufacturer (USA)
4. SNGN 120408T 02020 from mixed ceramics (Al_2O_3 + TiCN) MC2 of Kennametal Hertel manufacturer (USA)
5. SNMA 120408 from uncoated cemented carbide H10S of Baildonit manufacturer (Poland)

Laser assisted machining was carried out with a 2,6 kW CO_2 laser (2.6kW Trumpf, type TLF2600t). Distance angle between heated area by laser beam and zone of machining was 30 degree in case of MMC (Fig.3a) and 90 degree for Si_3N_4 (Fig.3b).

Machined surface roughness was measured for each trial of experiment. The surface roughness of the turned surface was measured by a Hommel T500 system. The stylus was traced over the machined surface in the direction of feed motion. The arithmetic mean deviation *Ra* (μm) with cut-off length equaled to 0.8 mm (measuring length was 4.8 mm) was applied as criterion for evaluating the surface finish. The machined surface finish was characterized by the average arithmetic roughness which was determined by the averaging of the arithmetic roughness values for the machined surface roughness profiles at 6 equally angular spaced radial sections on a turned surface. Additionally, images of machined structure topography 3D and power spectrum density of the roughness profile were made on the profile meter Hommel Tester T8000.

Cut zone was heated up by CO_2 laser beam to the temperature Θ_2 = 1100-1500°C in case of Si_3N_4 turning.

The selected machining parameters for MMC were: cutting speed (v_c=100m/min), depth of cut (a_p=0,1mm), feed rate (f=0,04mm/rev), work piece diameter (d=50mm), laser beam diameter (d_l=2mm), laser power (P=300÷1000 W). The temperature was controlled by two Raytek Company pyrometers.

The example of laser assisted turning of ceramics Si_3N_4 was illustrated in figure 2.

Figure 2. View of ceramics cutting zone area heated by laser.

The scheme of ceramics Si_3N_4 turning set-up was presented in figure 3a and the scheme of MMC turning in LAM conditions was presented in figure 3b.

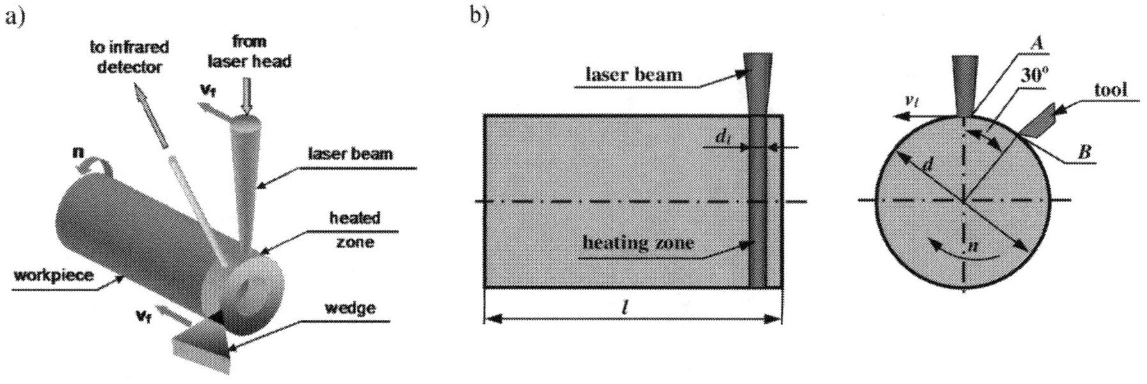

Figure 3. The experimental set-up. a) ceramics turning with LAM b) MMC laser assisted turning, designations: A - heating area by laser beam, B – zone of machining, d - workpiece diameter, d_l - laser beam diameter

3. Research results and analysis

3.1. Metal matrix composites surface analysis

The machined surface roughness is one of the most important quality indicators of hard to cut materials like ceramic and metal matrix composites. For the assessment of the surface quality the average roughness values Ra and Rz were chosen. Figure 4 shows the influence of laser power on surface roughness during turning of Al/SiC MMC.

Figure 5 shows a comparison of the surface profiles of the machined surface by conventional cutting and LAM. The profiles show the difference of surface finish achieved. The significant differences in machined surface roughness have been observed for laser assisted turning and conventional turning. It can be explained that if the cutting zone wasn't heated by laser beam, the SiC reinforcement was pulled out during cutting, cracks and pits were formed on the cutting surface and this caused poor surface finish. Surface roughness produced by LAM is much lower (for all laser power values) than that obtained by conventional turning (Fig.4). As can be seen, for 600 W of laser power, machined surface roughness (described by Ra and Rz parameters) is lower almost 30% than that obtained with conventional turning.

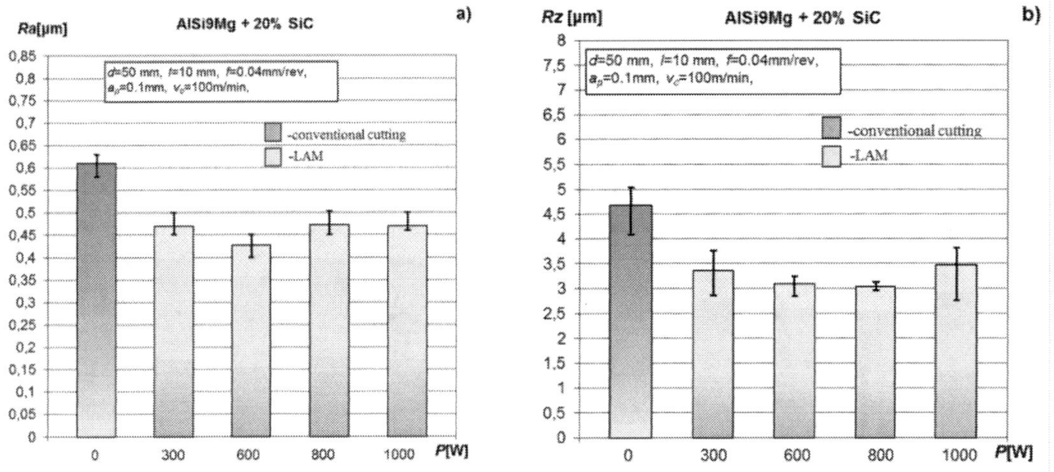

Figure 4. Average surface roughness after conventional turning and LAM of MMC a) *Ra* parameter, b) *Rz* parameter

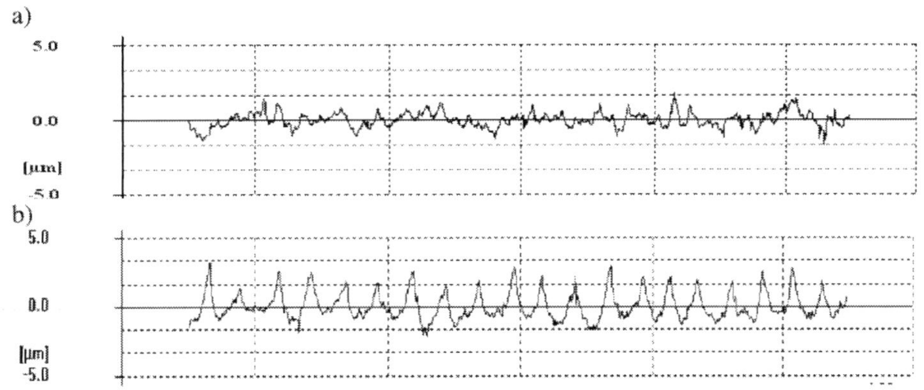

Figure 5. Comparison of machined surface roughness profiles for the same cutting conditions; a) laser assisted turning Ra=0,39µm, Rz=2,71µm b) conventional turning Ra=0,57µm, Rz=4,2µm

3.2. Ceramic surface roughness.

The Si_3N_4 ceramics surface texture after machining is changing together with the laser power density q growth. For the lowest investigated laser power density q the surface profile is determined. In the range of moderate laser power densities q the profile is mixed, however at the highest values of q it has random character (Fig. 6).

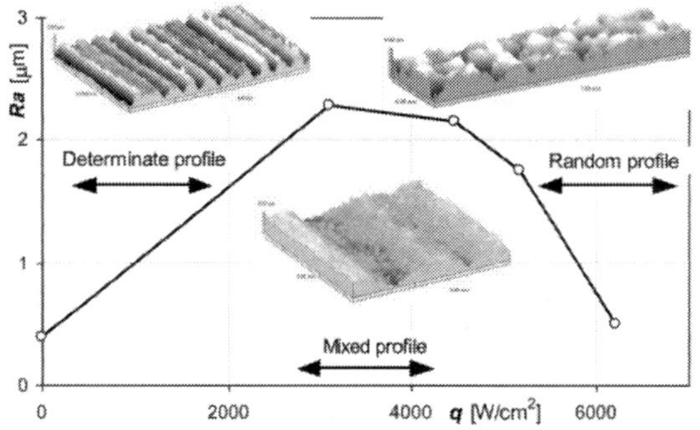

Figure 6. Surface roughness after conventional turning (q=0) and laser assisted turning (q>0) of the Si_3N_4 ceramics.

The machined surface roughness depends on machined surface temperature (Fig.7). In the (Θ_2 = 1200÷1300°C) lowest process temperatures Ra parameter values are comparable and not exceeding 1µm. The machined material become plastic in Θ_2 = 1400°C temperature. The quality of machined surface is decreasing due to plasticized chips welding phenomena.

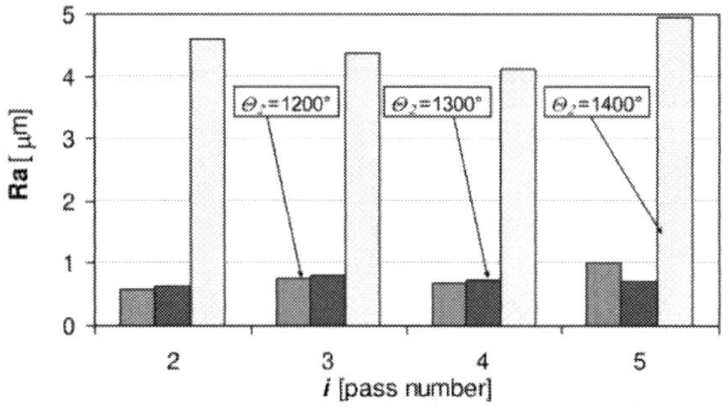

Figure 7. Roughness of Si_3N_4 machined surface. The applied parameters: v_c = 10m/min, a_p = 0,05mm, f = 0,04mm/rev.

Surface texture in the investigated feed range (f = 0,04mm/rev) has a random character without any privileged orientation (Fig.8). This is confirmed by a 3D profile chart (Fig.8a) and, power spectrum density (PSD) chart, which have dominant component different from the applied feed rate (Fig.8b).

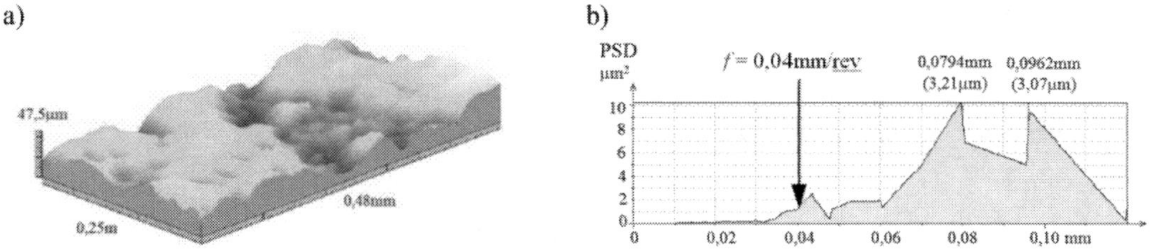

Figure 8. Machined Si_3N_4 ceramics surface texture (a) and power spectrum density of the roughness profile (b) after turning process within polycrystalline diamond (PCD) inserts. The applied parameters and obtained VB_c value: Θ_2 = 1400°C, v_c = 10m/min, a_p = 0,05mm, f = 0,04mm/rev, VB_c = 0,075mm

4. Conclusions

This paper presented an investigation of laser assisted turning effect on the machined surface roughness. The investigation was also focused on the tool wear phenomenon of sintered carbide during turning of Al/SiC metal matrix composites. The above results show that it is possible to benefit from the integration of laser technology in turning process. The results indicate that laser assisted hot turning improves machined surface roughness in comparison with conventional turning. The method of laser assisted machining MMCs seems to be very promising.

Generation of finished surface machined in a LAM process, in the range of various temperatures shows that machined surface texture has random character with significant disturbances in the cutting edge kinematic – geometric projection. It is probably caused by changes in decohesion mechanism, from brittle fracture to the plastic flow of the ceramics.

The tool-work piece temperature control system during LAM process was very useful, which was proved by carried out investigations.

5. References

[1] Ding X., Liew W.Y.H., Liu X.D., (2005) Evaluation of machining performance of MMC with PCBN and PCD tools. Wear 259, pp.1225-1234

[2] Kilickap E., Cakir O., Aksoy M., Inan A., (2005) Study of tool wear and surface roughness in machining of homogenized SiC-p reinforced aluminum metal matrix composite. Journal of Materials Processing Technology, Vol.164-165, pp.862-867

[3] Kishawy H. A., Kannan S., Balazinski M., (2005) Analytical modeling of tool wear progression during turning particulate reinforced metal matrix composites. Annals of the CIRP Vol.54/1, pp.55-58

[4] Chang C.-W., Kuo C.-P., „An investigation of laser-assisted machining of Al2O3 ceramics planning", International Journal of Machine Tools and Manufacture 47 (2007), pp.452–461.

[5] Jankowiak M., „Influence of Si3N4 ceramics laser heating conditions on wear of wedges and roughness of machined surface in hybrid turning", in Proceedings of 9th International Conference TECHNOLOGY 2005, Bratislava, pp. 391-395

[6] Jankowiak M., Przestacki D., Nowak I. "Effect of CO2 laser beam applied during machining of SiC particle reinforced aluminum matrix composites", 5th International PhD Conference on Mechanical Engineering, September 6-8, 2007, Pilsen, Czech Republic, pp.225-228

[7] Przestacki D., Jankowiak M., Mazur P., „Tool wear and machined surface roughness in hybrid turning of SiC reinforced aluminium matrix composite", XI Międzynarodowe Seminarium Naukowe Studentów i Młodych Inżynierów Mechaników, Gdańsk, 24-26 kwietnia 2008

Applications of Morphological Operations in Surface Metrology and Dimensional Metrology

Shan Lou, Xiangqian Jiang[1], Paul J. Scott
EPSRC Centre for Innovative Manufacturing in Advanced Metrology, University of Huddersfield, Huddersfield HD1 3DH, UK

E-mail: x.jiang@hud.ac.uk

Abstract. In contrast to the widely used mean-line based evaluation techniques, the capabilities of morphological methods are not fully recognized in practice. Morphological operations, e.g. dilation, erosion, closing and opening, are useful tools in surface metrology and dimensional metrology. This paper presents a variety of novel applications of morphological operations in association with several of existing critical cases to demonstrate their usability and capability. These applications include scanning process analysis, real mechanical surface reconstruction, freeform surface deviation evaluation, open surface and roundness filtration, form approximation, contact phenomenon simulation, establishment of uncertainty zone for continuous surface reconstruction and stratified functional surface evaluation.

1. Introduction

Mathematical morphological is a mathematical discipline established in the 1960's [1]. It studies the geometrical structure of an image by matching it with the structuring element at various locations in the image. By varying the size and the shape of structuring elements, geometrical information of the different parts of the image and their interrelation can be extracted [2]. These useful tools were first widely used in image processing as nonlinear operations and then extended to other disciplines, including geometrical metrology.

In contrast to the widely used mean-line based evaluation techniques, e.g. the Gaussian filter, morphological methods are not universally adopted. Nonetheless they are of great value if not consciously recognized in practice. The initial application of morphological operations in geometrical metrology is the scanning of workpieces using tactile probes. The traversing process over the workpiece surface is in actual a hardware implementation of the envelope filter proposed by Von Weingraber [3], which is performed by rolling a ball over the workpiece surface. The envelope method, also known as the E-System, in comparison to the mean-line based filters (called as the M-System) is a function oriented approach for the functional prediction of workpieces in that it is related to geometrical properties of workpiece surfaces. Later morphological operations were introduced into the E-System, resulting morphological filters, which offer more tools and capabilities [4].

This paper proposes a variety of novel applications of morphological operations coupled with several of existing cases aiming to demonstrate their usability and capability for surface metrology and dimensional metrology.

---

[1] To whom any correspondence should be addressed.

2. Mathematical morphology operations

Morphological filters are based on four basic morphological operations, namely dilation, erosion, opening and closing. They form the foundation of mathematical morphology [5].

Dilation combines two sets using the vector addition of set elements. The dilation of A by B is:

$$D(A,B) = A \oplus \check{B},$$

where $A \oplus B = \{c \mid c = a + b, a \subseteq A \ \& \ b \subseteq B\}$ and $\check{B}$ is the reflection of B through the origin of B. Figure 1 presents an example of dilating a square by a disk. The dilation of the light colour square by a disk results in the dark colour square with round corners.

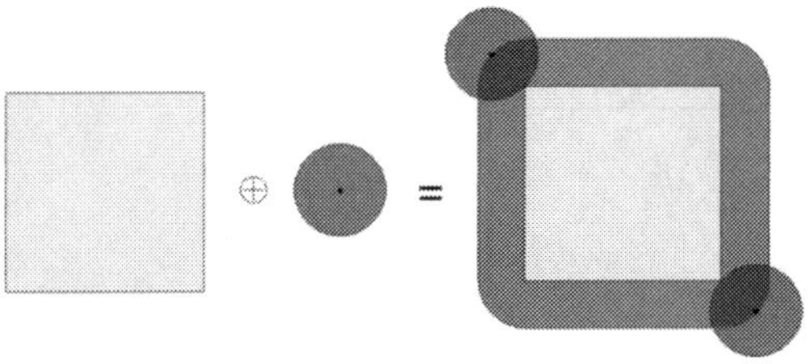

Figure 1. Dilation of a square by a disk

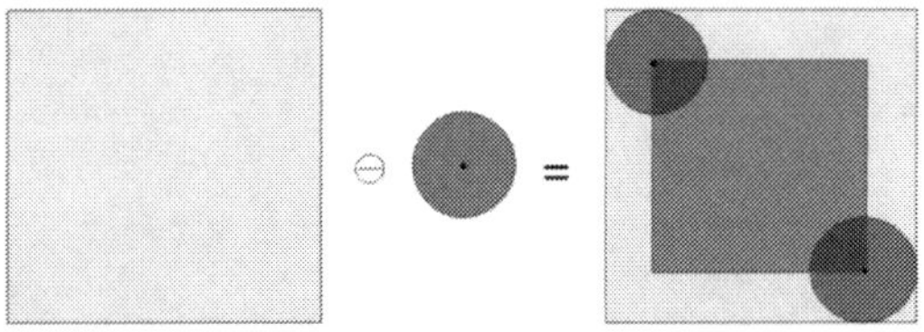

Figure 2. Erosion of a square by a disk

Erosion is the morphological dual to dilation. It combines two sets using the vector subtraction of set elements. The erosion of A by B is:

$$E(A,B) = A \ominus \check{B},$$

where $A \ominus B = \overline{\bar{A} + B}$. An example of erosion is illustrated in Figure 2. The erosion of the light colour square by a disk generates the dark colour square.

Opening and closing are dilation and erosion combined pairs in sequence. The opening of A by B is given by applying the erosion followed by the dilation,

$$O(A,B) = D(E(A,B),\check{B}) .$$

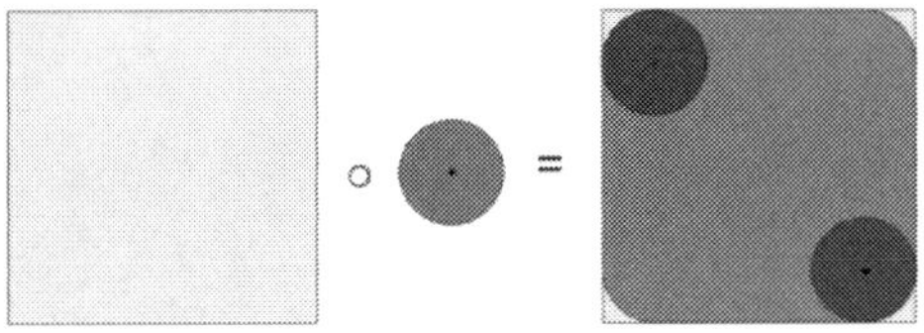

Figure 3. Opening of a square by a disk

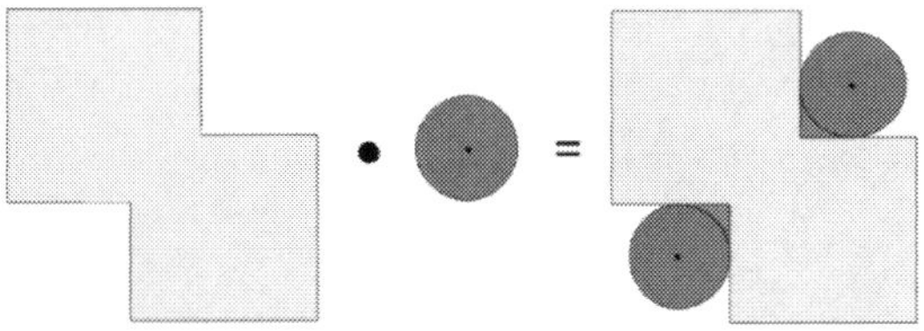

Figure 4. Closing of a union of two squares by a disk

In Figure 3, the opening of the light colour square by a disk generates the dark colour square with round corners.

Closing is the morphological dual to opening. The closing of A by B is given by applying the dilation followed by the erosion,

$$C(A,B) = E(D(A,B),\check{B}).$$

In Figure 4, the closing of the light colour shape (union of two squares) by a disk results in the union of the light colour shape and the dark colour areas.

3. Applications of morphological dilation and erosion operations

3.1. Surface scanning

The scanning of the workpiece surface using a tactile probe, e.g. the analog probe or the touch trigger probe, is a very common practice in geometrical measurement and a hardware implementation of morphological dilation operation [6]. The workpiece surface as the input set is dilated by the structuring element, the probe tip to generate the morphological output, the measured surface, which is also called the traced surface. Figure 5 illustrates the scanning process of a tactile probe. The scanning measurement is conducted by traversing the tip over the surface. The tip centre data are recorded at each sampling position and these sampled data form a discrete presentation of the measured surface. In ISO 3274 [7], the traced surface profile is defined as "*locus of the centre of a stylus tip which features an ideal geometrical form (conical with spherical tip) and nominal dimensions with nominal tracing force, as it traverses the surface within the intersection plane*".

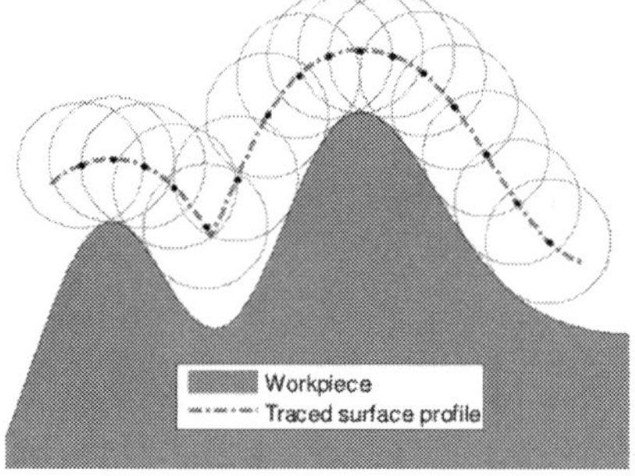

Figure 5. Scanning the probe tip over the workpiece surface

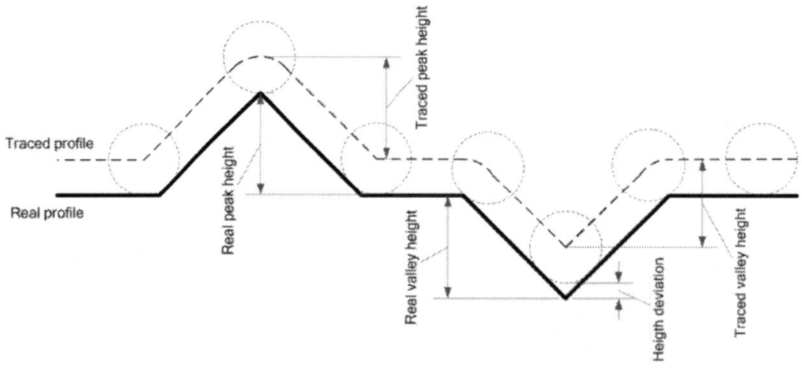

Figure 6. Tip mechanical filtering effects

In common practice, the probe tip employed for scanning are used to be small in size. However the tip size still influences the precision measurement of workpiece surfaces. Figure 6 demonstrates the effect of the probe tip traversing over the workpiece surface. By comparing the traced profile with the real workpiece profile in the figure, it is evident that the probe tip tends to round off peaks on the profile making it broader, nevertheless the peak height remains constant. The valleys on the profile are smoothed by the tip becoming narrow, meanwhile the valley height is reduced as well [8]. This effect introduces distortion into measurement of workpiece surfaces and is called as the mechanical filtration effect of tips. For the measurement of workpiece surfaces, especially for the freeform shaped workpieces, the distortions caused by the tip mechanical filtration effect appreciably influences the precision of measurement. Thus the correction to the traced surface is desired in order to restore to the real workpiece surface. However the traced surface is unable to be perfectly reconstructed to the real surface, but only to an approximate one, i.e. the real mechanical surface.

3.2. Real mechanical surface reconstruction

ISO 14406 [9] presents the definition of mechanical surface: "*boundary of the erosion, by a sphere of radius r, of the locus of the centre of an ideal tactile sphere, also with radius r, rolled over the real surface of a workpiece.*" Figure 7 demonstrates the reconstruction process. Use an ideal sphere with the same size to the probe tip to roll over the traced profile, i.e. the dilated profile by the probe tip (which is already presented in Figure 5), the locus of the sphere centre is treated as the real mechanical surface. Rolling the ball from the below of the traced surface is in essence an erosion operation.

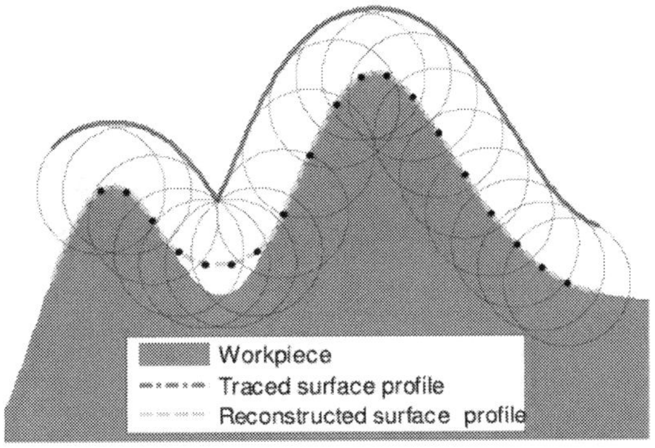

Figure 7. Reconstruction of mechanical surface

It is obvious that the morphological erosion operation is unable to perfectly reconstruct the original real surface of the workpiece. It was found that morphological operations can only reconstruct those portions of the surface where their local curvatures are larger than that of the probe tip [6, 10]. This indicates that the real mechanical surface differs from the real surface at the locations where the local surface curvature is small than the tips. Thus the reconstructed real mechanical surface varies with the probe tip size. Figure 8 presents such an example. Large probe tips tend to reduce and smooth the surface irregularities, while small tips enable the reconstructed surface more approximate to the real surface. The smaller the tip is, the closer the real mechanical surface approximates to the real surface.

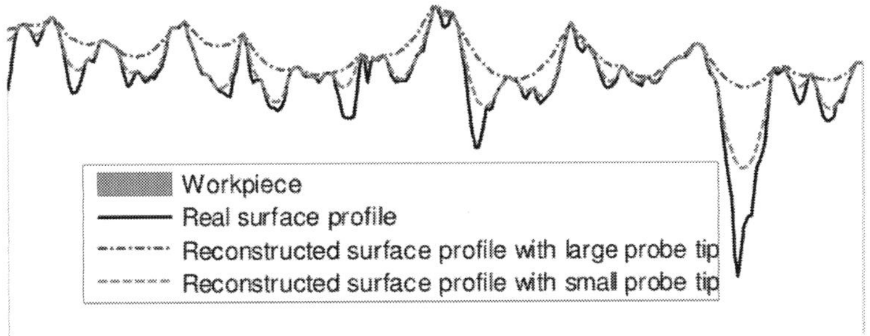

Figure 8. Reconstructed real mechanical surfaces vary with the tip size

In industry, the implementation of the reconstruction of the real mechanical surface varies from the application requirements. For surface texture instruments, for instance, the profilometer and the atomic force microscope, the reconstruction is usually performed by morphological image processing

Journal of Physics: Conference Series **483** (2014) 012020 doi:10.1088/1742-6596/483/1/012020

techniques [11, 12]. Figure 9 presents a basic method to compute the dilation operation with the disk structuring element [13]. The disk ordinates are computed from the disk centre to the two ends. These ordinates are placed over the profile ordinates with the disk centre ordinate over the target profile point. The ordinate where the mapping pair of the profile ordinate and the disk ordinate gives the maximum value determines the height of the disk centre. This procedure is repeated for all the profile ordinates to obtain the whole envelope.

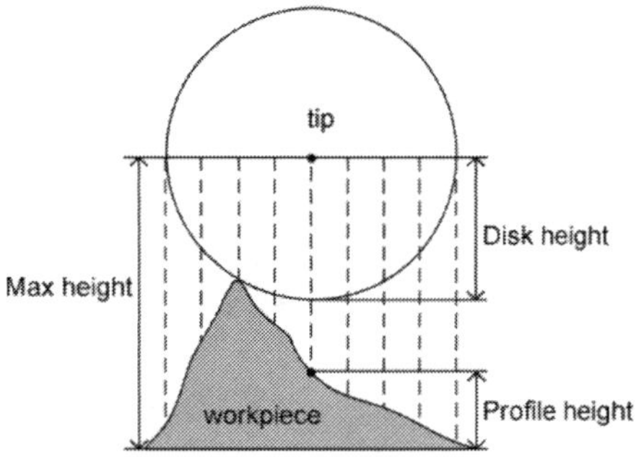

Figure 9. Computation of the dilation operation with the disk structuring element

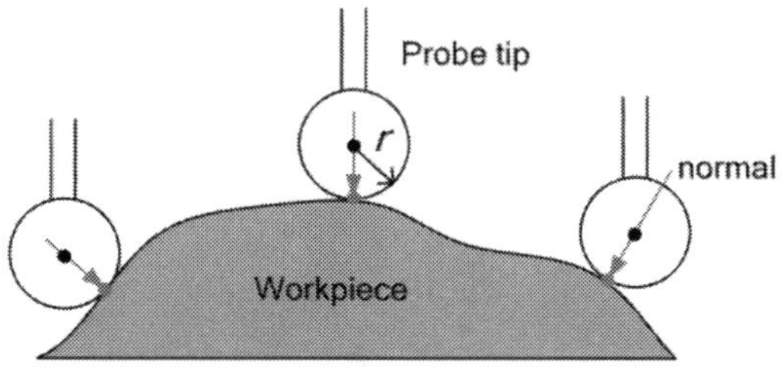

Figure 10. The radius compensation of CMM measurement

In dimensional metrology, for example, coordinate measurement, the reconstruction is usually implemented by the probe radius compensation. Compared to surface texture instruments, sampling of coordinate measurement machine (CMM) is usually less dense and the probe tip much bigger. As Figure 10 illustrates a couple of sampling positions, the contact points of the probe tip to the workpiece surface are obtained by compensating the tip radius in the direction of the surface normal at the contact point. The normal vectors for compensation are achieved either by guessing from the measured tip centre data in the case that the surface is densely scanned and nominal data is not available [14, 15], or by using the nominal vector at the matching point on the nominal surface model of the workpiece, e.g. the CAD model [16, 17]. Although dimensional metrology and surface metrology employ different routes, both of them are essentially morphological reconstruction to the real surface.

3.3. Freeform surface deviation evaluation

Usually the evaluation of freeform surfaces is challenging, especially for tactile measurement, due to their complex form. By convention, the evaluation routine for tactile measurement is constructed in the following sequence. The workpiece surface is measured using the scanning probe. The measured data are therefore compensated by tip radius along probing vectors, usually normal vectors extracted from the nominal surface model of the workpiece (e.g. CAD Model) at the corresponding point [18, 19]. The compensated points are then compared to the corresponding nominal points to capture the deviation. Although this method is reasonable and efficient, it is problematic in terms of two facts. For one thing, the extraction of the normal vectors are based on the nominal surface model and thus affected by the matching of measured data and nominal data. The misalignment between the coordinate frame of the nominal model and that of the real workpiece, which usually not precise in practice for freeform surface measurement because referencing datums are seldom available for establishing a fine alignment, will yield incorrect compensation vectors. For anther, this method assumes that the shape of the workpiece is same to that of the nominal surface model. Thus any deformation of the actual workpiece relative to its nominal shape will falsify the compensation vector [18].

Instead of compensating probe radius in the direction of the approaching vectors, an alternative method to evaluate the deviation of freeform surfaces is to dilate the nominal surface model of the workpiece using the structuring element equivalent to the probe tip and compare the dilated data with the measured data. As presented in Section 3.1 that the traced surface is a dilation of the real workpiece surface by the probe tip, the differences between the actual dilated surface and the nominal dilated surface could thereafter be viewed as the deviation of the freeform surface. Refer to Figure 11 and Figure 12 as a demonstration of the profile example. In contrast to the traditional radius compensation approach, this method brings in a couple of merits. First, this method does not require calculating compensation vectors, which are sensitive to the misalignment of the real workpiece and its nominal model. Second, the dilation method is not affected by the deformation of the workpiece relative to the nominal model. Third, the mechanical filtration effect of the probe tip is not taken into account for the deviation evaluation, which is reasonable in that the deviation caused by the tip effect should not contribute to the deviation evaluation of the workpiece.

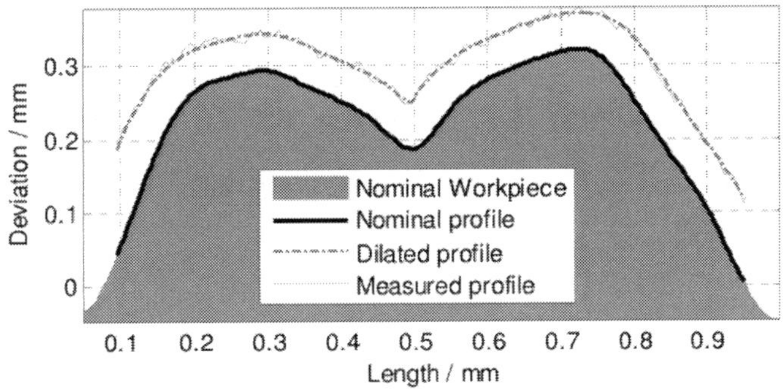

Figure 11. Evaluation of freeform surface profile using the dilation operation

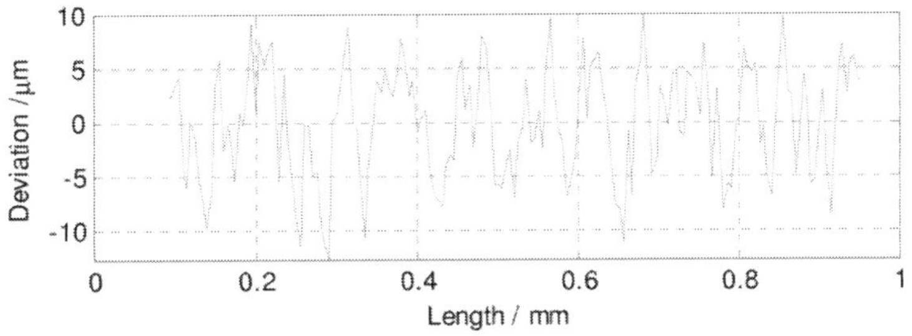

Figure 12. Deviation of freeform profile by subtracting the nominal dilated profile from the traced profile

4. Application morphological closing and opening operations

4.1. Closing and opening filters

Closing and opening filters are two envelope filter, whose output envelopes the input surface. They differ from the obsolete envelope filter proposed by Von Weingraber [3] in that the early envelope filter is essentially equivalent to the dilation operation, usually offset by ball radius.

Figure 13 and Figure 14 illustrate two examples of applying the closing operation and the opening operation on an open profile with the disk structuring element respectively. The closing filter is obtained by placing an infinite number of identical disks in contact with the profile from above along all the profile and taking the lower boundary of the disks [20]. On the contrary the opening filter is archived by placing an infinite number of identical disks in contact with the profile from below along all the profile and taking the upper boundary of the disks.

It is obviously revealed in the figures that the closing filter suppresses the valleys on the profile which are smaller than the disk radius in size, meanwhile it remain the peaks unchanged. Conversely, the opening filter suppresses the peaks on the profile which are smaller than the disk radius in size, while it retains the valleys. The selection of the disk radius depends on the size of physical features on the surface of workpiece. Except circular structuring elements, other most commonly used structuring elements recommended by ISO 16610 are flat structuring elements, for instance, the horizontal line segment for profile data.

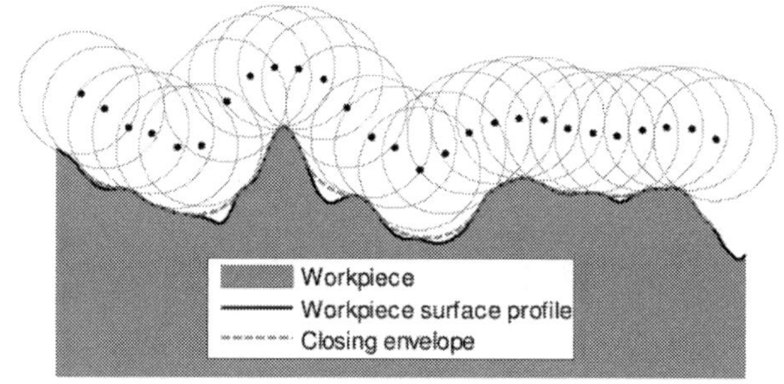

Figure 13. The closing envelope of an open profile by a disk

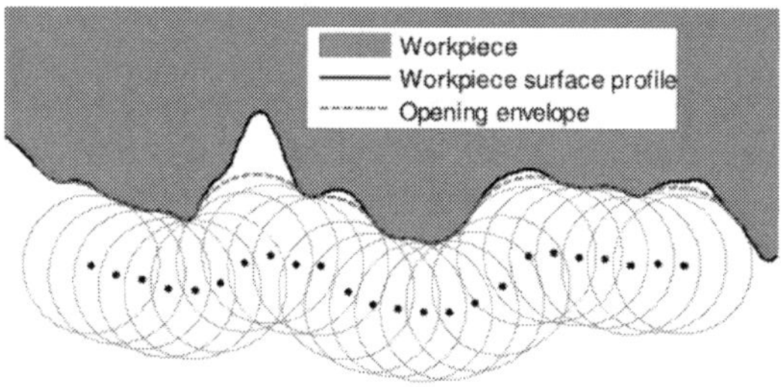

Figure 14. The opening envelope of an open profile by a disk

4.2. Form approximation

It has been illustrated that morphological envelopes could be utilized to approximate the form of functional surfaces for conformable interfaces [21], for instance a soft gasket in contact with a solid block in order to provide sealing function. The long wavelength component of the block surface could be tolerated by the compatibility of the gasket material while the middle wavelength components result in highly localized contacts. See Figure 15. The morphological closing envelope with the circular disk structuring element is used to approximate the conformable gasket surface such that the avoid areas between the conformable surface and the rigid surface can be obtained to characterize the sealing or load distribution. The radius of the circular structuring element should be chosen based on the compression and bending properties of the conformable component.

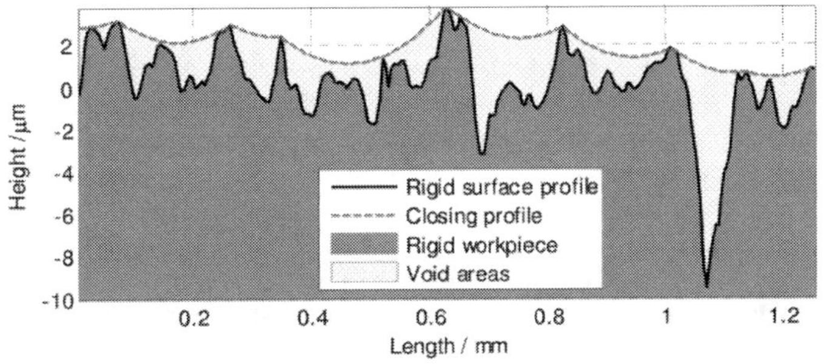

Figure 15. Approximation of the form of a conformable surface profile

4.3. Contact points

In many engineering applications, the contact between two surfaces is non-conforming, i.e., the contact area is very small when compared with the geometry of the bodies in contact. Even in situations between conforming contacts, the contact between the asperities that compose the surface topography is known to be non-conforming [22]. To investigate the contact phenomenon, the closing operation can be employed, whereby the interaction is simulated by rolling a ball with a given radius, which is sized to simulate the largest reasonable radius at a contact e.g. peak curvature, upon the underlying surface. The contact points of the rolling ball against the rolled surface are then captured. This serves as an indication of surface summits and surface portions which are in real contact.

In physics, the contact points are those points on the surface which are in contact with the rolling ball. From a point of view of mathematical morphology, the contact points are those points on the surface that remain constant with the morphological closing operation. Therefore these points might be captured by computing the closing envelope using appropriate algorithms [23]. Figure 16 illustrates such an example. The experimental profile is applied by the morphological closing filter with disk radius 5 mm. The contact points are those points on the profile remain constant before and after the closing operation, as illustrated by the dots in the figure.

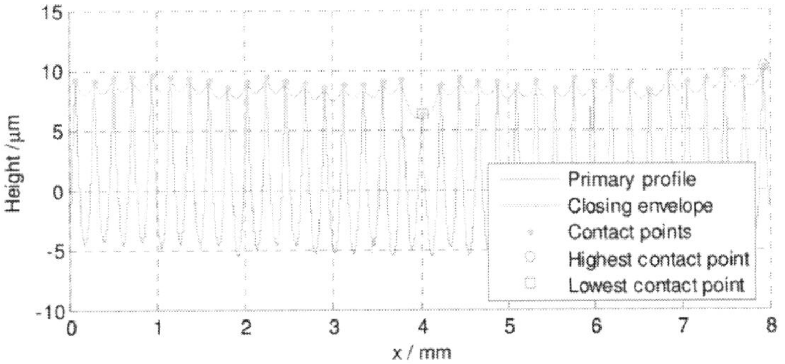

Figure 16. The contact points of the profile and the rolling ball with radius 5 mm.

4.4. Uncertainty zone for continuous surface reconstruction

The workpiece surface is the set of features that physically exist and separate the entire workpiece from the surrounding medium. The inspection of geometrical information of the workpiece surface is conducted by measuring the surface at certain sampling interval, using either the contact measurement instruments (e.g. CMM) or non-contact ones (e.g. interferometer). Either of them generates a series of sampled points, which form a discrete representation of the original surface. It should be noticed that the sampled points in this scenario differ from those presented in the preceding cases in that they are supposed to be the contact points on the real workpiece surface, instead of the tip centre points for tactile measurement. For non-contact measurement, the sample data are all "contact points". It may be desired to reconstruct the original continuous workpiece surface from the discrete samples. In the theory of signal processing, the Nyquist theorem indicates that an infinitely long band-limited signal could be perfectly reconstructed without loss of information from the discrete data sampled at regularly spaced intervals if that interval is smaller than half of the minimal wavelength comprised by the original signal. In mathematical morphology, there is no theorem equivalent to the Nyquist theorem in that a universal equidistant sampling scheme can be found without loss of information, however there are a number of morphological sampling theorems to limit the amount of information lose [24].

Figure 18 illustrates an example of determining the uncertainty zone for the reconstruction of the original surface from a sequence of sampled points taken by a circular disk structuring element. The morphological sampling theorem takes the prerequisite that the surface profile Z under the examination remains unchanged after applying the opening and closing operation by a particular structuring element SE (e.g. a disk) of a given size (e.g. the disk radius), i.e. $C(Z,SE) = Z = O(Z,SE)$. If the original surface Z is sampled with a sampling interval strictly less than the size of SE, yielding a sampled surface Z_s, the original profile is supposed to lie in the region constructed by the opening envelope $O(Z_s)$ and the closing envelope $C(Z_s)$. This region defines the uncertainty zone in which the original profile lies, i.e. $C(Z_s,SE) \le Z \le O(Z_s,SE)$ [9].

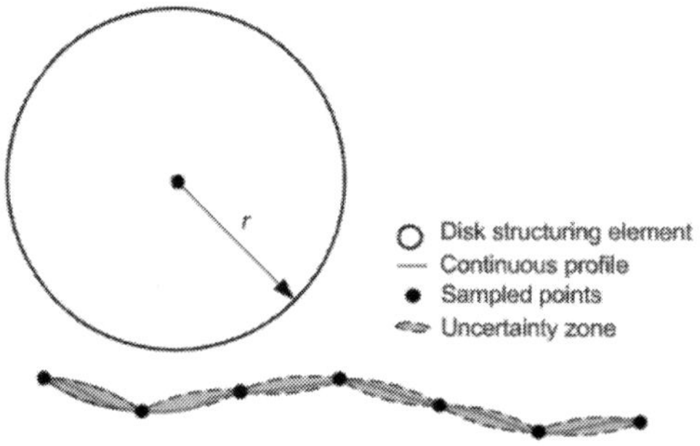

Figure 18. Uncertainty zone for continuous surface profile reconstruction with the disk

5. Applications of alternating symmetrical filters

5.1. Stratified functional surface evaluation

In engineering, surfaces with stratified functional properties are very common, for instance, the inner surface of cylinder liners for automotive engines. These kinds of surfaces are composed of deep valleys superimposed by plateaux. The plateaux support force, bearing and friction while the valleys serve as lubricant reservoirs and distribution circuits. The traditional method for the analysis of these surfaces is performed by applying the two-stage Gaussian filter, the so-called Rk filter. However there are several drawbacks of this method [25]. Firstly, it was derived from the empirical foundation with a significant assumption: surface contains a relative small amount of waviness. It is ambiguous and confusing. Secondly, running-in and running-out sections are generated from the Gaussian filter. These sections truncate the profile and only 20%-60% of the measurement data are used in evaluation. Thirdly, the form component needs to be removed from the profile before the Gaussian filter could apply to the data.

In contrast, morphological filters very suited for this kind of surfaces. Using morphological filters, the profile does not need to be pre-processed to remove the form. The roughness profile can be obtained over the complete measurement length if the end effects are cared for, therefore the roughness profile does not have running-in and running-out sections being "removed". Figure 19 presents such an example. The experimental profile was extracted from a plateau honed surface. The morphological alternating symmetrical filter, combination of first the closing filter and then the opening filter, with disk radius 5 mm, is employed to generate the reference line. As illustrated in Figure 19, the special alternating symmetrical filter reference line basically follows the form of the closing envelope, which is suitable for surfaces where valley features play a dominant role. The closing filter suppresses all the valleys on the original profile that are smaller than the disk radius and the opening filter removes all the peaks on the resulting closing envelope accordingly. The roughness profile is obtained by subtracting the reference line from the original profile.

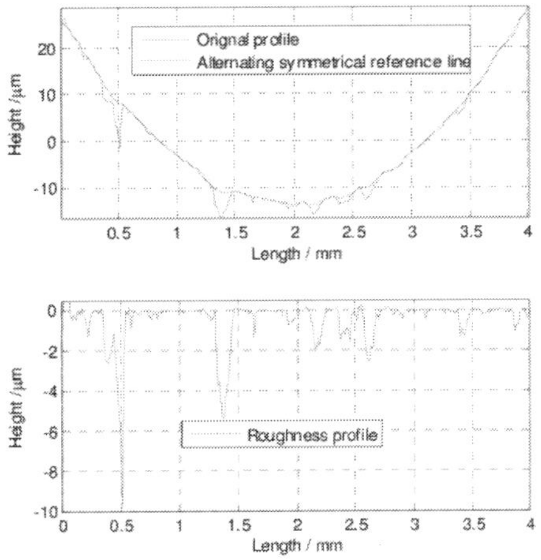

Figure 19. Roughness profile resulting from the alternating sequential filter with disk radius 5 mm

5.2. Roundness filtration

Unlike open profiles measured on planar surfaces, roundness profiles, such as those obtained using a rotating-spindle stationary probe roundness instrument, are closed profiles in shape. The roundness data should also be partitioned into different wavelength regimes to better understand process parameters and functional performance.

Conventionally, the data measured by roundness instruments are radius suppressed, which means the data only reflects the radial deviation while it is insensitive to the radius itself. In such a situation, the Gaussian filter and the spline filter have been successfully employed to decompose roundness data [26, 27]. Morphological filters can be applied for roundness filtration, however they should work on the unsuppressed roundness profiles, which means the component's radius and radius deviations are compounded together. Both the state of art roundness instruments and the CMM can provide such kind of data.

Figure 20 presents an example of applying the morphological alternating symmetrical filter (closing followed by opening) on a roundness profile measured from a cylinder part by a CMM. The cylinder is about 10 mm in radius and the roundness data is filtered with a disk of radius 1 mm. It should be noticed that for convenience of visualization both the roundness profile and the morphological envelope are radius suppressed (reduced by 9 mm).

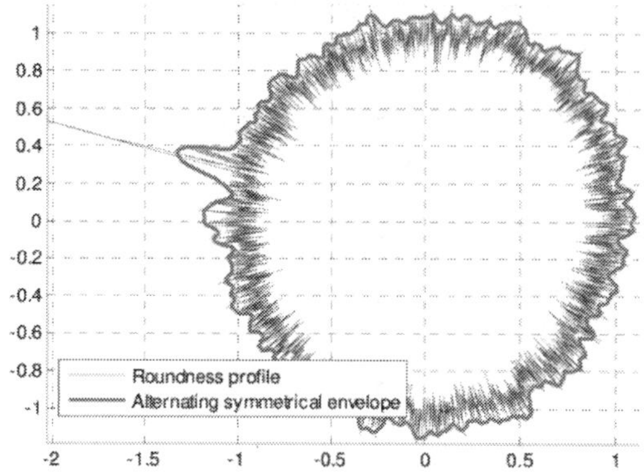

Figure 20. The morphological envelope obtained by the alternating symmetrical filter with disk radius 1 mm (Part radius suppressed)

6. Conclusion

Morphological operations, namely dilation, erosion, closing and opening, are useful tools in geometrical metrology. In this paper, a variety of novel applications of morphological operations in additional with several of existing cases are presented to demonstrate their usability and capability.

The traversing process using tactile probes and the reconstruction of real mechanical surfaces are morphological dilation and erosion respectively. The deviation evaluation of freeform surfaces could be conducted by dilating the nominal surface by tip radius and then compared to the measured data.

Closing and opening filters are superimposed results from combining dilation and closing operations in sequence. The closing filter could suppress the valleys on surfaces and retains the peaks, vice versa for the opening filter. Morphological envelopes could approximate the form of the functional surface, e.g. the conformable surface. The closing operation with the circular structuring element can be employed to simulate the contact phenomenon of two mating surfaces. To reconstruct the original continuous surface from the discrete samples, morphological closing and opening envelopes are able to provide an uncertainty zone for the reconstruction of original surfaces with limited amount of information loss.

The alternating symmetrical filter, as the combination of the closing filter and the opening filter, is an optimal alternative to the two-stage Gaussian filter for the evaluation of stratified functional surface. It can be also used in the roundness filtration.

Acknowledgements

The authors gratefully acknowledge the UK's Engineering and Physical Sciences Research Council (EPSRC) funding of the EPSRC Centre for Innovative Manufacturing in Advanced Metrology (Grant Ref: EP/I033424/1) and the European Research Council under its programme ERC-2008-AdG 228117-Surfund.

References

[1] Matheron G and Serra J 2002 The birth of mathematical morphology *Proc. 6th Intl. Symp. Mathematical Morphology* 1-16.

[2] Heijmans H J A M 1995 Mathematical morphology: a modern approach in image processing based on algebra and geometry *SIAM Review* 371-36

[3] Von Weingraber H 1956 Über die Eignung des Hüllprofils als Bezugslinie für die Messung der Rauheit *CIRP Ann. Manuf. Technol.* **5** 116-28

[4] ISO 16610-40 2010 *Geometrical Product Specification (GPS)-Filtration, Part 40: Morphological profile filters Basic concepts*.

[5] Serra J 1982 *Image Analysis and Mathematical Morphology* Academic Press New York.

[6] Krystek M 2004 Morphological filters in surface texture analysis *XIth international colloquium on surfaces* 43-55
[7] ISO 3274 1996 *Geometrical Product Specifications (GPS) – Surface texture: Profile method – Nominal characteristics of contact (stylus) instruments*
[8] Dagnall H 1998 *Exploring Surface Texture* Taylor Hobson
[9] ISO 14406 2003 *Geometrical Product Specifications (GPS) – Extraction*
[10] Roger S, Dietzsch M, Gerlach M, Jeβ S 2005 "Real mechanical profile" – the new approach for nano-measurement *J. Phys.: Conf. Ser.* **13** 13-9
[11] David J K and Fransiska S F 1993 Envelope reconstruction of probe microscrope images *Sur. Sci.* **294** 409-19
[12] Villarrubia J S 1996 Scanned probe microscope tip characterization without calibrated tip characterizers. *J. Vac. Sci. Technol. B.* **14** 1518-21
[13] Shunmugam M S and Radhakrishnan V 1974 Two-and three dimensional analyses of surfaces according to the E-system *Proc. Instn. Mech. Eng.* **188** 691-9
[14] Mayer J R R, Mir Y A, Trochu F, Vafaeesefat A and Balazinski M 1997 Touch probe radius compensation for coordinate measurement using kriging interpolation *Proc. Instn. Mech. Engrs. Part B* **211** 11-8.
[15] Wozniak A, Mayer J R R and Balazinski M 2009 Stylus tip envelop method corrected measured point determination in high definition coordinate metrology *Int. J, Adv. Manuf. Tech.* **42** 505-14
[16] Liang S R and Lin A C 2002 Probe-radius compensation for 3D data points in reverse engineering *Comput. Ind.* **48** 241-51
[17] Yin Z, Zhang Y and Jiang S 2003 Methodology of NURBS surface fitting based on off-line software compensation of errors of a CMM *Precis. Eng.* **27** 299-303
[18] Ristic M, Anisworth I and Brujic D 2001 Contact probe radius compensation using computer aided design models *Proc. Instn. Mech. Engrs. Part B* **215** 819-34
[19] Jeong J and Kim K 1999 Generation of tool paths for machining free-form pockets with islands using distance maps *Int. J. Adv. Manuf. Technol.* **5** 311-6
[20] Scott P J 2000 Scale-space techniques *Proceedings of the X International Colloquium on Surfaces* 153-61
[21] Malburg C M 2003 Surface profile analysis for conformable interfaces *J. Manuf. Sci. Eng., Trans. ASME* **125** 624-27
[22] Bruzzone A A G, Costa H L, Lonardo P M and Lucca D A 2008 Advances in Engineered Surfaces for Functional Performance *CIRP Ann. Manuf. Technol.* **57** 750-69
[23] Lou S, Jiang X and Scott P J 2012 Algorithms for morphological profile filters and their comparison *Precis. Eng.* **36** 414-23
[24] Haralick R M, Zhuang X, Lin C and Lee J 1988 The digital morphological sampling theorem *IEEE T. Acoust. Speech* **37** 2067-90
[25] Jiang X 2010 Robust solution for the evaluation of stratified functional surface *CIRP Ann. Manuf. Technol.* **59** 573-76
[26] Krystek M 1996 Discrete L-spline filtering in roundness measurements *Measurement* **18** 129-38.
[27] Zeng W, Jiang X and Scott P J 2011 Roundness filtration by using a robust regression filter *Meas. Sci. Technol.* **22** 035108.

Journal of Physics: Conference Series **483** (2014) 012021 doi:10.1088/1742-6596/483/1/012021

Comparison of parameters describing stratified surface topography

P Pawlus[1], **R Reizer**[2], **A Lenart**[3]
[1]Rzeszow University of Technology, W. Pola 2, 35-959 Rzeszow, POLAND
[2]University of Rzeszow, Institute of Technology, ul. Rejtana 16A, 35-959 Rzeszow, POLAND
[3]Pratt & Whitney AeroPower Rzeszów, Hetmanska 120, 35-078 Rzeszow, POLAND

E-mail: ppawlus@prz.edu.pl

Abstract. There are a lot of methods of two-process surface topography description. Some parameters can be computed from material ratio graph. There are included in ISO standards. The methods described in ISO standards 13565-2 (Pk, Ppk, Pvk, Pmr1 and Pmr2) and 13565-3 (Pmq, Pvq and Pmq) are compared in this paper. Profiles with given values of Pmq, Pvq and Pmq parameters and wavelengths were modelled. For these profiles, material ratio curve and Pk, Ppk, Pvk, Pmr1 and Pmr2 parameters were calculated. As a result, dependencies among parameters from ISO 13565-3 and ISO 13565-2 standard were found.

1. Introduction

Most engineering surfaces have approximately Gaussian height distributions. However two-process texture is more important from functional point of view. The precise description of manufactured 2-process surface is a problem of a great practical importance.

Parameters of two-process surfaces are often calculated using material ratio curve. One of the first methods originates from French automobile industry [1]. According to it, it is possible to calculate the height of: the run-in zone, the operating zone and the lubricant zone. Trautwein [2,3] defined the point which separated the core roughness region and the deep valleys region. It resulted of the cross-cut of a tangent to the curve in its middle part with the regression straight line describing the deep valleys region. The method described in German standard DIN 4776 (and then in ISO 13565-2) corresponded to Trautwein's description of the bearing length curve, but was more synonymous. The following parameters can be determined: the reduced peak heights Rpk, the reduced valleys depth Rvk, the core roughness depth Rk, and material ratios determined by the straight line separating the core roughness from the material side (Rmr1) and free from material side (Rmr1) [1]. Zipin criticised this method [4], since it can be used not only for two-process but also for one-process surfaces. Some parameters found in German standard are non-synonymous [5, 6], for example the slope of material ratio curve in its middle part affects the so-called oil capacity [6]. Nielsen found that honing process can be controlled by Rk parameters [7]. Augustyn [8] also expressed his doubts regarding the synonymously of determining parameters using Trautwein method. His proposal was to define the point of the passage from the core roughness to the deep valleys region as the point of maximum curvature of the normalised curve. Michalski and Pawlus [9] defined the limits of the characteristic profile regions in similar manner. They used different formula for approximation of the normalised material ratio curve.

ISO committee incorporated also other alternative way in characterising two-process surface [10, 11, 12, 13]. It is based on the cumulative distribution plot on normal probability graph analysis. The intersection point on normal probability graph of abscissa Rmq defines the separation of plateau and base textures and is an important feature of the model. The proposed plateau roughness Rpq, valley roughness Rvq and Rmq are three parameters characterising plateau honed surface. This method was used previously to study running-in process [14]. In this approach, two regions of roughness exist: peak (plateau) and valleys. Contrary to parameters from Rk family, parameters Rpq, Rvq and Rmq can be calculated only for stratified textures. This method is conceptually simpler and more elegant than method described previously (ISO 13565-2), but the practical difficulty of the parameters Rpq, Rvq and Rmq computations exists [15, 16]. Parameters from standard ISO 13565-3 can be obtained in different way [17].

Figure 1 shows the parameters included in ISO 13565-2 standard. A2 is the so-called oil capacity. Figure 2 presents parameters described in ISO 13565-3 standard. The plateau depth Pd (ordinate of plateau-to valley transition) is also included.

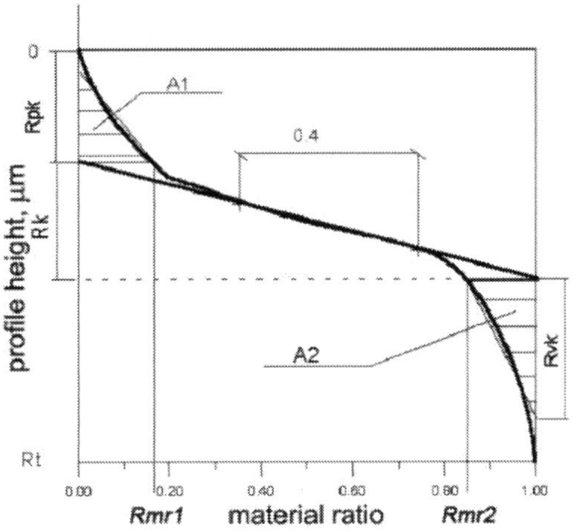

Figure 1. Parameters included in ISO 13565-2 standard.

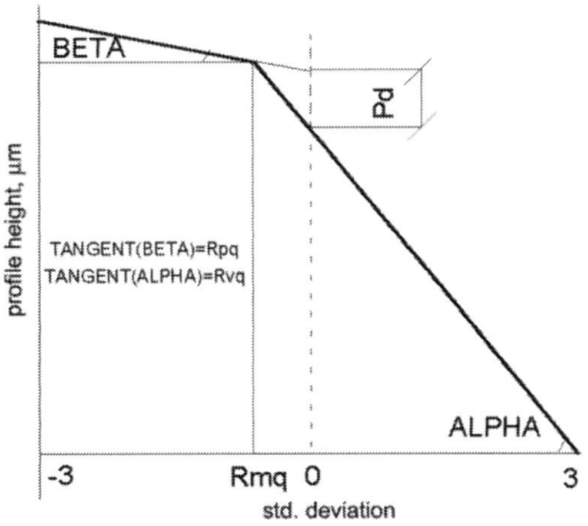

Figure 2. Parameters included in ISO 13565-3 standard and plateau depth Pd.

Rosen et al. [18] analysed cylinder surfaces and compared parameters from ISO 13565-2 and 13565-3 standards using regression and correlation analysis methods. The special tests were done in order to compare Cummins and Rk methods [19]. Measured cylinder surfaces from VOLVO engines were analysed. The obtained results [18, 19] suggest that optimum parameter set describing 2-process cylinder surface topography has not been found yet.
Usually the shape of the profile ordinate distribution can be described by parameters Rsk and Rku. However these parameters are sensitive on the local peaks and valleys existence. They were used with regard to their statistical significance. The so-called emptiness coefficient can be alternative to Rsk. The Ra/Rq and Rda/Rdq parameters are the alternatives to Rsk or Rp/Rt.

2. Simulation of two-process profiles

Because of sound theoretical basis of probabilistic method, it is possible to computer generate stratified cylinder profile when the values of parameters Ppq, Pvq and Pmq (of primary profile) are known. This can be done be superimposition of two profiles of Gaussian ordinate distribution when their standards deviation of height and depth of plateau Pd (transition characteristic from one process to another) are known. Each profile of normal ordinate distribution is characterized by 2 parameters: standard deviation of height and horizontal parameter – correlation length (the distance, at which autocorrelation function slowly decays to 0.1 value).
The aim of this research is to find the effect of changing of one parameter from Rk family: Ppq, Pvq and Pmq on parameters from the Rk family: Pk, Ppk, Pvk, Pmr1 and Pmr2 when the other parameters from Rq group were constant.
Each two-process profile is characterised by the following 5 parameters: Ppq, Pvq, Pmq, CLp (correlation length of plateau profile part) and CLv (correlation length of valley profile part). Parameters Ppq, Pvq, CLp and CLv as well as plateau depth were input parameters. The Pd parameter depends on Ppq, Pvq and Pmq parameters.
Three cases were analysed. In the first of them the correlation length of valley part was larger than of valley part: CLp = 8 μm, CLv = 40 μm – it is the case typical for plateau honed cylinders. The second case was opposed: horizontal parameter of valley part was smaller than that of plateau part: CLp = 40 μm, CLv = 8 μm – typical for surfaces from elements after a low wear process (when wear was within the limit of the original surface topographies). In third case correlation lengths of plateau and valley part were the same: CLp = CLv = 40 μm.
In each three cases first the Pmq parameter increased from 40% to 95% with step of 5%, when Ppq and Pvq parameters were the same. Calculation were done for various Pvq/Ppq ratios: 4.5, 9 and 18. Then for constant Pmq parameter (84.3%) and Pvq parameter (of the following values: 0.45 μm; 0.9 μm and 1.8 μm) the Ppq parameter increased from 0.04 μm to 0.4 μm with step of 0.01 μm. Finally, for constant Pmq parameter (84.3%) and Ppq parameter (0.1 μm) va;ues the Pvq parameter changed from 0.45 μm to 1.9 μm with step of 0.05 μm. The Pmq parameter of 84.3% corresponds to standard deviation of 1 and is typical for plateau honed cylinder profile.

3. Results and discussion

First two process profiles of larger values of correlation length of valley part comparing to plateau part were studied. Increase of the Pmq parameter (when the Ppq and Pvq parameters are the same) causes decrease of the Pk parameter; higher for smaller values of Pmq. The effect of Pvq parameter on the values of the core height Pk is evident for smaller value of the Pmq material ratio; for higher Pvq the Pk is larger. Increase of the Pmq parameter causes also decrease of Pvk; the Pvk value is proportional to the Pvq value. Differently to the Pk parameter, larger changes of the Pvk were found for higher values of the Pmq ratio. The effect of Pmq increase on value of Ppk parameter is negligible. Increase of the Pmq causes increase of Pmr1 and Pmr2 material ratios; this proportionality was more evident for the Pmr2 parameter. For the small Pmq parameter increase of Pvq causes decrease of Pmr2. Figure 3 presents the relation between Pmq parameter and parameters from the Sk family.

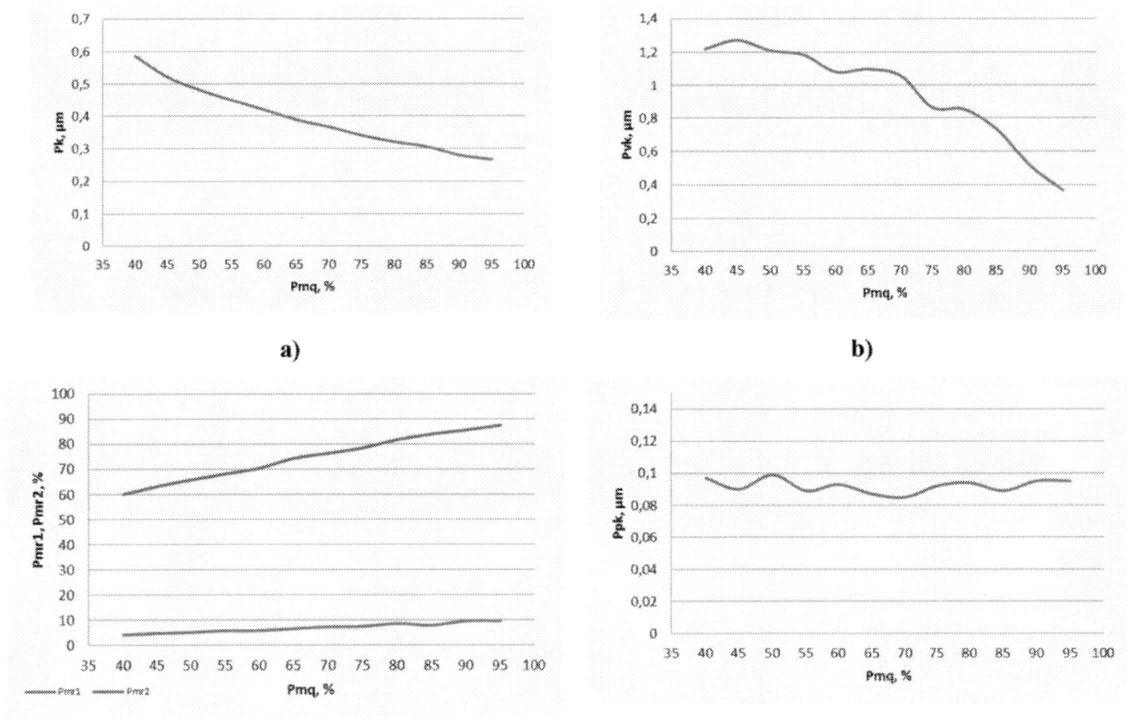

Figure 3. The effect of Pmq parameter on the Pk (a), Pvk (b), Pmr1, Pmr2 (c), Ppk (d), CLp = 8 µm, CLv = 40 µm, Ppq = 0.1 µm, Pvq = 0.9 µm.

Increase of the Ppq parameter (when Pmq and Pvq were constant) causes increase in the Pk parameter, for higher Pvq value Pk is larger for high values of Ppq. Similar, increase of the Ppq parameter causes also increase of the Ppk parameter. The tendency was found that increase of the Ppq parameter caused decrease of the Pvk parameter; this trend is more visible for smaller values of Pvq. Due to increase of Ppq, material ratios Pmr1 and Pmr2 increased; this trend is more visible for smaller values of the Pvq parameter. Dependence between Ppq parameter and parameters Pk, Ppk, Pvk, Pmr1 and Pmr2 is presented in Figure 4.

Increase of the Pvq value (when Pmq and Ppq are constant) causes increase of the Pvk parameter. The tendency was found that Pmr1 and particularly Pmr2 parameters decreased due to Pvq increase. The influence of Pvq change on the Ppk and Pk parameters is small, with tendency of the Pk increase due to Pvk increase (see Figure 5).

For profiles of the same correlation length of plateau and valley parts the observed previously effects of increase of the Pmq parameter on the Pk, Pvk, Pmr1 and Pmr2 were confirmed. The small dependence between Pmq material ratio and the Ppk parameter was also confirmed for high Pvq values; for smaller ones the tendency was found that increase of the Pmq caused also increase of Ppk.

The characters of dependencies between change of the Ppq parameter and the Pk, Ppk, Pmr1 and Pmr2 parameters found previously were confirmed, similarly to trend of decrease in Pvk due to increase of Ppq.

Increase of the Pvq caused increase of Pvk parameter. The effect of Pvq change on the Ppk value was small. The tendency obtained previously that Pvq increase causes decrease of Pmr1 and Pmr2 was confirmed.

Journal of Physics: Conference Series **483** (2014) 012021 doi:10.1088/1742-6596/483/1/012021

Generally, the tendencies of changes of parameters of two-process profiles with smaller values of horizontal length of plateau part comparing to valley part were confirmed for profiles of similar correlation lengths of 2 profile parts. In addition, the trend of increase of Ppk due to increase of the Pmq for small Pvq/Ppq ratio was found (see Figure 6).

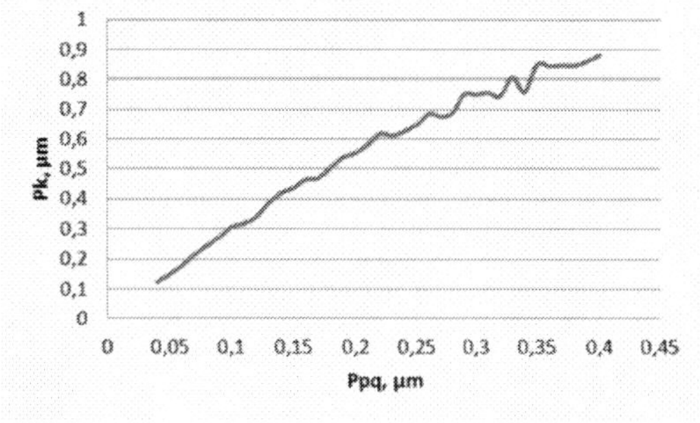

a)

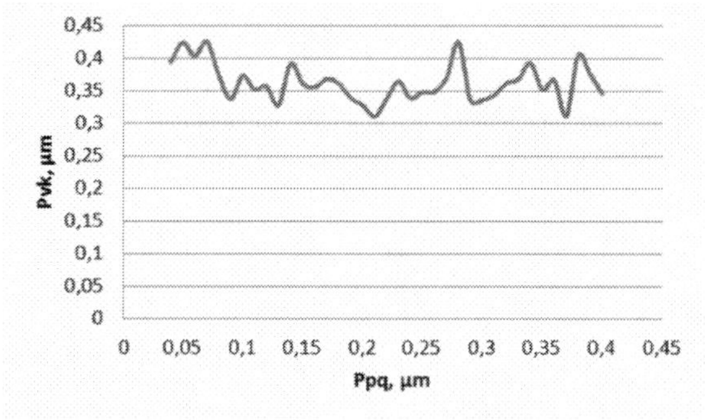

b)

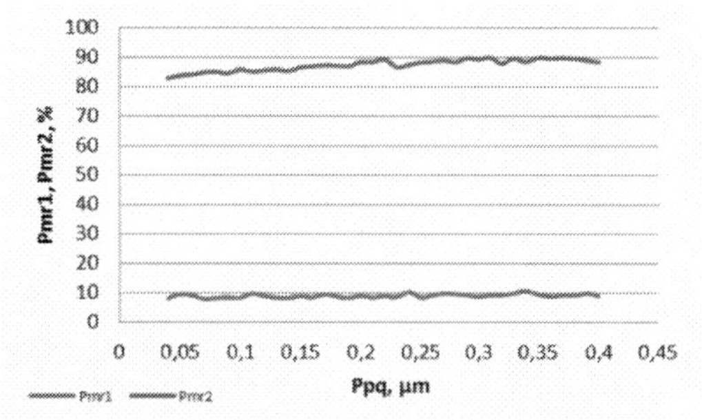

c)

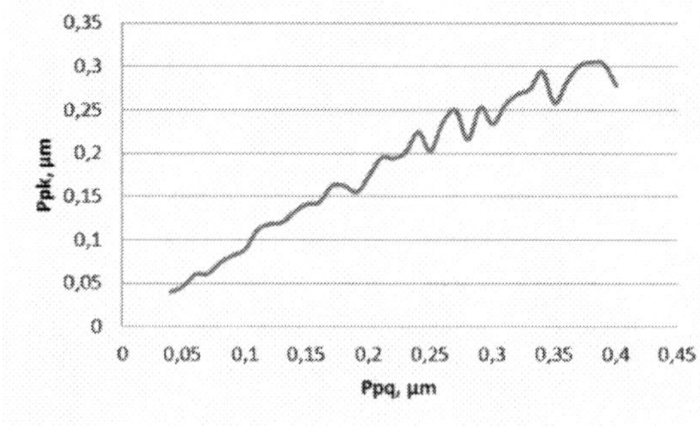

d)

Figure 4. The effect of Ppq parameter on the Pk (a), Pvk (b), Pmr1, Pmr2 (c), Ppk (d), CLp = 8 µm, CLv = 40 µm, Pvq = 4.5 µm, Pmq = 84%.

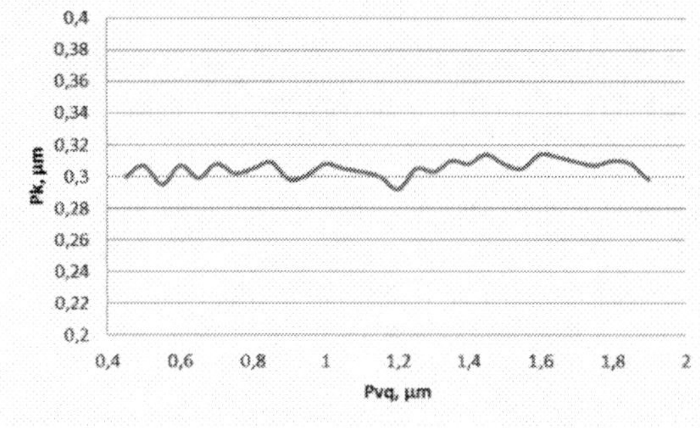

a)

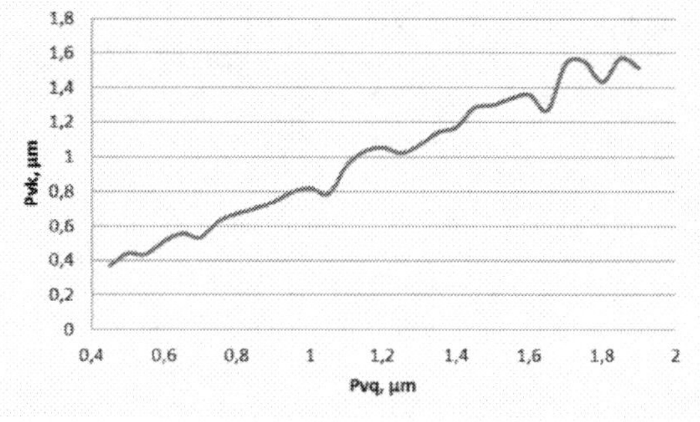

b)

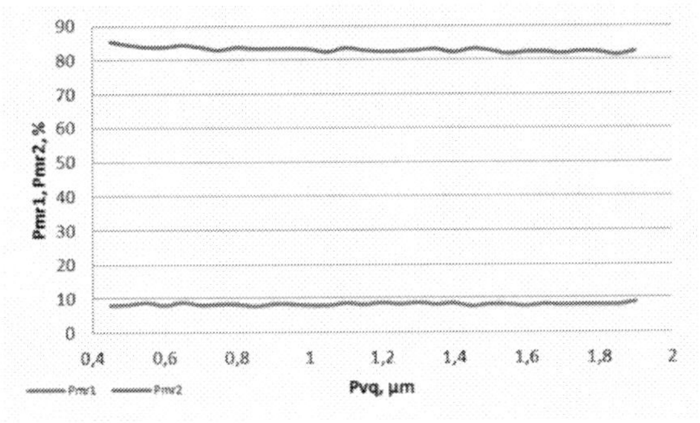

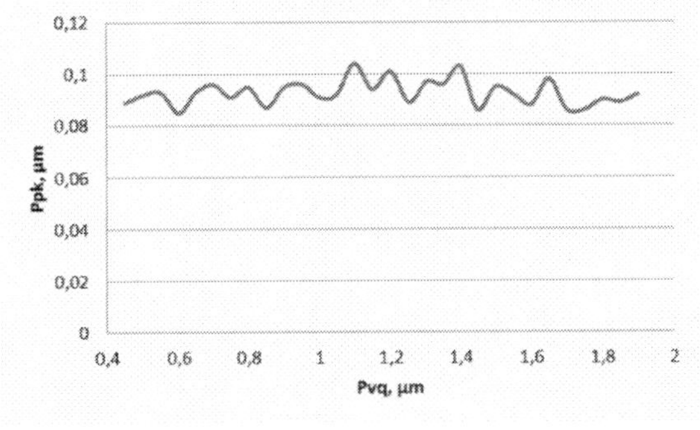

c) **d)**

Figure 5. The effect of Pvq parameter on the Pk (a), Pvk (b), Pmr1, Pmr2 (c), Ppk (d), CLp = 8 µm, CLv = 40 µm, Ppq = 0.1 µm, Pmq = 84%

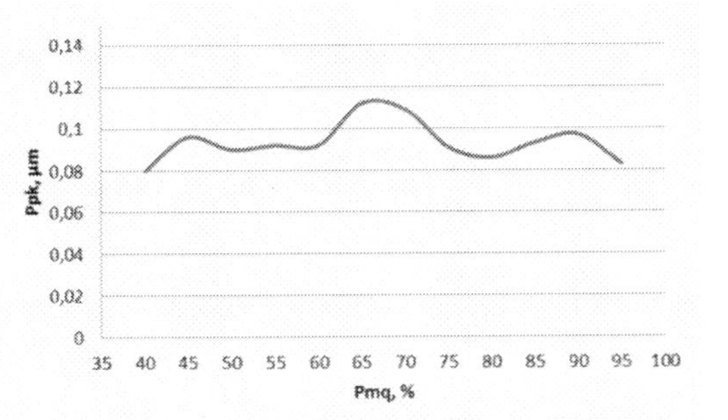

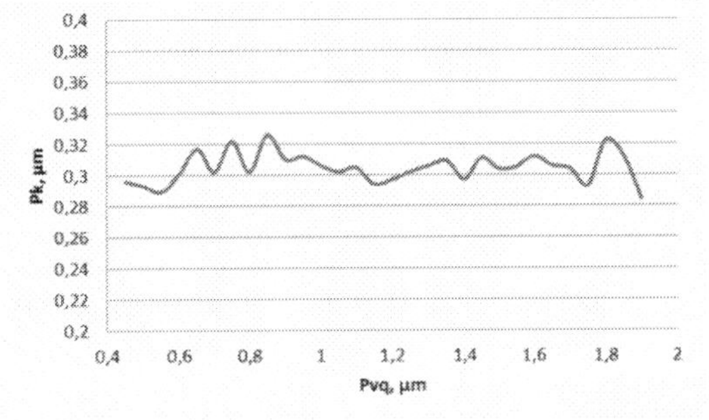

a) **b)**

Figure 6. The effect of Pmq parameter on Ppk parameter for Ppq = 0.1 µm, Pvq = 0.45 µm (a), the effect of Pvq parameter on Pk parameter, Pmq = 84%, Ppq = 0.1 µm (b), CLp = Clv = 40 µm.

The effects of the Pmq parameter increase and Pvk change on values of the parameters from the Sk family found for profiles of the same correlation lengths of plateau and valley parts were confirmed also for profiles of smaller correlation length of valley part comparing to plateau zone. The character of dependencies between increase of the Ppq parameter on parameters from the Sk group found for two analysed previously profiles sets was also confirmed.

Similar results were obtained in other similar research [22]. Figure 7 presents examples of modeled profiles.

The effect of changing parameters from the Sq group on other parameters was also studied. When Ppq/Pvq was the same, the increase of Pmq caused decrease of the statistical height parameters height (Pa, Pq) and of Rq/Rt ratio. In this case Psk, Pp/Pt, Pda/PΔq, Pa/Pq decreased and Pku increased.

When Pmq was the same, Ppq/Pvq decrease caused decrease of Pp/Pt, Psk, Pk/Pt, Pda/PΔq, Pa/Pq, increase of Pku and increase of Pvk/Pt.

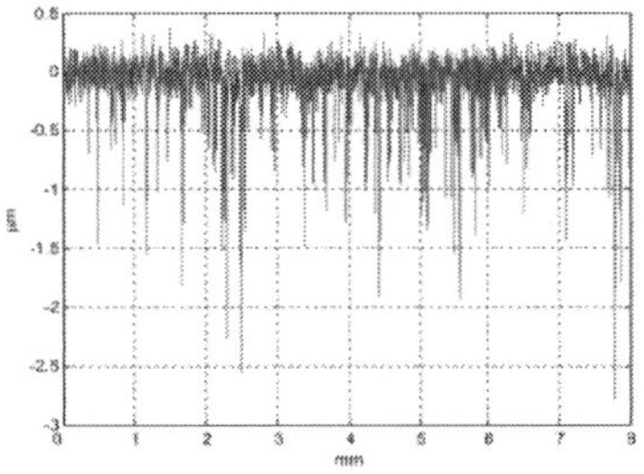

a)

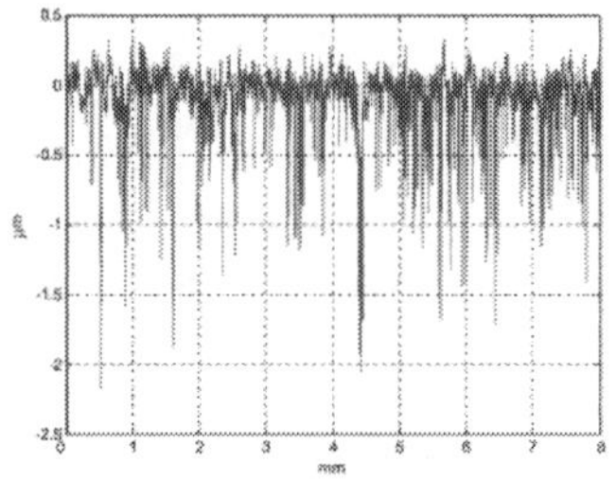

b)

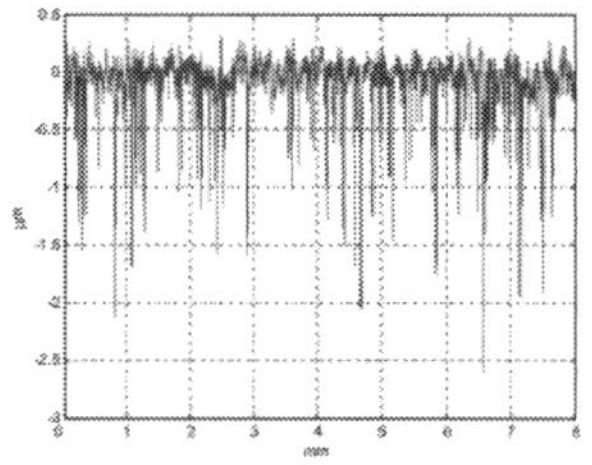

c)

Figure 7. Two-process profiles characterized by the following parameters: Ppq = 0.1 μm, Pvq = 0.9 μm, Pmq = 84%, CLp = 8 μm, CLV = 40 μm (a), CLp = 40 CLV = 8 μm (b), CLp = 40 μm, CLV = 40 μm (c).

In addition, the values of linear correlation coefficient were computed (see Table 1). Only part of findings obtained previously was confirmed. It was found that Pmq parameter is correlated only with Pmr1 and Pmr2 material ratios, Ppq with Ppk and Pk and Pvq with Pvk. Parameters from ISO 13565-3 standard are not inter-correlated contrary to those from ISO 13565-2 standard (only Pvk is independent parameter).

It is evident from this research that plateau profile part from ISO 13565-3 standard is connected with core and peak parts from ISO 13565-2 standard.

Table 1. Values of linear correlation coefficient between analysed parameters.

	Pmq	Ppk	Ppq	Pvk	Pvq	Pk	Pmr1	Pmr2
Pmq	1							
Ppk	0.109	1						
Ppq	0.085	0.977	1					
Pvk	-0.21	0.023	0.081	1				
Pvq	0.001	0.081	0.101	0.87	1			
Pk	-0.073	0.95	0.972	0.148	0.151	1		
Pmr1	0.757	0.32	0.271	-0.411	-0.239	0.102	1	
Pmr2	0.769	0.43	0.345	-0.612	-0.43	0.211	0.845	1

4. Conclusions

Increase of the Pmq parameter causes decrease of the Pq parameter, with larger changes obtained for smaller Pmq values. The effect of the Pmq parameter change on the Ppk parameter is negligible. Increase of the Pmq parameter causes decrease of the Pvq parameter; the changes are higher for larger Pmq material ratios. Due to Pmq increase, the material ratios Pmr1 and Pmr2 also increases.

Increase of the Ppq parameter causes increase of Ppk, Pk, Pmr1 and Pmr2 parameters and decrease of the Pvk parameter,
Increase of the Pvq parameter causes increase of the Pvk parameter. The Pvk parameter is the most independent parameter from the Rk group (ISO 13565-2 standard).
The obtained dependencies were found for various horizontal parameters of plateau and valley profile parts. They were partly confirmed by correlation analysis of parameters. For profiles, for which correlation length of plateau part was not smaller than that of valley part the tendency that due to Pmq parameter increase the Ppk parameter increased was additionally found.

5. References

[1] Bohm H J 1992 *International Journal of Machine Tools and Manufacture* **32** pp 109-113
[2] Trautwein R 1974 *MTZ* **35** pp 339-340
[3] Trautwein R 1977 *MTZ* **38** pp 565-566
[4] Zipin R B 1990 *Precision Engineering* **12** pp 106-108
[5] Michalski J, Pawlus P 1992 *Proceedings of the International ASME Conference: Application of Signals, Data, Systems Methodologies to Engineering Problems* Alexandria Egypt pp 3-13
[6] Malburg M C, Grant M B 1992 *The characterization of two-process surface texture using the cumulative Gaussian probability distribution* ISO Paper
[7] Nielsen H S 1988 *Precision Engineering* **10** pp 209-213
[8] Augustyn W 1985 *Elaboration of the method describing surface topography of cylinder liners* PhD Dissertation (Wroclaw: Wroclaw University of Technology) (in Polish)
[9] Michalski J, Pawlus P 1993 *Wear* **161** pp 135-143
[10] Malburg M C, Raja J 1993 *CIRP Annals* **42** pp 637-640
[11] Zipin R B 1983 *Applied Surface Science* **15** pp 334-360
[12] Whitehouse D J 1985 *Proceeding of the Inst. Mech. Engrs.* **199** pp 263-270
[13] Sanna Reddy H, Raja J, Chen K 1997 *Transactions of the 7th International Conference on Metrology and Properties of Engineering Surfaces* Gothenburg Sweden pp 111-118
[14] Williamson J P B 1967-68 *Proc. Inst. Mech. Engrs London* **182** pp 21-30
[15] Graboń W, Pawlus P 2010 *10th International Symposium on measurement and Quality Control* Osaka Japan
[16] Codgell J D 2008 *Precision Engineering* **32 pp** 34-46
[17] Jakubiec W, Brylski M 2004 *Proceedings of 9th International Conference on Metrology and Properties of Engineering Surfaces* (Halmstad: Halmstad University Sweden) pp 77-84
[18] Rosen B G, Andeberg C, Ohlsson R 2004 *XI International Colloquium on Surfaces* Chemnitz, Germany
[19] Anderberg C, Pawlus P, Rosén B G, Thomas T R 2009 *Journal of Materials Processing Technology* **209** pp 1936-1942
[20] Szczepański M 2012 *Comparative analysis of parameters from standards ISO 13565-2 and ISO 13565-3* MSc Dissertation (Rzeszow: Rzeszow University of Technology)(in Polish)

Journal of Physics: Conference Series **483** (2014) 012022 doi:10.1088/1742-6596/483/1/012022

Deposition uniformity inspection in IC wafer surface

W C Li, Y T Lin, J J Jeng and C L Chang
Mechanical and Systems Research Laboratories, Industrial Technology Research Institute, 195, Sec. 4, Chung Hsing Rd., Hsinchu, Taiwan 300, R.O.C.

E-mail: jeecool@itri.org.tw

Abstract. This paper focuses on the task of automatic visual inspection of color uniformity on the surface of integrated circuits (IC) wafers arising from the layering process. The oxide thickness uniformity within a given wafer with a desired target thickness is of great importance for modern semiconductor circuits with small oxide thickness. The non-uniform chemical vapor deposition (CVD) on a wafer surface will proceed to fail testing in Wafer Acceptance Test (WAT). Early detection of non-uniform deposition in a wafer surface can reduce material waste and improve production yields. The fastest and most low-priced inspection method is a machine vision-based inspection system. In this paper, the proposed visual inspection system is based on the color representations which were reflected from wafer surface. The regions of non-uniform deposition present different colors from the uniform background in a wafer surface. The proposed inspection technique first learns the color data via color space transformation from uniform deposition of normal wafer surfaces. The individual small region statistical comparison scheme then proceeds to the testing wafers. Experimental results show that the proposed method can effectively detect the non-uniform deposition regions on the wafer surface. The inspection time of the deposited wafers is quite compatible with the atmospheric pressure CVD time.

1. Introduction

Non-contact visual inspection based on image analysis techniques has become important for qualitative evaluation of surface quality in manufacturing. Defects that appear as local anomalies, such as stains, scratches, and wear, on material/product surfaces generally lack quantitative measures to define. They are also difficult to identify if the surrounding background contains complex patterns. Automated visual inspection has been successfully applied to a wide variety of material surfaces found in industry, such as printed circuit boards, PCBs [1-3], semiconductor wafers [4-6], liquid crystal display (LCD) panels [7-9], solar cells [10, 11], and textile fabrics [12, 13]. For PCB inspection, embedded fiducial markers on the boards allow the image under inspection to be registered with respect to the golden template for comparison. For semiconductor wafer inspection, the surface shows repetitive die patterns in the image so that the defect can be identified when it breaks the periodicity on the surface, but the uniformity inspection of a deposited wafer surface was less mentioned in technical literature. This paper focuses on the task of automatically visual inspection of color uniformity on the surface of integrated circuits (IC) wafers arising from the layering process. The non-uniform chemical vapor deposition (CVD) on a wafer surface will proceed to fail testing in Wafer Acceptance Test (WAT). Early detection of non-uniform deposition in a wafer surface can reduce material waste and improve production yields.

Journal of Physics: Conference Series **483** (2014) 012022 doi:10.1088/1742-6596/483/1/012022

2. Non-uniform deposition inspection in IC wafer surface

This section presents the proposed defects detection scheme for deposited wafer images. Section 2.1 first describes the HSI color feature extraction and normal data training procedure in the IC wafer surface. Section 2.2 further discusses the defect identification scheme in the deposited wafer surface.

2.1. Color features extraction from HSI color space

HSL and HSV are the two most common cylindrical-coordinate representations of points in an RGB color model. The two representations rearrange the geometry of RGB in an attempt to be more intuitive and perceptually relevant than the Cartesian (cube) representation. A third model, common in computer vision applications, is HSI, for hue, saturation, and intensity. The hue-saturation-intensity color model has human-intuitional advantages in image processing [14] such as color image enhancement, segmentation, fusion, color-based object detection, recognition, traffic signal detection and skin detection. The HSI color space, which describes more exact color than RGB model describes for human interpretation. Though the HSI model is non-uniform in perception, it is still one of the most popular color models for image processing. In practical applications, the non-uniformity in perception can be solved by some techniques. For example, in color image segmentation the deformed boundaries can be used to enclosed the target cluster so that the perception uniformity is not so critical. The algorithm of HSI color model is as follows:

$$I = \frac{R+G+B}{3}, \tag{1a}$$

$$H = \begin{cases} \theta, & \text{if } B \le G \\ 360° - \theta, & \text{if } B > G \end{cases} \quad \text{with } \theta = \cos^{-1}\left\{ \frac{\frac{1}{2}[(R-G)+(R-B)]}{\left[(R-G)^2 + (R-B)(G-B)\right]^{1/2}} \right\}, \tag{1b}$$

$$S = 1 - \frac{3}{R+G+B}\min(R,G,B), \tag{1c}$$

where I and S are in the range of [0, 1] and H is in the range of [0, 360°]. Figure 1(a1) shows a normal chemical deposited wafer surface image, and (a2)-(a4) present its HSI color model images individually. Figure 1(a3) and (a4) present non-uniformity color properties from wafer surface. Figure 1(b1) shows a defect deposited wafer image and its HSI color model images individual shown in (b2)-(b4). Defect region can be found on top of wafer with triangle shape. The triangle defect region shows higher hue value than normal wafer in H color band (shown in (a4) and (b4)). Because of the non-uniformity color properties in wafer surface, the defect detection strategy can not only define an arbitrary threshold to distinguish defect region from normal background.

In this research, the normal deposited wafer images are first grabbed from camera and selected the region of interesting (ROI) to calculate statistical data as the training databases. The training databases can be constructed a group of thresholding from individual color model image. The group of thresholding is defined as the measurement of uniformity for testing wafer images. The inspected wafers also have to convert HSI color model and calculate regional statistics. The deposited wafer surface inspection scheme is described in figure 2.

2.2. Deposited wafer inspection

Figure 3 describes of training ROI and definition of parameters in wafer image. The region of interesting is given by of size $R_W \times R_H$. In this paper, we define that the patch as "cell" in the region of interesting. The cell width w_{cell} and height h_{cell} is defined by equation (2).

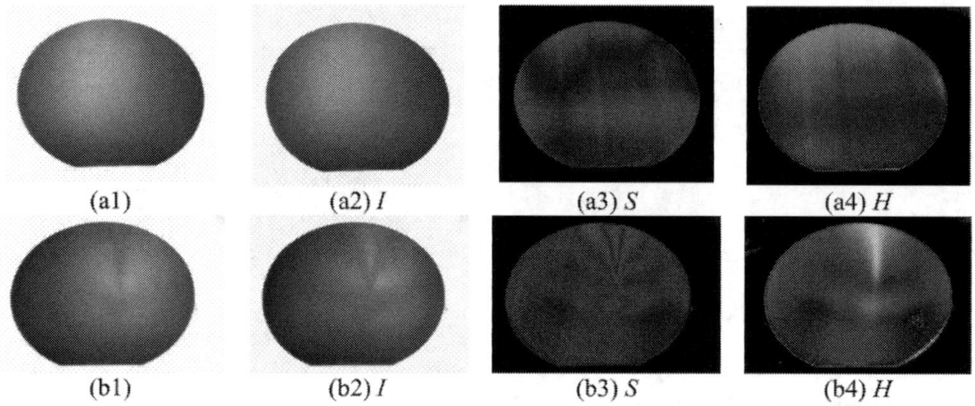

Figure 1. Deposited wafer image: (a1) normal wafer image; (a2)-(a4) the individual HSI color model images; (b1) defect wafer image; (b2)-(b4) the individual HSI color model images.

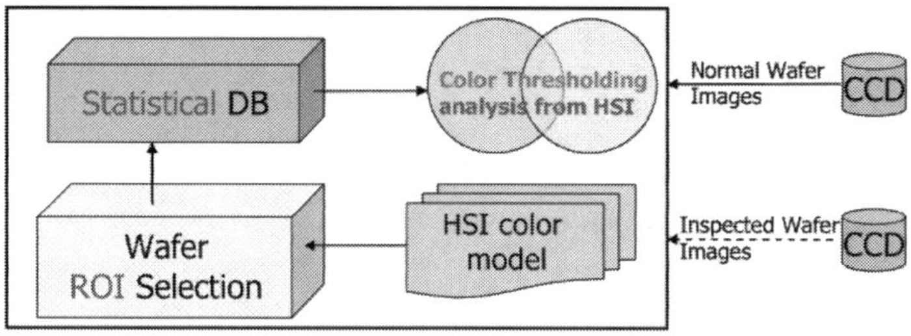

Figure 2. Deposited wafer surface inspection scheme.

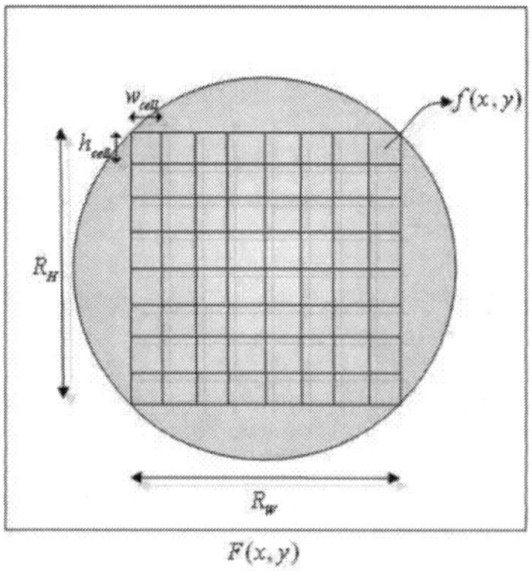

Figure 3. Description of inspection area in wafer image and size of cells.

$$w_{cell} = \frac{R_W}{N_W},\ h_{cell} = \frac{R_H}{N_H} \tag{2}$$

where N_W is the number of segment at R_W, and N_H is the number of segment at R_H. Therefore, the average $\mu(x, y)$ of each cell at left-top position (x, y) is calculated by

$$\mu(x,y) = \frac{1}{w_{cell} \times h_{cell}} \sum_{i=1}^{w_{cell}} \sum_{j=1}^{h_{cell}} f(x+i, y+j),$$
$$x = c \times w_{cell}, y = c \times h_{cell}, c = 0, 1, \ldots, N_W \times N_H$$

and the standard deviation $\sigma(x, y)$ is calculated by

$$\sigma(x,y) = \left\{ \frac{1}{w_{cell} \times h_{cell} - 1} \sum_{i=1}^{w_{cell}} \sum_{j=1}^{h_{cell}} \left[f(x+i, y+j) - \mu(x,y) \right]^2 \right\}^{1/2}$$

Since every cell in the region of interesting is calculated its average and standard deviation, the total average $\hat{\mu}$ and total standard deviation $\hat{\sigma}$ can be given by

$$\hat{\mu} = \frac{1}{N_W \times N_H} \sum_{x=1}^{N_W \times N_H} \sum_{y=1}^{N_W \times N_H} \mu(x,y)$$
$$\hat{\sigma} = \left\{ \frac{1}{N_W \times N_H - 1} \sum_{x=1}^{N_W \times N_H} \sum_{y=1}^{N_W \times N_H} \left[\mu(x,y) - \hat{\mu} \right]^2 \right\}^{1/2} \tag{3}$$

To segment the defect cell from the background in the $F(x, y)$ image, a simple thresholding process is used. If the discriminant measure $\mu(x, y)$ of each cell is significantly larger or less than threshold, the cell $f(x, y)$ in the original image is classified as a defect one. Otherwise, it is defect-free. The simple statistic process control (SPC) is applied to set up the upper and lower limit for $\mu(x, y)$. The control limit (i.e., threshold) used to identify defect points is given by

$$\begin{aligned} U_I &= \hat{\mu}_I + k \times \hat{\sigma}_I \\ L_I &= \hat{\mu}_I - k \times \hat{\sigma}_I \\ U_S &= \hat{\mu}_S + k \times \hat{\sigma}_S \\ L_S &= \hat{\mu}_S - k \times \hat{\sigma}_S \end{aligned} \tag{4}$$

where U_I and U_S is upper control limit in intensity and saturation image, and L_I and L_S is respectively lower control limit. $\hat{\mu}_I$ and $\hat{\mu}_S$ is the total average of measured value $\mu(x, y)$ for all cells in intensity and saturation image. $\hat{\sigma}_I$ and $\hat{\sigma}_S$ is the total standard deviation of cells. In this research, H color model image presents high color variation which may cause to mis-detected in normal image surface with lightly non-uniform. Therefore, we only concerned about I and S color model for wafer inspection. The parameter k is a predetermined multiplication of the standard deviation from the mean. In the experiment, the control constant k is empirically given by 2.0 for all test images. A defect cell will be set to 1 (marked with red rectangle) if the measured value is larger than U_I or U_S or less than L_I or L_S, whereas a defect-free cell will be set to 0 (do nothing) in the original image. That is,

$$f(x,y) = \begin{cases} NG(1) & \text{if } \mu(x,y) > U_I \text{ or } \mu(x,y) < L_I \\ & \text{or } \mu(x,y) > U_S \text{ or } \mu(x,y) < L_S \\ Pass(0) & \text{otherwise} \end{cases} \tag{5}$$

3. Experimental results

The proposed algorithms were implemented on an Intel Core 2, 3.4GHz CPU personal computer. The mean computation time of the proposed method was 0.5s for a 1280×1024 image (which exclude the training time). The control constant k of the statistical control limit is set at 2 for defect detection. Figure 4 shows the detection results of four normal wafer images and four defect wafer images. Figure 4(a2)-(d2) show the individual detection results of (a1)-(d1), and they are indexed with the green "PASS" symbol on the right-top of panel. They are no false alarms are generated. Figure 4(e1)-(h1) present four defect wafer images and their detection results which shown in figure 4(e2)-(h2). The individual defect cells are marked with red rectangles and indexed with red "NG" symbol to notice users.

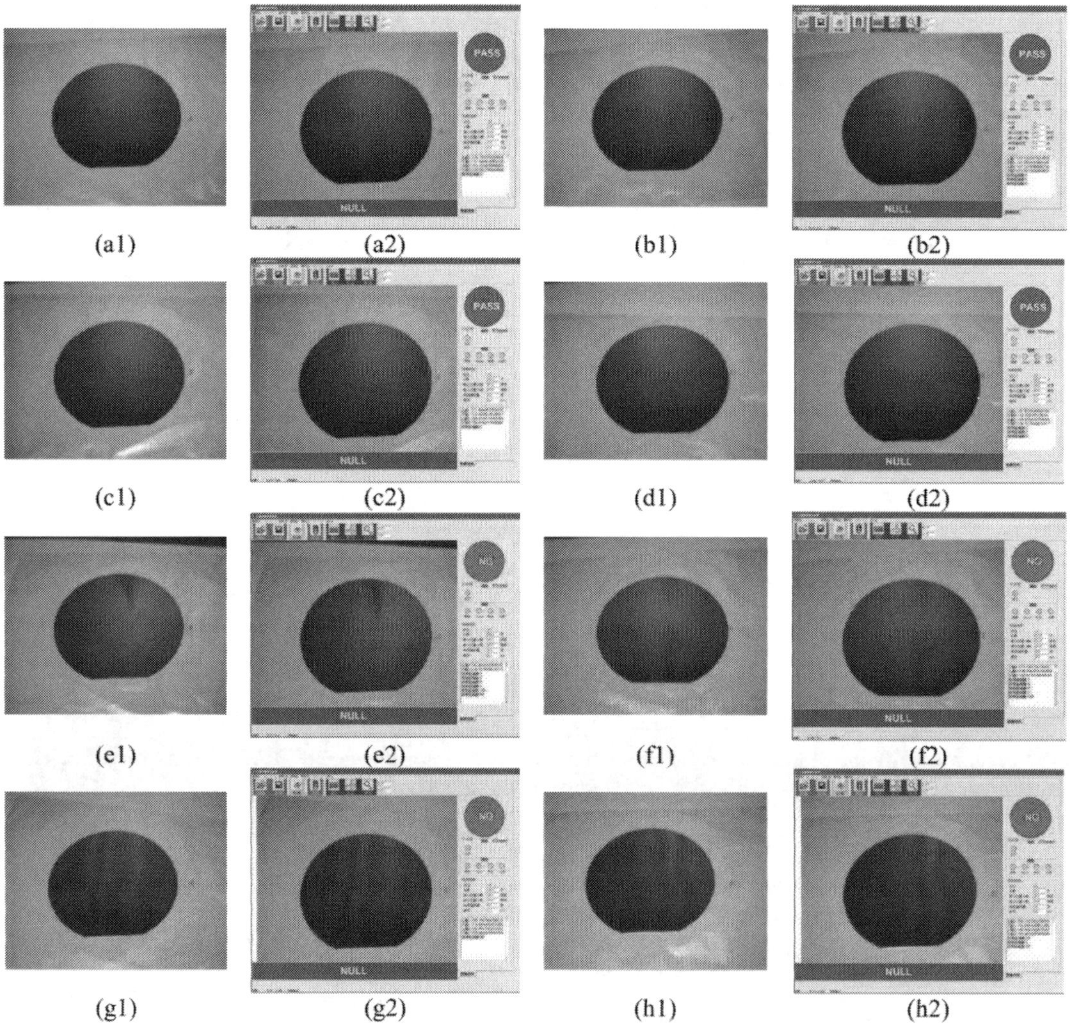

Figure 4. Detection results of deposited wafers: (a1)-(d1) four normal wafer images and individual detection results shown in (a2)-(d2); (e1)-(h1) four defect wafer images; (e2)-(h2) the individual detection results and marked defect cells with red rectangles.

Journal of Physics: Conference Series **483** (2014) 012022 doi:10.1088/1742-6596/483/1/012022

4. Conclusion

This paper has proposed a defect detection scheme based on regional statistical data technique to identify non-uniform defects in deposited IC wafer images. The individual small region statistical comparison scheme then proceeds to the testing wafers. Experimental results show that the proposed method can effectively detect the non-uniform deposition regions on the wafer surface. The inspection time of the deposited wafers is quite compatible with the atmospheric pressure CVD time.

5. References

[1] Moloney C R 1991 *IEEE Pacific Rim Conference on Communications, Computers and Signal Processing* pp. 811–814.

[2] Yeh C H and Tsai D M 2001 "Wavelet-based approach for ball grid array (BGA) substrate conduct paths inspection," *International Journal of Production Research*, Vol. 39, pp. 4281–4299.

[3] Leta F R, Feliciano F F and Martins F P R 2008 "Computer vision system for printed circuit board inspection," *ABCM Symposium Series in Mechatronics*, Vol. 3, pp. 623–632.

[4] Su C T, Yang T and Ke C M 2002 "A neural-network approach for semiconductor wafer post-sawing inspection," *IEEE Trans. on Semiconductor Manufacturing*, Vol. 15, pp. 260–266.

[5] Shankara N G and Zhongb Z W 2005 "Defect detection on semi-conductor wafer surfaces," *Microelectronic Engineering*, Vol. 77, pp. 337–346.

[6] Yeh C H, Wu F C, Ji W L and Huang C Y 2010 A wavelet-based approach in detecting visual defects on semiconductor wafer dies", *IEEE Trans. on Semiconductor Manufacturing*, Vol. 23, pp. 154–164.

[7] Oh J H, Kwak D M, Lee K B, Song Y C, Choi D H and Park K H 2004 Line defect detection in TFT-LCD using directional filter bank and adaptive multilevel thresholding," *Key Engi-neering Materials*, Vol. 270-273, pp. 233–238.

[8] Zhang Y and Zhang J 2005 Fuzzy recognition of the defect of TFT-LCD," *Proc. of SPIE*, pp. 233–240.

[9] Tsai D M and Tsai H Y 2010 Low-contrast surface inspection of mura defects in liquid crystal displays using optical flow-based motion analysis," *Machine Vision and Applications*, Vol. 22, pp. 629–649.

[10] Seiler D 2010 "Dual exposure speeds solar cell inspection," *Photonics Spectra*, Vol. 44, pp. 238–248.

[11] Zhang W-J, Li D and Feng Y 2010 "Investigation of visual inspection method for silicon solar cell," *Journal of Computer Applications*, Vol. 30, pp. 249–252.

[12] Cho C S, Chung B M and Park M J 2005 "Development of real-time vision-based fabric inspection system," *IEEE Transactions on Industrial Electronics*, Vol. 52, pp. 1073–1079.

[13] Kumar A 2008 "Computer-vision-based fabric defect detection: a survey," *IEEE Transactions on Industrial Electronics*, Vol. 55, pp. 348–363.

[14] Chien C-L and Tseng D-C 2011 "Color image enhancement with exact HSI color model," *International Journal of Innovative Computing, Information and Control*, Vol. 7, pp. 6691-6710.

Acknowledgments

This material is based upon work supported by the Industrial Technique Research Institute, Mechanical and Systems Research Laboratories under Grant No. C301AA4110. My thanks and appreciations also go to my colleague in developing the project and people who have willingly helped me out with their abilities.

Journal of Physics: Conference Series **483** (2014) 012023 doi:10.1088/1742-6596/483/1/012023

The effect of thickness and loading force on wear behavior of HfO_2

B-C He, W-E Fu, Y-Q Chang and H-C Liou
Centers for Measurement Standards, Industrial Technology Research Institute
No.321, Sec.2, Kuangfu Rd., Hsinchu, 30011, Taiwan

E-mail: hopc@itri.org.tw

Abstract. The effect of annealing treatment on hafnium oxide (HfO_2) thin films with two thicknesses (20 and 50 nm) has been observed. The surface morphology and tribological properties of HfO_2 thin films were measured through atomic force microscopy (AFM). It is found that annealing treatment and increasing thickness of thin film would promote the formation and growth of island-like grains, which contributes a raise in surface roughness. During scratch test, plowing behavior dominated the deformation mechanism in the form of lumps along the edge of grooves by AFM-3D images. Besides, slower scratch speed led to larger deformation energy and caused a serious fracture. The annealing-induced crystallization resulted in reduced penetration depth and coefficient of friction (COF). The varied COF with respect to different normal forces reflected substrate effect. The thicker HfO_2 films exhibited better wear resistance regardless of annealing conditions. It is because that increase in film thickness accompanied growth of polycrystalline structure.

1. Introduction

There are many applications based on hafnium oxide (HfO_2) because of its unique chemical and physical properties such as high chemical stability, hardness [1], and thermal stability. In semiconductor industry, it is increasingly difficult to reduce the gate leakage and the tunneling current in the SiO_2 and SiON layer for the state-of-the-art and future nanotechnology nodes as the device feature size gets smaller and smaller. Therefore, HfO_2 based gate dielectric materials becomes the main components allowing improvement in the miniaturization of advanced integrated circuits, particularly complementary metal oxide semiconductor devices [2-4]. Although detailed studies exist on the electrical characterisation of HfO_2, its mechanical properties still need to be investigated. While the thickness of thin films has shrunk from micrometer to nanometer range for achieving the need of miniaturization of electronic devices, it would accompany with the varied structure of material, which changes morphology and nano-mechanical properties such as surface roughness, wear resistance, and fatigue [5,6]. Furthermore, annealing treatment usually accompanies with changes of thin films structure like phase transition and relaxation [7,8]. Consequently, cyclic thermal stress and wire bonding-induced stress during the semiconductor processing could possibly damage the films, and thus lower the structure reliability, if nano-mechanical properties of HfO_2 thin films are not fully understood.

Nanoindentation technique is frequently used to measure the nano-mechanical properties of thin films such as hardness and Yong's modulus [9,10], it is difficult to remove substrate effect while film shrinks to nanoscale [11,12]. Nanoscratch technique can avoid this difficulty encountered in nanoindentation and is well suited for characterizing practical adhesion failure of thin films and coatings. The wear resistance of thin film can also be accessed through elastic deformation and pileup [13]. Under controlled scratch length and scratch speed, the residual penetration depth, coefficient of friction (COF) can be measured and assessed for the surface hardness and wear characterization under varied normal force and annealing temperatures.

In previous results, the thermal treatment-induced crystallization can enhance surface hardness of HfO_2 thin films by nanoscratch technique [14]. In this study, the effect of not only thermal treatment but also films thickness on the surface hardness of HfO_2 thin films was discussed. Furthermore, the crystallization of thin films can be reflected on surface morphology. For these reasons, the morphologies of as-deposited and annealed HfO_2 thin films with thickness of 20 and 50 nm were scanned by atomic force microscope (AFM). Nanoscratch experiments were also executed by AFM in lateral force measurement mode, in which a diamond coated tip was used to scratch HfO_2 thin films surface. The deformation mechanism was assessed by measured groove morphologies and cross-section scratch profiles. In addition, the coefficient of friction (COF) was evaluated using a combination of normal and lateral forces. The reliability of HfO_2 thin films deposited on Si wafer was assessed for semiconductor applications based on the evaluation of surface hardness and wear resistance.

2. Experiments

The method of preparing HfO_2 thin films has been described in previous study [14]. Firstly, the HfO_2 layers were grown on 8-inch (1 0 0) p-type Si wafers by using atomic layer deposition (ALD, Polygon 8200, ASM, USA) method that makes atomic scale deposition control possible due to the characteristics of self-limiting and surface reactions. Detail illustration of ALD process can be found elsewhere [15]. During the process, alternating surface saturating reactions between hafnium tetrachloride ($HfCl_4$) and water (H_2O) precursors at 300 °C were performed. Since no cleaning process before the deposition was applied to the bare Si wafer, a native SiO_2 thin layer existed between HfO_2 and Si wafer. The SiO_2 interface was not significantly affected by the ALD process. The as-deposited HfO_2 thin film was then cut into 20 mm × 20 mm samples for the subsequent annealing process. The annealing process was performed by rapid temperature annealing (RTA) method for 2 minute at 900 °C with a heating rate of 50 °C /sec under an argon environment. There has a potential crystalline phase transition and changes in crystallite sizes of HfO_2 thin films under the selected annealing temperatures [16].

For the nanoscratch tests, AFM (Dimension Icon, Bruker, USA) was used to perform the experiments. A nanoscale scratch on the HfO_2 thin film surface was obtained using a diamond-coated Si tip (DT-NCHR, Nanosensors, USA). This coating on tip features with extremely high wear resistance due to the unsurpassed hardness of diamond. The tip radius of curvature is ~100 nm on a 10-μm-tall triangular pyramid probe. As far as mechanical specification is concerned, the probe includes a spring constant of ~42 N/m and a resonance frequency of ~330 kHz. The scratches were carried out with constant normal loads of 8, 24, 40, and 56 μN. The length of each groove was 2 μm. The scratching direction was perpendicular to the direction of cantilever length. At the same time, the lateral force was measured in situ through lateral force mode [17].

The surface morphology were probed using a SiN tip (SCANASYST-AIR, Bruker, USA) in peak force tapping mode to prevent additional damage to the groove. The pyramid shape SiN tip had a typical tip radius of 2~12 nm with a spring constant of ~0.4 N/m and a resonance frequency of 70 kHz. The surface hardness of the as-deposited and annealed HfO_2 thin films was characterized by the analysis of the scratch depths.

3. Results and Discussions

3.1. Surface property analysis

Figure 1 shows the morphology of HfO_2 thin films. It is obvious that the as-deposited HfO_2 thin films with 20 nm in thickness exhibited a flattest surface with low island density and minimum root mean square (RMS) surface roughness of 0.459 nm. The surface became rougher due to an increase in island density after 900 °C annealing. According to previous research [18], there has a crystallization process from amorphous to polycrystalline structure with appropriate thermal treatment. As thin films thickness increased to 50 nm, more islands emerged which brings significant increase in roughness. It is because the HfO_2 thin films structure is closely dependent on the film thickness. The nucleations of other growth directions are increased and the grain growth is expanded as the film thickness is increased [19]. The defects formed at the film growth at low temperature may provide nucleation sites for crystalline growth. Besides, the increase of the stress with the film thickness can contribute to the change in structure, which means relatively lower crystallization energy is required. It is also found that annealing treatment promote more crystallization of 50 nm thin films and can be proved by rougher morphology.

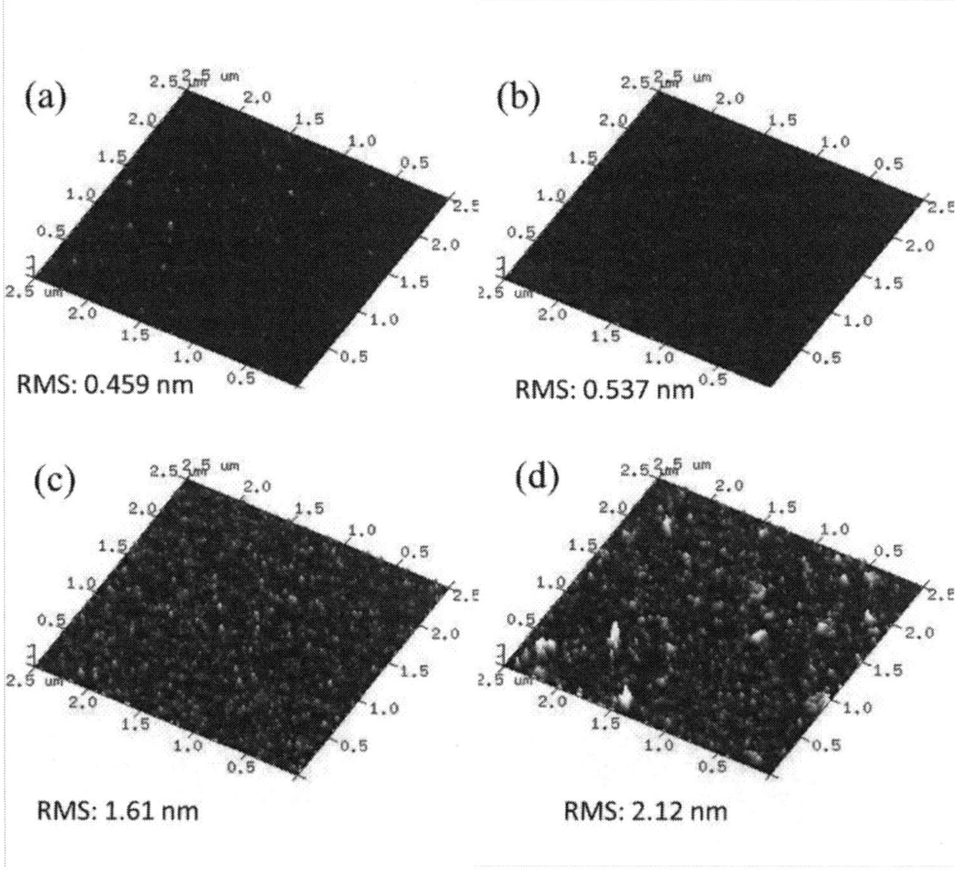

Figure 1. AFM-3D surface morphology of (a) as-deposited and (b) 900 °C-annealed HfO_2 thin films with 20 nm in thickness, and (c) as-deposited and (d) 900 °C-annealed HfO_2 thin films with 50 nm in thickness.

The surface hardness of the HfO_2 thin films can be evaluated through the volume of the removed material using scratch technique. Figure 2(a) shows the AFM image of scratch topographies for as-deposited HfO_2 thin films with 20 nm in thickness under different normal forces. It was found that lumps appeared along the two sides of scratch groove and became significant as normal force increased. The formation of lumps was a plastic deformation mechanism that the surface material underwent a plowing behavior by AFM tip as a result of pileups on the groove edges. Furthermore, the degree of plastic deformation is proportional to the volume of lump [20]. Hence, the plastic deformation became significant accompanying with lager normal force from increased lump. It is worth noting that there was no agglomerated continuous ribbon-like debris and only a small amount of wear plate-like debris around the grooves. It means that HfO_2 thin films exhibited good scratch resistance and high fracture strength under normal force range from 8 to 56 μN. It also indicates that the material removal mechanism is primarily based on plowing behavior, but not cutting behavior and brittle fracture [21]. Figure 2(b) shows AFM images of scratch topographies for HfO_2 thin films with 20 nm in thickness after 900 °C annealing. It is found that the plowing behavior still dominated on annealed samples and less debris was observed comparing with as-deposited one. However, the scratch grooves became narrower under all normal forces after annealing treatment. In addition, the scratches grooves were more unobvious on 50 nm thin films, as seen in figure 2(c) and (d). It indicated both thermal treatment and increase in film thickness can enhance the wear resistance. In addition, scratch speed effect on tribological properties was discussed. Figure 3 shows the scratch topographies for 20-nm-thickness thin films with 0.2 μm/s scratch speed. Compared with Figure 2 in which scratch speed of 2 μm/s was executed for all the scratches, the larger plate-like debris was observed on the scratch groove. This seems that AFM tip supplied more deformation energy to the surface materials at slower scratch speed and caused a serious fracture.

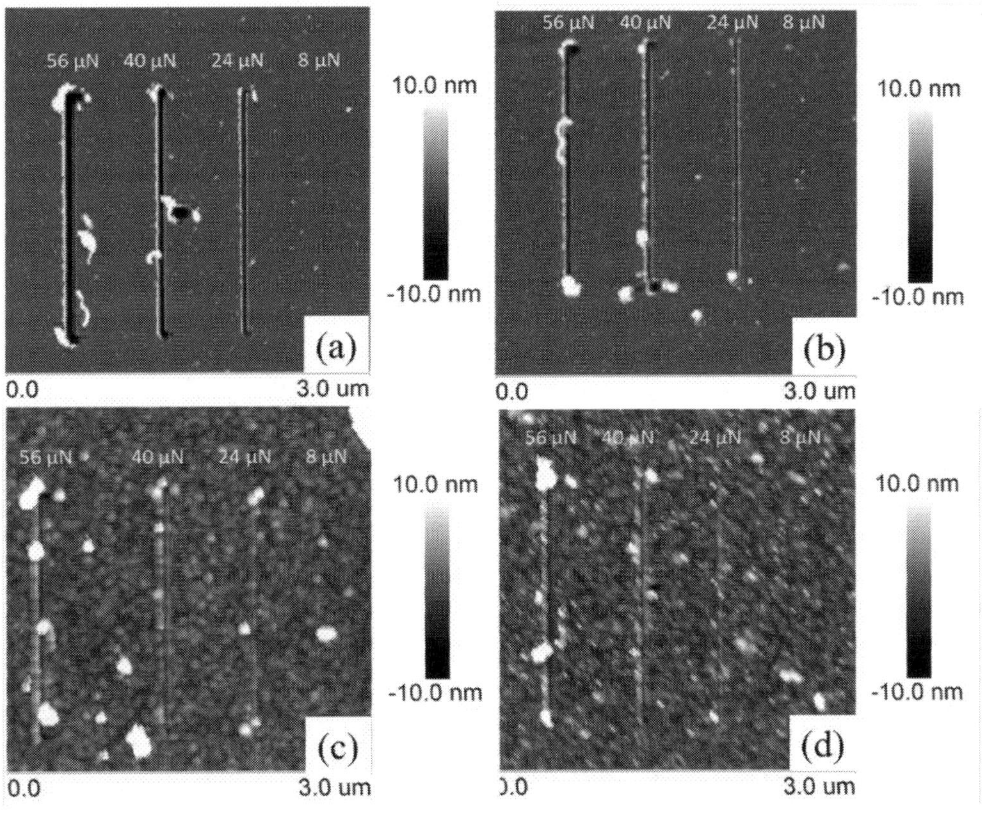

Figure 2. AFM-2D images of scratches on 20-nm-thick HfO_2 films under (a) as-deposited and (b) 900 °C- annealed condition and 50-nm-thick HfO_2 films under (c) as-deposited and (d) 900 °C-annealed condition.

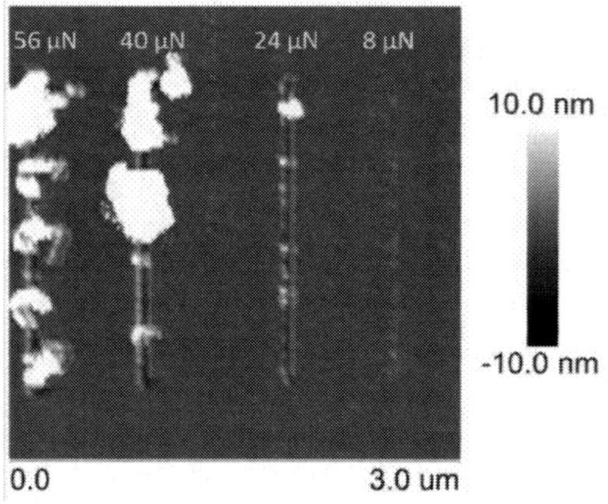

Figure 3. AFM-2D images of scratches on 900 °C annealed HfO_2 thin films (20 nm in thickness) with 0.2 μm/s scratch speed.

3.2. Cross-section analysis of scratches

Figure 4 shows the cross-section profiles of scratch depth under 56 μN normal force. Compared with minimum scratch depth (~2.5 nm) in this experiment, the roughness is small enough to be ignored. In 20 nm samples, the groove depth obviously decreased after annealing treatment. From previous study [14], the crystallization process from amorphous to polycrystalline structure would increase surface hardness. In addition, amorphous structure force contributed to relative significant pileups due to plowing behavior. There has a further decrease in groove depth while HfO_2 thin films thickness increase from 20 to 50 nm, which proves that the thicker HfO_2 thin films exhibited better surface hardness either in as-deposited or annealed state due to polycrystallization.

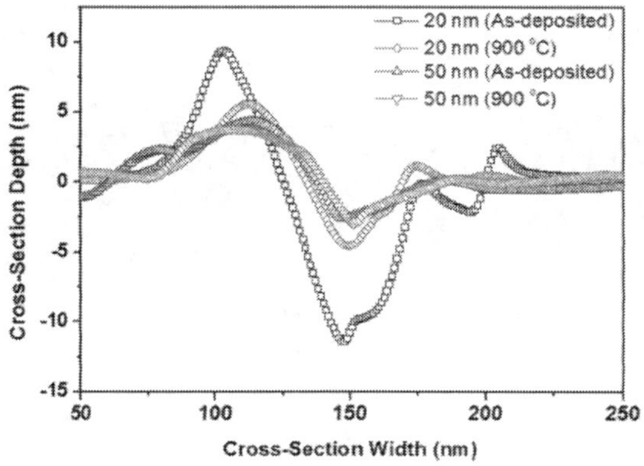

Figure 4. Cross-section profiles of AFM scratches at 56 μN normal forces on as-deposited and 900 °C-annealed HfO_2 thin films with 20 and 50 nm in thickness.

Journal of Physics: Conference Series **483** (2014) 012023 doi:10.1088/1742-6596/483/1/012023

3.3. Coefficient of friction

The COF is the relation between normal and lateral force [21,22]. In figure 5, the COF for annealed surface is smaller than as-deposited one for both 20 and 50 nm samples under all normal forces, indicating that crystallization by annealing treatment can enhance the wear resistance [14,22]. In addition, the COF increased at maximum normal force for as-deposited samples with both 20 and 50 nm thickness, which was due to substrate effect [23]. This effect reduces the hardness while hard thin films (HfO_2) was deposited on soft substrate (Si), and also lowers the wear resistance. However, there has no same tendency on annealed samples and COF even decreased under maximum normal force. It is due to the formation of $HfSi_xO_y$ layer, which possesses higher hardness than HfO_2 thin films [14]. Therefore, the annealed samples can keep and even exhibit higher wear resistance. It also can be observed that thicker films has lower COF. The results can be contributed to higher degree of crystallization of thicker films than thinner ones.

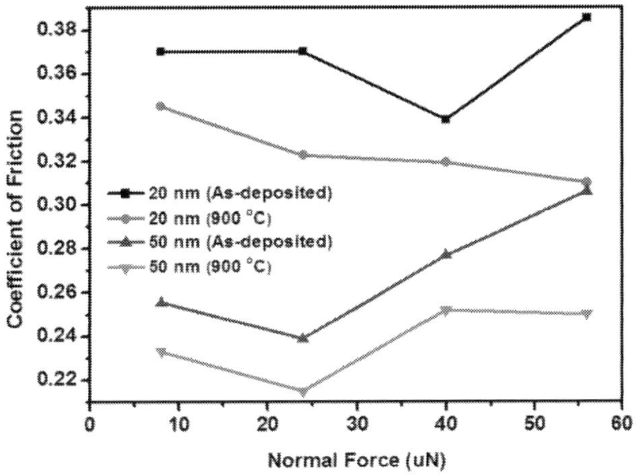

Figure 5. The normal force with respect to COF.

4. Conclusion

The structure of mechanical behaviors of HfO_2 thin films with different films thickness, normal forces, and thermal treatments were investigated through morphology and nanoscratch tests with AFM system. The groove morphology, depth cross-section, and COF were considered for evaluating the phase transition, material removal mechanism, surface hardness, and wear resistance. From AFM images, rougher surface can be observed through annealed and thicker samples due to crystallization of HfO_2 thin films. Lump along the scratch groove without the ribbon-like debris denoted that plowing behavior dominated the plastic deformation. Besides, low scratch speed made significant fractures due to large deformation energy. It was found that the scratch depth can be affected by annealing treatment and film thickness, which means that the structure transformation from amorphous to polycrystalline phases can enhance the surface hardness of the HfO_2 thin films. Compared with as-deposited and thinner sample, lower COF and scratch groove of annealed and thicker ones under the same normal forces condition were also attributed to the annealing-induced crystallization which increased the wear resistance.

Acknowledgements
The financial support provided by the Bureau of Standards, Metrology and Inspection (BSMI) Nanometer-Scale Metrology Project is gratefully acknowledged.

5. Reference

[1] Wang J, Li H P and Stevens R 1992 *J. Mater. Sci.* **27** 5397
[2] Wilk G D, Wallace R M and Anthony J M 2001 *J. Appl. Phys.* **89** 5243.
[3] Geng H and Hwaiyu G 2005 *Semiconductor Manufacturing Handbook* (New York: McGRAW-HILL)
[4] Locqute J, Marchiori C, Sousa M, Fompeyrine J and Seo J W 2006 *J. Appl. Phys.* **100** 051610
[5] Paldey S and Deevi S C 2003 *Mater. Sci. Eng., A* **342** 58
[6] Purswani J M and Gall D 2008 *J. Appl. Phys.*
[7] Pearson A J, Wang T, Jones R A L and Lidzey D G 2012 *Macromolecules* 45 1499
[8] Gu D, Zhu M, Celler G K, and Baumgart H 2009 *Electrochem. Solid-State Lett.* **12** H113
[9] Wang Z M, Wang J Y, Jeurgens L P H, Phillipp F and Mittemeijer E J 2008 Acta Mater. **56** 5047
[10] Welzel U, Ligot J, Lamparter P, Vermeulen A C and Mittemeijer E J 2005 J. Appl. Crystallogr. **38** 1
[11] Bhushan B 1995 *Handbook of Micro/nano Tribology* (CRC Press)
[12] Moody N R, Gerberich W W, Burnham N and Baker S P 1998 *Mater. Res. Soc. Symp. Proc.* **522** 39
[13] Doerner M F and Nix W D 1986 *J. Mater. Res.* **1** 601
[14] He B C, Fu W E, Chang Y Q and Liou H C 2013 *Thin Solid Films* **544** 212
[15] Bhattacharya A K and Nix W D 1988 Int. J. Solids Struct. **24** 1287
[16] Jardret V, Zahouani H, Loubet J L and Mathia T G 1998 *Wear* **218** 8
[17] Fu W E and Chang Y C, 2011 Appl. Surf. Sci. **257** 7436
[18] SpigaS, Driussi F, Lamperti A, Congedo G and Salicio O 2012 *Appl. Phys. Express* **5** 021102
[19] Cho M-H, Roh Y S, Whang C N, Jeong K, Nahm S W, Ko D-H, Lee J H, Lee N I and Fujihara K 2002 *Appl. Phys. Lett.* **81** 472
[20] Lysaght P S, Barnett J, Bersuker G I, Woicik J C, Fischer D A, Foran B, Tseng H H and Jammy R 2007 *J. Appl. Physiol.* **101** 024105
[21] Fu W E, Chang C W, Chang Y Q, Yao C K and Liao J D 2012 *Appl. Surf. Sci.* **258** 8974
[22] Nie X, Zhang P, Weiner A M and Cheng Y T 2005 *Nano Lett.* **5** 1992
[23] Li Q, Lee C, Carpick R W and Hone J 2010 *Phys. Status Solidi B* **247** 2909

Qualification and validation the metrology module located on CN machine tool -EMCO PC Mill 155 –

S. Merghache[1], A.Ghernaout[2]

[1]Department of Mechanics, Faculty of the Engineer's Sciences, University of Tlemcen, University Center of Tissemsilt, BP 119 Tlemcen, 13000 Algeria.
[2]Department of Mechanics, University of Tlemcen, BP 119 Tlemcen, 13000 Algeria.

Abstract. The contribution of machine tools in the blooming of several industrial domains is not to be demonstrated any more. However, during the last decade, the economic context imposed on this industry new standard of performance as regards the quality, the productivity, the costs and the delay of the production. The evolution of digitally operated modern machines drives integration of measurement and control in the cycle of manufacturing. The probe system brings an innovative solution to the improvement of CN machine tools efficiency and significantly expand its range of applications to improve the manufacturing process. The purpose of this article is to present our approach on the integration and the qualification of a metrology module associated with a software acquisition and analysis in relation to the digital control, a solution which we fulfilled and validated within our laboratory.

1. Introduction

The economic profitability of the automation of means of production requires a rigorous and automatic control of the manufactured products. Indeed, the quality of these products requires so far, means of automatic surveillance of tools and measuring the sides in the course of production. The evolution of the digitally operated modern machines drives to an integration of the measurement and the control directly in the cycle of manufacturing Figure 1. Probing on machine tools has greatly expanded its range of applications to improve manufacturing process by allowing [1, 3, 4, 6, 12]:

- To avoid pursuing an expensive manufacturing (complex parts, not corresponding),
- To measure before finishing and optimizing the last pass.
- To optimize the adjustment parameters of a part holder.
- To optimize the balance of manufactured surfaces in an acceptable gross.
- To optimize the corrections of tools and taking of the origins of a digital control.
- To correct the geometrical defects of the trajectories of a machine.
- To minimize the defect of the shape of a surface.
- To verify or measure a geometrical or the functional specification of a mechanical part.

Met & Props 2013 Bldg. 16, 321, Kuang-Fu Road Sec. 2Hsinchu, Taiwan 30011, R.O.C.

This list is not restrictive but it shows that such an organization is likely to make considerable gains in quality control, operating machinery, production and transit time of the parts in the workshop. The general problem so far is the geometrical identification of the real surfaces of a made part and the real trajectories of a machine. This identification has for objective to guarantee that the manufactured surfaces really coincide with the robust model defined by specifications. We present in our article the configuration which is materialized in our laboratory, the integration of a probe associated with a chain acquisitions data, software acquisition and analysis in relation to the control [2, 6].

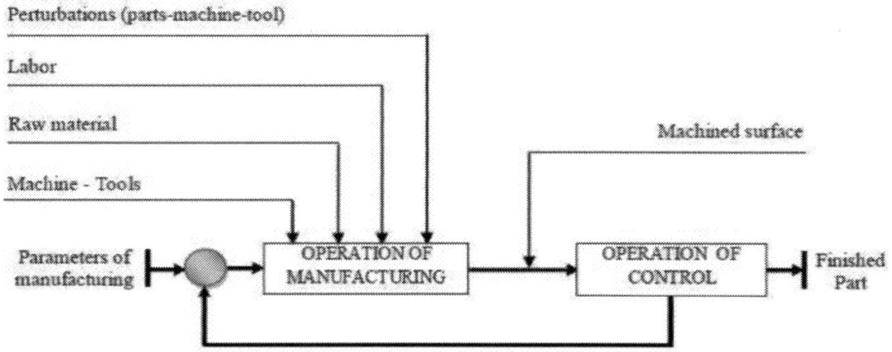

Figure 1. Buckle of integration of the measure and the control in the manufacturing. [12].

2. Integration of probe system on CN machine tool «EMCO PC Mill 155»

2.1. The Description of EMCO PC Mill 155 machine

2.1.1. Generalities

For more than fifty years, EMCO develops and builds machines to manufacture metals, and since 1980, the company is also represented on the market with CN machine tools, in particular in the sector of training. The machine «EMCO PC Mill 155" is intended for the reaming and the drilling of machined metals and machined plastics [13].

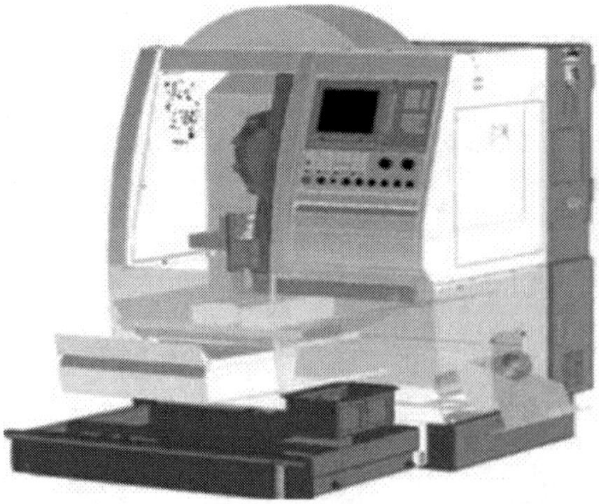

Figure 2. CN Machine tool EMCO PC Mill 155. [13]

Our digitally operated machine tool represents 3, already possesses an articulated mechanical structure and a sensor of position on each of its axes. Thus she can become a machine to measure in or a CMM if she allows assuring all the following functions:

- To arrange a sensor.
- To allow the acquisition of coordinates of measurement.
- Allow the storage of the data.
- Assure the treatment of the result.

The set of these functions is not in priori assured but the current DCN authorizes them. The basic function which allows this use is "the stop on interruption". The machine which possesses this function and which can be endowed with a sensor of metrology can become a machine to measure. In the absence of resident sub programs assuring the storage of the information and their processing, it is always possible to the user to create files and necessary subroutines [5, 7].

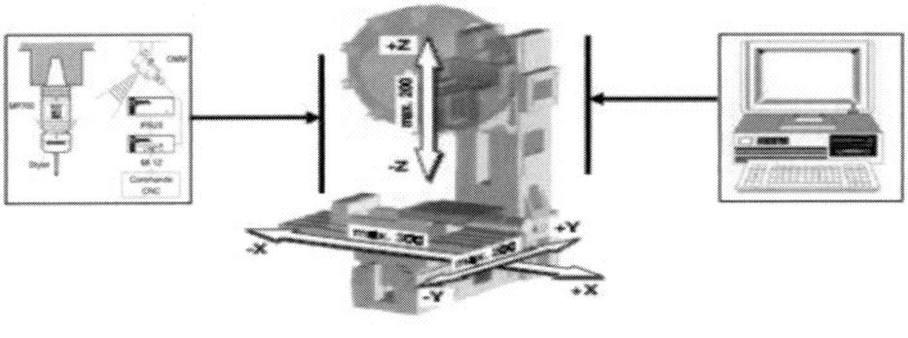

Figure 3. Transformation the CN machine tool of to the CMM.

2.2. Adopted Solution

2.2.1. The choice of the probe system

The choice of the system of probe depends naturally on its capacity to reach the point of measurement but also the shape and the requirements of the manufactures part as well as the nature of the machine which is used [8]. The probe system from Renishaw brings an innovative solution to the improvement of the efficiency on the machine tools:

- Saving of time.
- Reduction of rubbishes.
- Preservation of the competitiveness.

Thus we chose the system of probe MP700 of Renishaw Table.1, because it is an equipment of high precision designed specially to work in the environments the most hostile machine tools. This system was developed to supply:

- Repeatability improved in all the directions of probe.
- A low strength of release combined with a low variation of pre -race to supply a high precision, even during the use with long stilets.
- A decoupled life expectancy (10 million releases).
- A more precise and faster measure.
- An elimination of the defects of reset.
- A big immunity in the vibrations of machine tools.
- An impact resistance and in false release thanks to the use of a digital filtration, to the multiple ways. [8]

Table 1. Control probes for CN machine tool.

Machine	Compact	Small	Medium	Large
CN machine tool				
vertical	OMP40	OMP40	OMP60/RMP60	RMP60
horizontal	OMP40	OMP60	OMP60/RMP60	RMP60
high precision	OMP40	OMP400	OMP400/MP700	**MP700**
manual machines	Probe "job contact" (JCP)			

This system contains a probe MP700, the optical module machine, interface unit of probe MI12 and a power supply unit PSU3 [9].

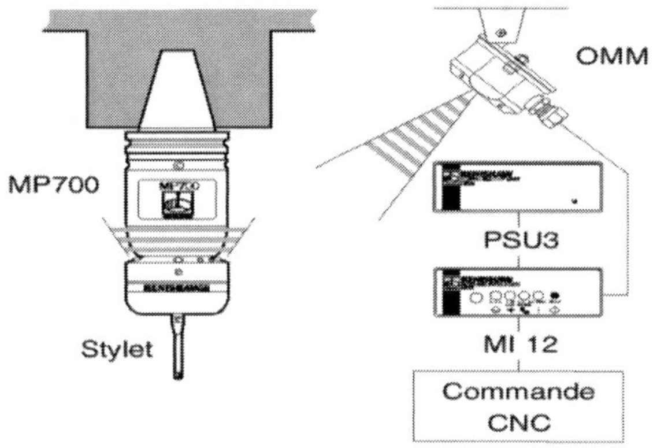

Figure 4. Probe system MP700 "variant OMM / MI 12".

2.2.2. Installation of the probe system MP700 on CN machine tool

For the good functioning of our system, the optical module machine (OMM), must have risen on the digitally operated machine tool so that the optical envelopes between the OMM and the sensing optical MP700 (OMP) module are positioned such as on the Figure 5. [9]

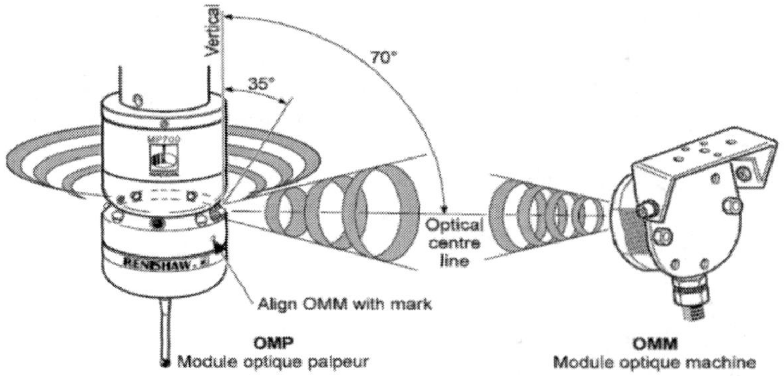

Figure 5. Requirements of working of the OMM.

3. Development of software for verification of the geometric specifications of the shape

For the purpose of controlled mechanical parts. We added a chain of acquisitions (MP700 probe system from Renishaw) and control software to our NC machine tool that exists in our atelier. Among all the functions of this software, we will limit ourselves in this article to those related to treatment of the measure, that is those who allow to determine the coordinates of points of contact between the probe and the surface to measure, and which allow, by a mathematical treatment of coordinates, To perform dimensional measurements and verify the characteristics of geometric tolerances of the parts. We expose an example of the righteousness, an example for the flatness and an example for the circularity.

3.1. General algorithm

A good definition of modules of the program and their interaction is one of the most important phases of the conception. The figure 6 illus trates, in generally the functional architecture of developed software. This figure represents all phases of our conception and our implementation [12].

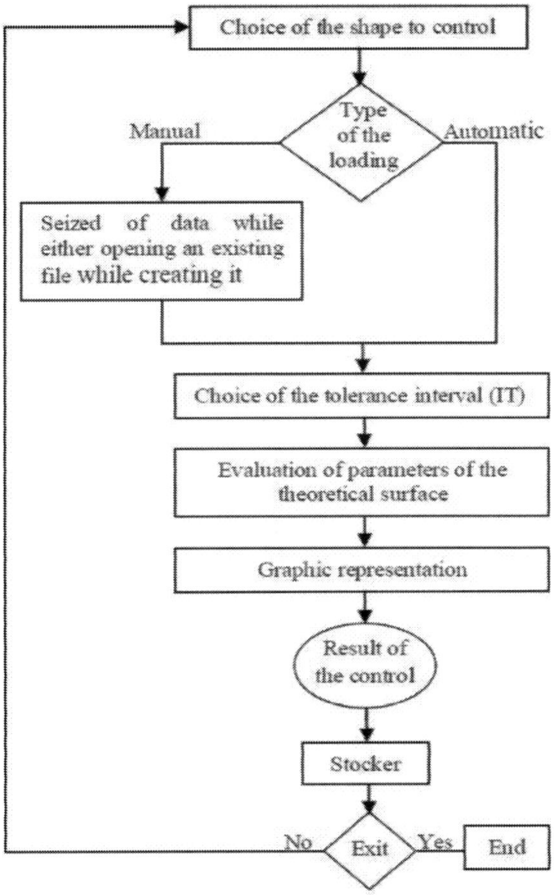

Figure 6. General Algorithm of surface control.

Journal of Physics: Conference Series **483** (2014) 012024 doi:10.1088/1742-6596/483/1/012024

3.2. Calculation of the parameters

The done measures are some submissive sizes to the difficult disruptions to synthesize and information that they contain is with difficulty usable. We try to show to get the different parameters of the theoretical shapes how here. All calculations are based on the principle of optimization by the method of least squares.

3.2.1. Straightness

In this case, the application of the method of least square allows obtaining from a cloud of point, the coefficients A and B of the characteristic equation of a straight line [10]. This last is gotten so that the distance that separates it of every experimental point is minimal. To see Figure 7.

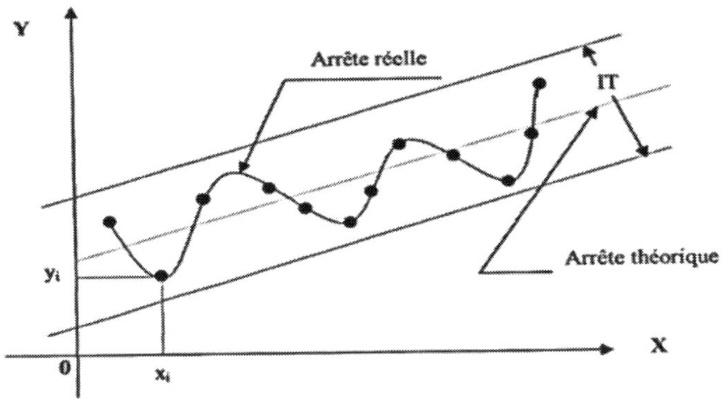

Figure 7. Reference theoretical for the Straightness.

x, y are the coordinates of the experimental statements that form the cloud of points and n represents their number

$$y = Ax + B \tag{1}$$

With

$$A = \frac{n\sum_{i=1}^{n} x_i y_i - \sum_{i=1}^{n} x_i \sum_{1}^{n} y_i}{n\sum_{i=1}^{n} x_i^2 - \left[\sum_{i=1}^{n} x_i\right]^2} \tag{2}$$

And

$$B = \frac{\sum_{i=1}^{n} y_i \sum_{i=1}^{n} x_i^2 - \sum_{i=1}^{n} x_i \sum_{i=1}^{n} x_i y_i}{n\sum_{i=1}^{n} x_i^2 - \left[\sum_{i=1}^{n} x_i\right]^2} \tag{3}$$

3.2.2. Flatness

For flatness we move from one dimensional to two-dimensional plane at the right in Figure 8.For this we must calculate the coefficients of the optimal characteristics, knowing that his equation in space is written as follows [11].

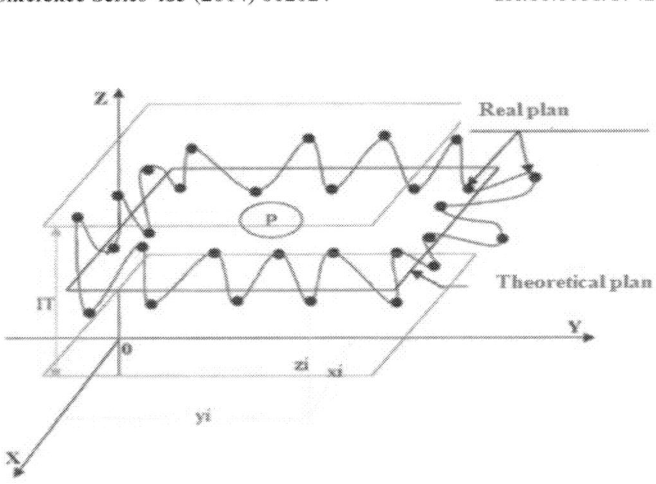

Figure 8. Reference theoretical for the flatness.

$$Z = Ax + By + C \tag{4}$$

It is therefore about determining coefficients A, B, and C according to relations:

$$A = \frac{[n\sum_{i=1}^{n} x_i y_i - \sum_{i=1}^{n} x_i \sum_{i=1}^{n} y_i] - B[n\sum_{i=1}^{n} x_i y_i - \sum_{i=1}^{n} x_i \sum_{i=1}^{n} y_i]}{\left[n\sum_{i=1}^{n} y_i^2 - \left[\sum_{i=1}^{n} y_i\right]^2\right] - \left[n\sum_{i=1}^{n} x_i y_i - \sum_{i=1}^{n} x_i \sum_{i=1}^{n} y_i\right]^2} \tag{5}$$

$$B = \frac{b_1 - b_2}{\left[n\sum_{i=1}^{n} y_i^2 - \left[\sum_{i=1}^{n} y_i\right]^2\right] - \left[n\sum_{i=1}^{n} x_i y_i - \sum_{i=1}^{n} x_i \sum_{i=1}^{n} y_i\right]^2} \tag{6}$$

$$C = \frac{\sum_{i=1}^{n} z_i}{n} - A\frac{\sum_{i=1}^{n} x_i}{n} - B\frac{\sum_{i=1}^{n} y_i}{n} \tag{7}$$

with

$$b_1 = \left[n\sum_{i=1}^{n} x_i^2 - \left[\sum_{i=1}^{n} x_i\right]^2\right]\left[n\sum_{i=1}^{n} y_i z_i - \sum_{i=1}^{n} y_i \sum_{i=1}^{n} z_i\right]$$

and

$$b_2 = \left[n\sum_{i=1}^{n} x_i y_i - \sum_{i=1}^{n} x_i \sum_{i=1}^{n} y_i\right]\left[n\sum_{i=1}^{n} x_i z_i - \sum_{i=1}^{n} x_i \sum_{i=1}^{n} z_i\right]$$

3.2.3. Circularity

As regards the circular shape she defined in a Cartesian mark by an equation of the shape [3,4]:

$$(x - a)^2 + (y - b)^2 = r^2 \tag{8}$$

Where, r represents the radius of the circle. We calculate the optimal radius of circle r_{opt}, of the way that difference par rapport to every radius raised by the measure laughed r_i is minimal, Figure 9.

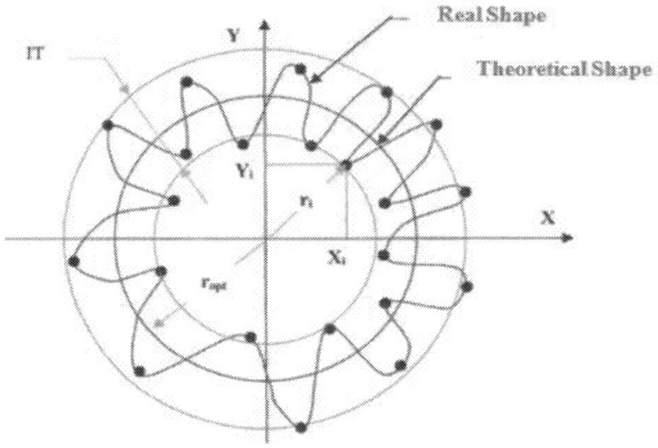

Figure 9. Reference theoretical for the circularity.

We know that a and b are the coordinates of the center of the circle, and n the number of points

$$b = \frac{b_1 - b_2}{2\left[n\sum_{i=1}^{n} y_i^2 - \left[\sum_{i=1}^{n} y_i\right]^2\right] - 2\left[n\sum_{i=1}^{n} x_i y_i - \sum_{i=1}^{n} x_i \sum_{i=1}^{n} y_i\right]^2} \tag{9}$$

$$a = \frac{\left[n\sum_{i=1}^{n} x_i y_i - \sum_{i=1}^{n} x_i \sum_{i=1}^{n}(x_i^2 + y_i^2)\right] - 2b\left[n\sum_{i=1}^{n} x_i y_i - \sum_{i=1}^{n} x_i \sum_{i=1}^{n} y_i\right]}{2\left[n\sum_{i=1}^{n} y_i^2 - \left[\sum_{i=1}^{n} y_i\right]^2\right] - 2\left[n\sum_{i=1}^{n} x_i y_i - \sum_{i=1}^{n} x_i \sum_{i=1}^{n} y_i\right]^2} \tag{10}$$

with

$$b_1 = \left[\sum_{i=1}^{n} x_i^2 - \left[\sum_{i=1}^{n} x_i\right]^2\right]\left[n\sum_{i=1}^{n} y_i(x_i^2 + y_i^2) - \sum_{i=1}^{n} y_i \sum_{i=1}^{n}(x_i^2 + y_i^2)\right]$$

$$b_2 = \left[\sum_{i=1}^{n} x_i y_i - \sum_{i=1}^{n} x_i \sum_{i=1}^{n} y_i\right]\left[n\sum_{i=1}^{n} x_i(x_i^2 + y_i^2) - \sum_{i=1}^{n} x_i \sum_{i=1}^{n}(x_i^2 + y_i^2)\right]$$

And the optimal radius of circle r_{opt}:

$$r_i = \sqrt{(x_i - a)^2 + (y_i - b)^2} \tag{11}$$

And

$$r_{opt} = \frac{\sum_{i=1}^{n} r_i}{n} \tag{12}$$

4. Experimentation

The tests are divided in to three phases:
The first tests to the level of manufacture laboratory by the system of probe MP700 from Renishaw that is installed on CN machine tool. For it as must do the following stages:

1. Choose the shape to control.
2. Grid the surface according to measurements of parts.
3. Put the piece on the marble of the machine.
4. Probing the part.
5. Raise coordinates of points probed entered into tables.
6. Validate the treatment of control.

The second test is done on our software with the same coordinates of points probed by the CN machine tool.
The third test is of compared results of measure of CN machine tool with results of measure of CMM.

5. Results and discussions

5.1. Results of measure

Numerous data essay have been reassembled in this article. Figure 10, Figure 11, And Figure 12 represents coordinates of points probing by our probe MP700, for the verification of the Straightness, the Flatness and circularity on our CN machine-tool.

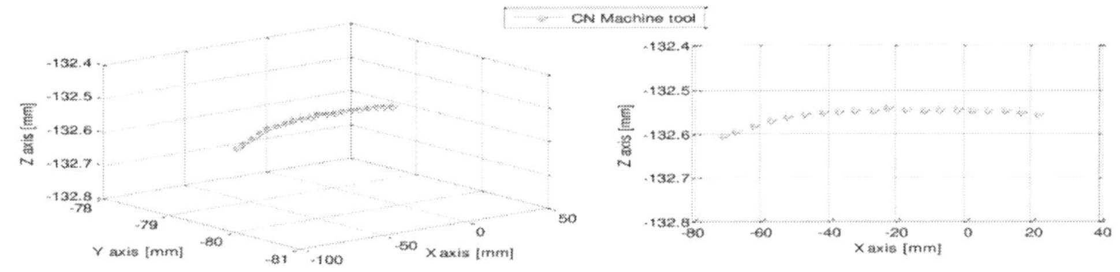

Figure 10. Cloud points « case of the Straightness ».

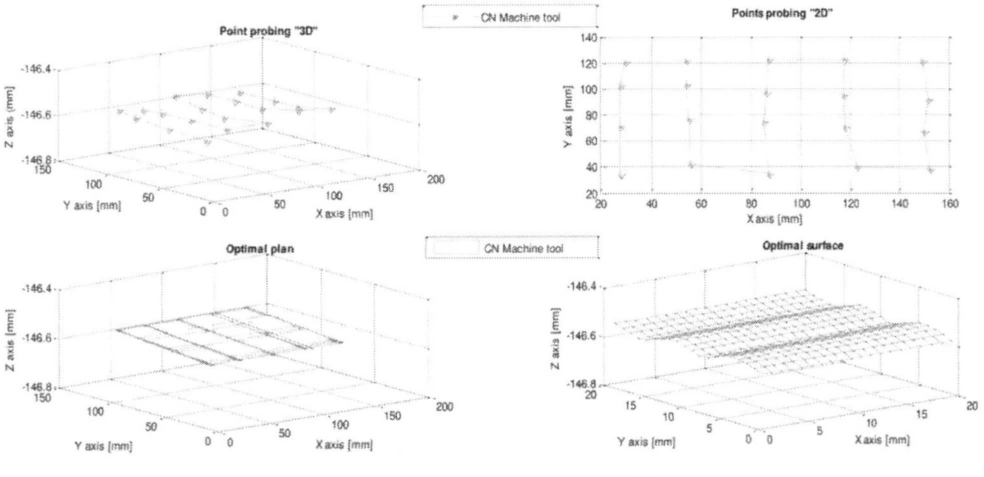

Figure 11. Cloud points « case of the flatness ».

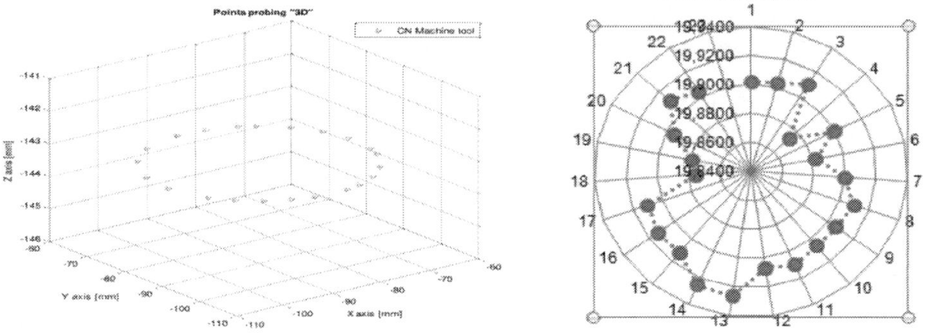

Figure 12. Cloud points « case of the Circularity ».

5.2. Comparison of the measure between CN machine tool and CMM

Figure 13, Figure 14 and Figure 15 represent the comparison of measure of the Straightness, the flatness and the circularity between CN machine tool and CMM that is the precision of the two machines, in 2D and 3D.

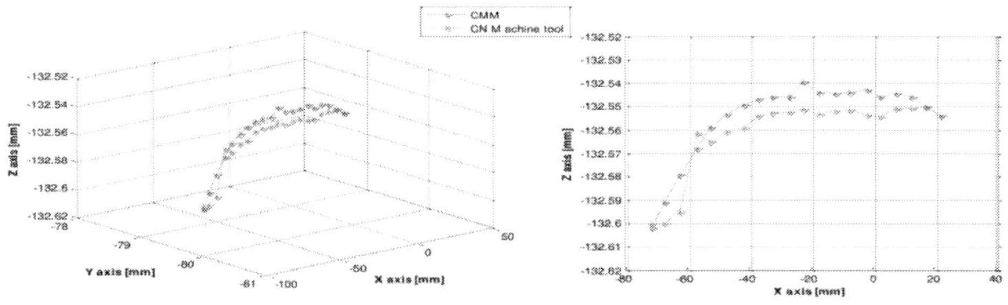

Figure 13. Comparison of the measure of the straightness between CMM and CN Machine tool.

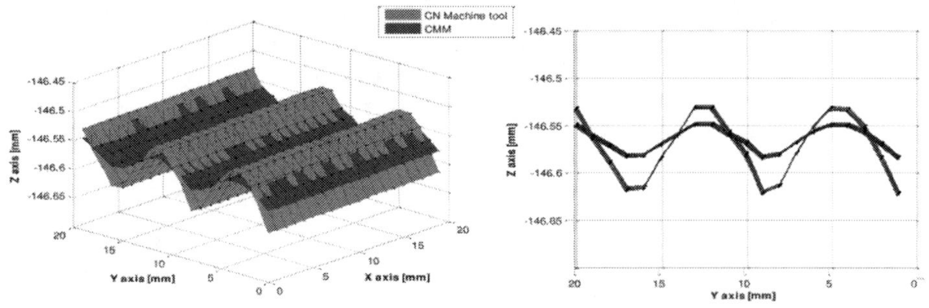

Figure 14. Comparison of the measure of the flatness between CMM and CN Machine tool.

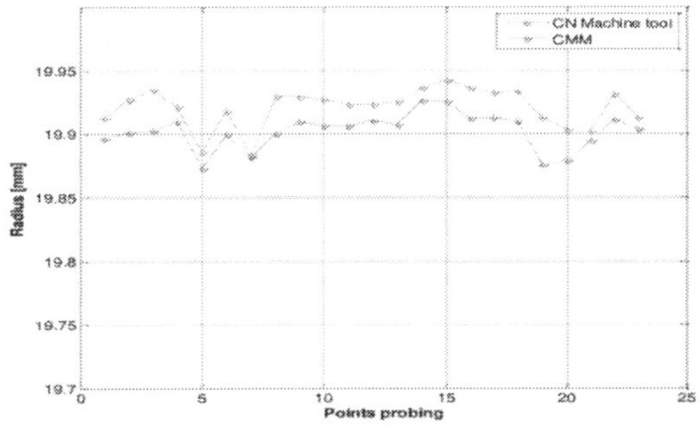

Figure 15. Cloud points « case of the flatness ».

5.3. Discussed results

One observes that the acceptance of the measure of the part is directly dregs with the interval of tolerance, that is to say when one changes the value of the tolerance interval, the nature of measure of the piece changes. One also notes that the different geometric specification measure gives the following shape shortcomings:

Table 2. form errors of the geometric specifications

	CMM			CN Machine Tool		
	Form errors min [mm]	Form errors max [mm]	Interval [mm]	Form errors min [mm]	Form errors max [mm]	Interval [mm]
Straightne	0.0004	0.0390	0.0386	0.0007	0.0471	0.0464
Flatness	0.0007	0.0486	0.0479	0.0005	0.0826	0.0821
Circularity	0.0004	0.0291	0.0287	0.0003	0.0381	0.0378

If one compares results of measure of the flatness between CN Machine tool and CMM, one gets a reduction of 41.66% of form errors respectively. On the other hand one can affirm that the form errors in the case of the measure of the circularity, is decreased of 24.07% between CN Machine tool and CMM. So in the case of the measure of the straightness by CN Machine tool and CMM gives a reduction form errors of 16.81%.

One also notices that the precision of the measure of a mechanical parts decreases for the two machines when the number of axes of the diplacement of the probe augment, is that owed to the quasi -static errors of axes Table 3. Therefore the measure of a geometric specification by CMM is more precise than the measure by the CN machine-tool.

Table 3. the variation of the errors forms with regard to axes of displacement.

Geometric specification	The variation of from errors between CMM and CN Machine tool [%]	Number of axis of the probe displacement
Straightness	16.81	1 axe [X]
Flatness	41.66	3 axes [X, Y, Z]
Circularity	24.07	2 axes [X, Y]

6. Conclusion

One conclusion the presence of a probe MP700 from renishaw on our CN machine tool "EMCO PC Mill 155" permits to do real operations of metrology, analogous to the one of machines to measure (CMM). However one can also achieve operations of measure that are very useful to improve the fruitfulness. So this system costs 20 less dear times that a CMM.

One also observes in all cases of measures to study a meaningful grain of the defect of shape between the measure by the two CN machine tool machine and CMM that to be-to-say the measure by CMM is more precise than the measure by CN machine tool is that owed to the almost-static mistakes. These mistakes of behavior are owed to the structure of the profile geometric-kinematics of the machine, combined to the mechanical, thermal solicitations and inertielleses developed during the operation of machining. If they are known, they can be made up for by the mediator of a command module interns external either that one can integrate to the CN machine tool.

4. References

[1] Xiu-Lan Wen, Jia-Cai Huang, Dang-Hong Sheng, Feng-Lin Wang, «Conicity and cylindricity error evaluation using particle swarm optimization», Automation Department, Nanjing Institute of Technology, Nanjing 211167, China, Precision Engineering 34 (2010) 338–344

[2] V.N. Narayanan Namboothiri, M.S. Shunmugam, « Form error evaluation using L 1 – approximation », Department of Mechanical Engineering, Indian Institute of Technology, Madras-600 036, India, Computer methods in applied mechanics and engineering 162 (1998) 133-149

[3] Hsin-Yi Lai, Wen-Yuh Jywe, Chao-Kuang Chen, Chien-Hong Liu, « Precision modeling of form errors for cylindricity evaluation using genetic algorithms », Department of Mechanical Engineering, National Cheng-Kung University, Tainan, Taiwan 701, Precision Engineering 24 (2000) 310–319.

[4] Xiangchao Zhang*, Xiangqian Jiang, Paul J. Scott, « Minimum zone evaluation of the form errors of quadric surfaces », Centre for Precision Technologies, University of Huddersfield, Huddersfield HD1 3DH, UK, Precision Engineering 35 (2011) 383–389.

[5] Fabricio Tadeu Paziani , Benedito Di Giacomo, Roberto Hideaki Tsunaki, « Robot measuring form errors », Department of Mechanical Engineering, University of Sao Paulo, 400, 13566-590 Sao Carlos, SP, Brazil, Robotics and Computer-Integrated Manufacturing 25 (2009) 168–177

[6] P. Bourdet, « Metrology tree dimensional and geometric of parts mechanics», University Paris VI-ENS of Cachan, mechanical technology License 1998/1999.

[7] B. Mery, «Machine to Numeric Command», the superior school of arts and professions the ENS of Cachan Paris, Edition Hermes 1997.

[8] RENISHAW, «system of probe MP700 RENISHAW », Guide of the operator H-2000-5133-06.A, RENISHAW S.A.S 15 Street Albert Einstein, Fields on Marne, 77437 Marne the Valley, France, Edition April 2003

[9] RENISHAW, « Manual of installation of the system of probe MP700 RENISHAW », Guide of the operator H-2000-5143-01-A, RENISHAW S.A.S 15 Street Albert Einstein, Fields on Marne, 77437 Marne the Valley, France, Edition 1997.

[10] M. Boumahrat, A. Gourdin, Edition O.P.U 1993, «applied numeric Method».

[11] Adamaneo, Hoang-Ky, Ouaklin, Edition O.P.U 1991, « Statistical ».

[12] S. Merghache, A. Ghernaout, A. Cheikh, « probing on NC machine-tools », International Seminary on Technologies of Mechanics, December 2009, Tlemcen, Algeria.

[13] EMCO MAIER, Catalogue for CN machine tool 2001, Réf. -N°. FR 4345.

Spikes removal in surface measurement

P Podulka[1], **P Pawlus**[1], **P Dobrzański**[1], **A Lenart**[2]
[1]Rzeszow University of Technology, W. Pola 2, 35-959 Rzeszow, POLAND
[2]Pratt & Whitney AeroPower Rzeszów, Hetmanska 120, 35-078 Rzeszow, POLAND

E-mail: ppawlus@prz.edu.pl

Abstract. Several cylinder surface topographies made from grey cast iron were measured by Talysurf CCI white light interferometer with and without use of spikes filter. They were plateau honed by abrasive stones. Measured area was 3.3 mm x 3.3 mm, height resolution was 0.01 nm. The forms were eliminated using polynomial of the 3rd degree. After it, spikes were removed using four methods. These approaches were compared. Parameters of the smaller and highest sensitivity on spikes presence were selected.

1. Introduction

The initial assessment of surface topography was made simply by running a fingernail across the surface. This technique survives to this day as tactile comparison. The first surface topography measuring instrument was the light section microscope developed in Germany early in the XX century. The subject of surface roughness measurement began when a simply profilometer was developed. The measurement process was achieved by drawing a stylus across a surface and recording vertical deviations of the surface. From the late 1970s analogue instruments were replaces by those supported by computers [1]. A number of profiling techniques have been developed for measuring the topography of rough surfaces. There are several types of optical techniques in addition to the contacting stylus method and atomic force microscopy. These include white light interferometer [2], confocal methods [3] and others [4, 5, 6]. Optical measurement suffers from the same intrinsic constrains as the stylus technique but also has additional problem that the scattered light from the surface does not entirely react normally to the surface which restricts optical use in some application [4]. During the period of development of traditional stylus type instrumentation for areal (3D) measurement huge progress was made (mainly in the USA) with regard to development of optical systems. Contrary to stylus methods which are robust but slow optical methods are fast but more sensitive to extraneous effects. Optical methods need isolation of devices from the external environment. Both thermal and vibrations changes influence reliability of the results. Very careful cleaning of the measured surface is necessary from the point of view of the industrial application. Sharp edges, inclusions, defects and other peculiarities of the surface can cause outliers and dropouts of data points in the topographical images measured with optical methods, more than with stylus technique.

Scanning white-light interferometry is gaining more and more importance in research and development as well as in manufacturing quality control [7, 8]. It is probably the most useful optical instrument at present in use for measuring surfaces, films and coatings [4]. White light interferometer has been shown to be susceptible to a skewing effect when the amplitude is less than the coherence length of the light source. This skewing effect leads to spikes in the profile data near steep edges

[1, 9]. Although new types of measurement interferometers are produced, the noise-like spikes presence is still a real problem.

Spikes in the data points should be detected and points interpolated to prevent singularities. There are various selective spike-removal filters. Application of morphological filters [10] and removal through detection of points of high slopes [11] can be the examples.

The fundamental aim of investigation is to compare different methods of spikes removal from the data points. The other aim is to select parameters of the smallest sensitivity on spikes presence.

2. Material and methods

Several cylinder surface topographies made from grey cast iron are the object of investigations. They were plateau honed by abrasive stones. Cylinder liners were measured by Talysurf CCI white light interferometer. Measured area was 3.3 mm x 3.3 mm, height resolution was 0.01 nm. Surfaces were measured in the same place with and without use of spikes filter. The forms were removed using polynomial of the 3rd degree. After it, spikes were removed using different methods. Areal surface topography parameters were then calculated using software Talymap Gold for surfaces with and without removed spikes.

The parameters from ISO 25178 standard [12] were analysed:

- height: Sq, Sa, Ssk, Sku, Sp, Sv and Sz,
- functional: Smr, Smc and Sxp,
- spatial: Str, Sal and Std,
- hybrid: Sdq and Sdr,
- functional volume : Vm, Vv, Vmp, Vmc, Vvc and Vvv,
- feature: Spd, Spc, S10z, S5p, S5v, Sda, Sdv and Shv.

Feature parameters were computed after a discrimination by segmentation using a Wolfpruning of 5% of the value of the Sz parameter. Also parameters from the Sk family were calculated: Sk, Spk, Svk, Sr1, Sr2, Sa2 and Sa1. Digital filtration was not used.

In addition, surfaces were measured in the same place (mechanical relocation method was applied) using spikes removal procedure of measuring equipment.

3. Methods of spikes removal

The first method used is based on the shape of material ratio curve. About 20 cylinder liners were analysed. When spikes are presented, the initial upper part of this curve forms a straight vertical line (see Figure 1 b).

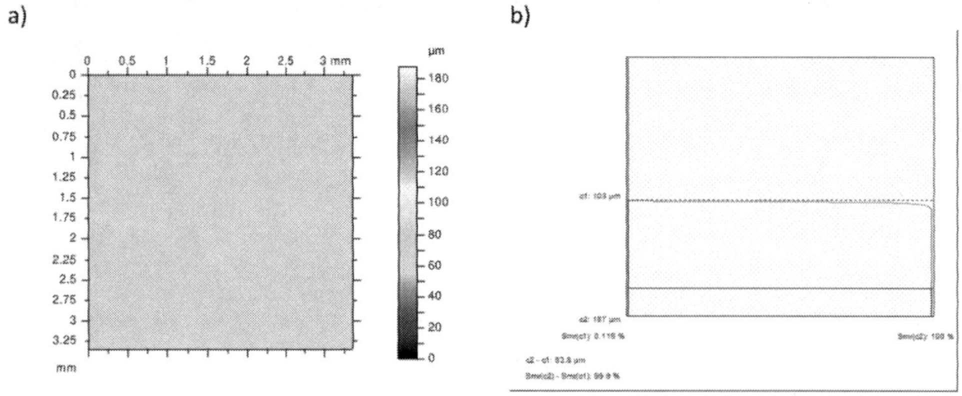

Figure 1. Contour plot of cylinder surface containing spikes (a), material ratio curve of this surface (b)

Journal of Physics: Conference Series **483** (2014) 012025 doi:10.1088/1742-6596/483/1/012025

For spikes removal, the point of material ratio curve shape changing should be determined and then surface ought to be truncated from ordinate of this point. After analysis of many surfaces, it was decided to threshold the surface from point of material ratio 0.13%, which means that when material ratio is smaller than 0.13%, the ordinate should be equal to height corresponding to this ratio.

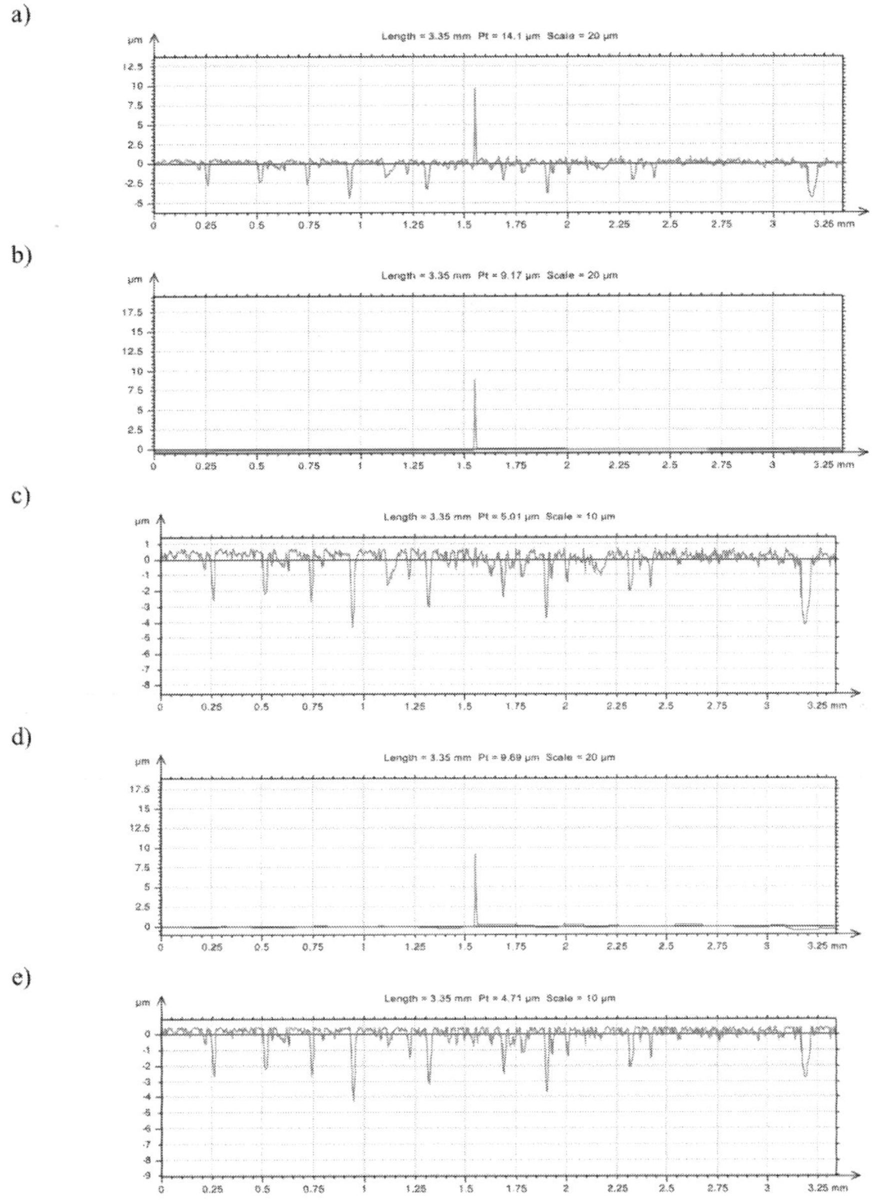

Figure 2. Cylinder profile containing spike (a), filtered profile using horizontal line of 0.2 mm length (b), residue profile without spike (c), filtered profile using horizontal line of 0.05 mm length (c), residue profile without spike (d)

Additionally, for random surface of Gaussian ordinate distribution height difference between material ratios 0.13-99.87% is equal to 6 standard deviation of surface amplitude [13]. It was also found previously that when height of areal cylinder surface is limited to material ratios between 0.13% and 99.87% areal (3D) surface height is similar to that of 2D profile [14].

In the other method closing filter was used with structuring element of horizontal plane shape. The filtered surface is the upper envelope. Residue surface (unfiltered minus filtered) doesn't contain spikes. The length of this shape should be selected carefully. When it is too small, the structuring elements penetrates into deep valley, so residue surface can be distorted. For too big length, large wavelengths are eliminated. The intention of the present authors was to remove spikes only. The length of structuring elements should be smaller than distance between two adjacent spikes. Therefore in selection of length of structuring elements the widths of valleys and distances between spikes should beconsidered. After taking these limitations into consideration it was decided to select length of structuring element as 0.2 mm. Figure 2 presents illustration of this method. When length of structuring element was too short (0.05 mm) the filtered profile was affected by some irregularities (for example wide valley in the right side of figures) and consequently the height of residue profile was underrated. When the length of structuring element is 0.2 mm, filtered profile contains mainly spike.

Third method based on the comparison of standard deviations. For each point of surface standard deviation of height was calculated for this point and its neighbors. Then standard deviation of all the points was averaged. Then for points lying above surface mean plane spikes were identified as points, for which vertical distance to mean plane was higher than multiplication of k_{sd} and averaged height standard deviation. Then the ordinates of spikes were replaced by average value of neighbors, not being spikes.

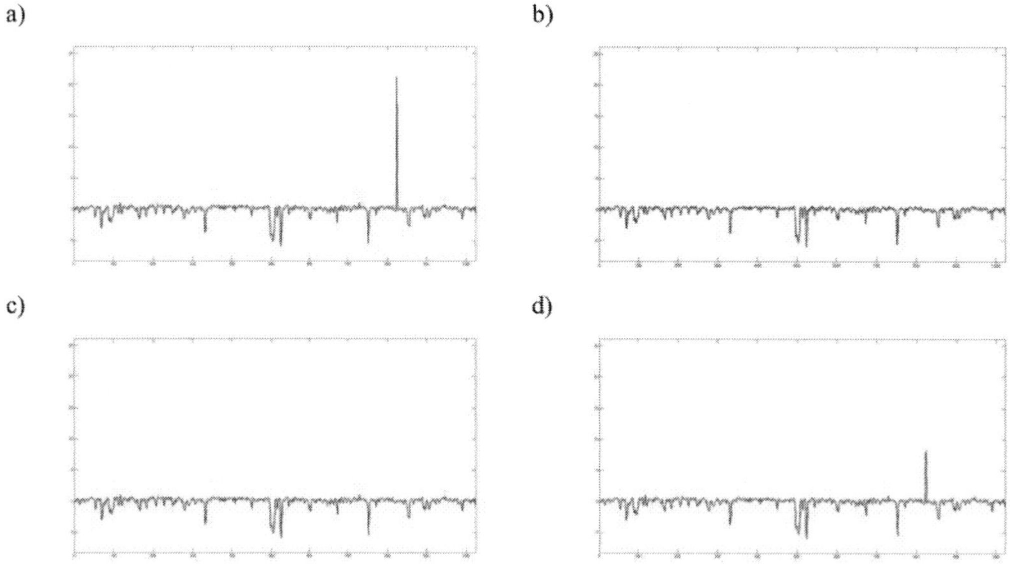

Figure 3. Profile containing spike (a), profile with removed spike for k_{sd} value of 2.5 (b), 10 (c) and 12 (d)

Selection of the k_{sd} value is important. If it is too small, height of some individual peaks can be reduced, when too large, only upper part of spikes can be removed. After analysis of many surfaces the k_{sd} value was set to be 10. Figure 4 shows example of this method implementation.

Fourth method is based on calculation of rms. slope. It is based on elimination of points of the highest slope. Slope was calculated for each surface point. Then average surface slope was computed. When slope of points lying above surface mean line was larger than selected k_{sl} value multiplied by average slope, these points were qualified as spike. For determined spikes, similar to the previous method, their ordinates were replaced by average ordinate value of neighbors, not being spikes.

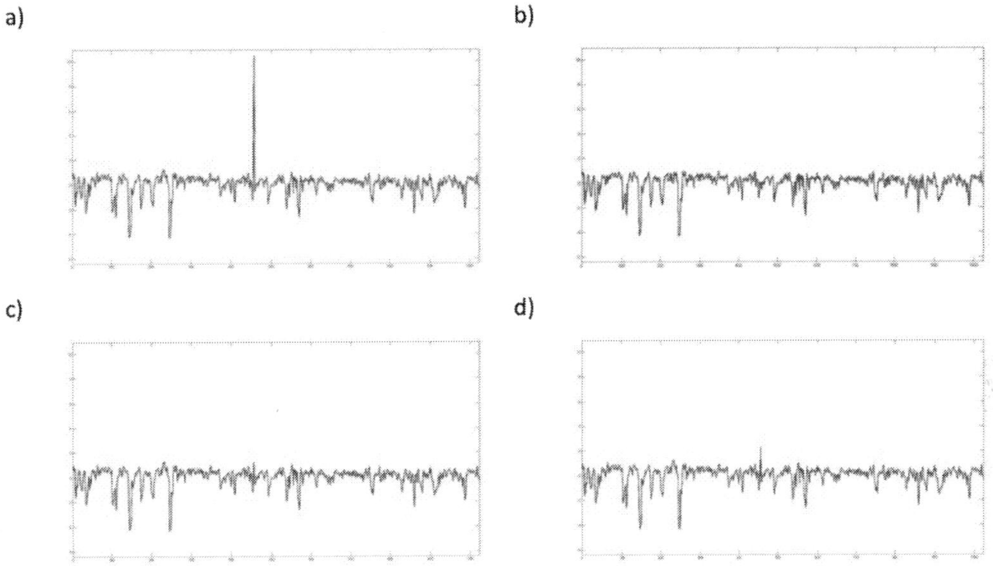

Figure 4. Profile containing spike (a), profile with removed spike for k_{sl} value of 2.5 (b), 7.5 (c) and 12 (d)

This approach seems to be of more importance than the previous approach based on standard deviation, because surface slope depends on both amplitude and spatial parameters of the surface. The k_{sl} value was selected to be 7.5. Figure 4 presents illustration of this method.

4. Results and discussion

Detailed analysis of one surface is given. Table 1 presents parameters of surface before and after removal of spikes using four methods. In order to compare the results with those obtained using spikes removal during measurement, the measured surface was initially thresholded (truncated) to material ratio range of 0-99.87%. In addition, Figure 5 presents contour plots and Figure 6 material ratio curves of measured surfaces.

It is evident that spikes removal caused decrease of the majority of height parameters describing peak surface part. The change of the Sp parameter is the largest (up to 99%) after using closing filter. In this case the height parameters are the smallest. Decrease of the S5p parameter is high, too. Decreases of the Spk parameter is smaller than 40%, because of its definition (Spk is reduced peak height). Maximum height also decreases as a result of spikes removal.

After application of closing filter the value of the Sk parameter was the smallest. Differences between parameters describing the valley part Sv and Svk were negligible. Surface correction caused changes of the Sk, Sq and Sa parameters of a few percents. Especially the small relative changes of the Sq parameter were confirmed for other analysed surfaces.

The skewness Ssk and kurtosis Sku decreased as a results of spikes removal, the largest changes were obtained after using closing filter. The functional Sxp parameter is stable on surface independently of spikes removal method, contrary to unstable parameters Smr and Smc. The spatial parameter Sal was constant. The Str parameter decreased due to surface correction. As predicted, surface main direction Std did not change as a results of spikes removal.

Sdq and Sdr parameters decreased after surface modification, which is related to amplitude decrease. Material volume parameters decreased as a results of peak removal, however changes of parameters Vv, Vvc and Vvv were rather small. Summit density Spd increased, after spikes removal is the smallest value was obtained for truncated surface. Peak density Spd increased, the highest value was found after surface thresholding. Feature parameters Sda, Sha, Sdv and Shv decreased due to spikes removal.

Table 1. Parameters of plateau-honed cylinder surface for various methods of spikes removal

	Surface with spikes	Application of closing filter with 0.2 mm length of horizontal plane	Peak part truncation, corresponding to material ratio of 0.13%	Algorithm based on standard deviation	Algorithm based on rms. slope
Sq, µm	0.806	0.768	0.771	0.77	0.77
Ssk	5.01	-2.52	-2.4	-2.42	-2.42
Sku	795	11.2	10.6	10.6	10.6
Sp, µm	109	1.09	2.48	1.64	1.55
Sv, µm	4.27	4.32	4.27	4.27	4.27
Sz, µm	114	5.41	6.75	5.91	5.82
Sa, µm	0.52	0.513	0.519	0.518	0.518
Smr, %	0.0000954	62.5	0.039	8.85	14.9
Smc, µm	109	0.488	1.86	1.02	0.934
Sxp, µm	2.42	2.44	2.42	2.42	2.42
Sal, mm	0.0164	0.0164	0.0164	0.0164	0.0164
Str	0.0235	0.0198	0.02	0.02	0.02
Std, °	117	117	117	117	117
Sdq	0.175	0.14	0.14	0.14	0.14
Sdr, %	1.35	0.962	0.96	0.953	0.954
Vm, mm^3/mm^2	9.72e-006	1.12e-005	1.31e-005	1.27e-005	1.25e-005
Vv, mm^3/mm^2	0.00069	0.000614	0.000634	0.000633	0.000634
Vmp, mm^3/mm^2	9.72e-006	1.12e-005	1.31e-005	1.27e-005	1.25e-005
Vmc, mm^3/mm^2	0.000437	0.000461	0.000471	0.000471	0.000471
Vvc, mm^3/mm^2	0.000496	0.00043	0.00045	0.000449	0.00045
Vvv, mm^3/mm^2	0.000194	0.000184	0.000184	0.000184	0.000184
Spd, $1/mm^2$	1.61	1042	754	927	946
Spc, 1/mm	4.24	0.0666	0.0724	0.0698	0.0695
S10z, µm	-	5.77	5.96	5.61	5.61
S5p, µm	85.7	1.83	2.17	1.83	1.83
S5v, µm	-	3.94	3.78	3.78	3.78
Sda, mm^2	10.6	0.0106	0.0148	0.0125	0.0131
Sha, mm^2	25	0.0079	0.0121	0.00946	0.00927
Sdv, mm^3	0.0122	7.07e-008	7.75e-008	6.8e-008	8.13e-008
Shv, mm^3	0.000285	5.41e-007	9.41e-008	6.56e-008	6.39e-008

Generally, application of algorithms based on standard deviation and rms. slope gave similar values of all the analysed parameters, height parameters and hybrid parameters were also similar to those obtained by truncation. However application of upper thresholds caused different values of parameters characterizing peaks than use of the other methods; it was probably caused by truncation of peak surface part. This should be treated with a care when peak parameters are important. Maybe this method should be changed by replacing spikes by ordinates calculated from neighboring points. The method based on closing filter gave usually the smallest values of height and hybrid parameters from all the analysed methods. When length of structuring element is too big, some narrow valleys can arise, they can be removed by subsequent thresholds.

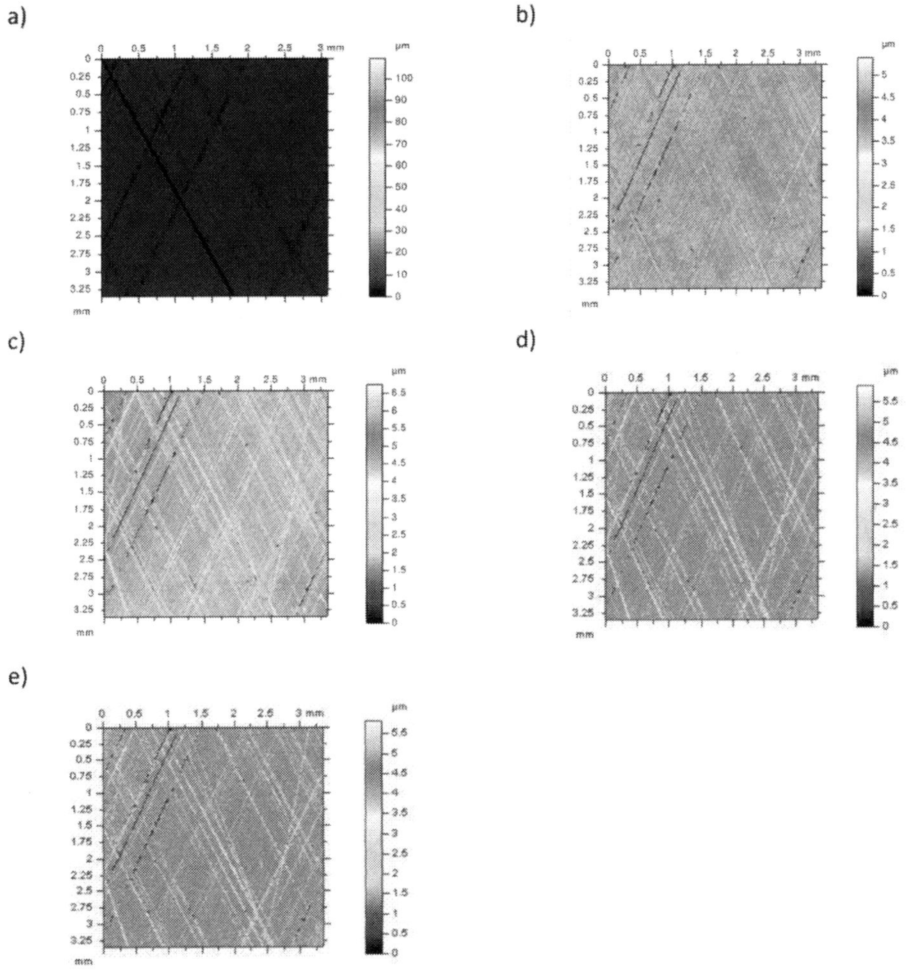

Figure 5. Surface containing spike (a), filtered using closing filter of 0.2 mm length of horizontal plane (b), after peak part truncation, corresponding to material ratio of 0.13 (c) and after using algorithms based on standard deviation (d) and rms. slope (e)

It was found from the measurement of surface with filter removing slopes that obtained parameters were close to those obtained by surface truncation.

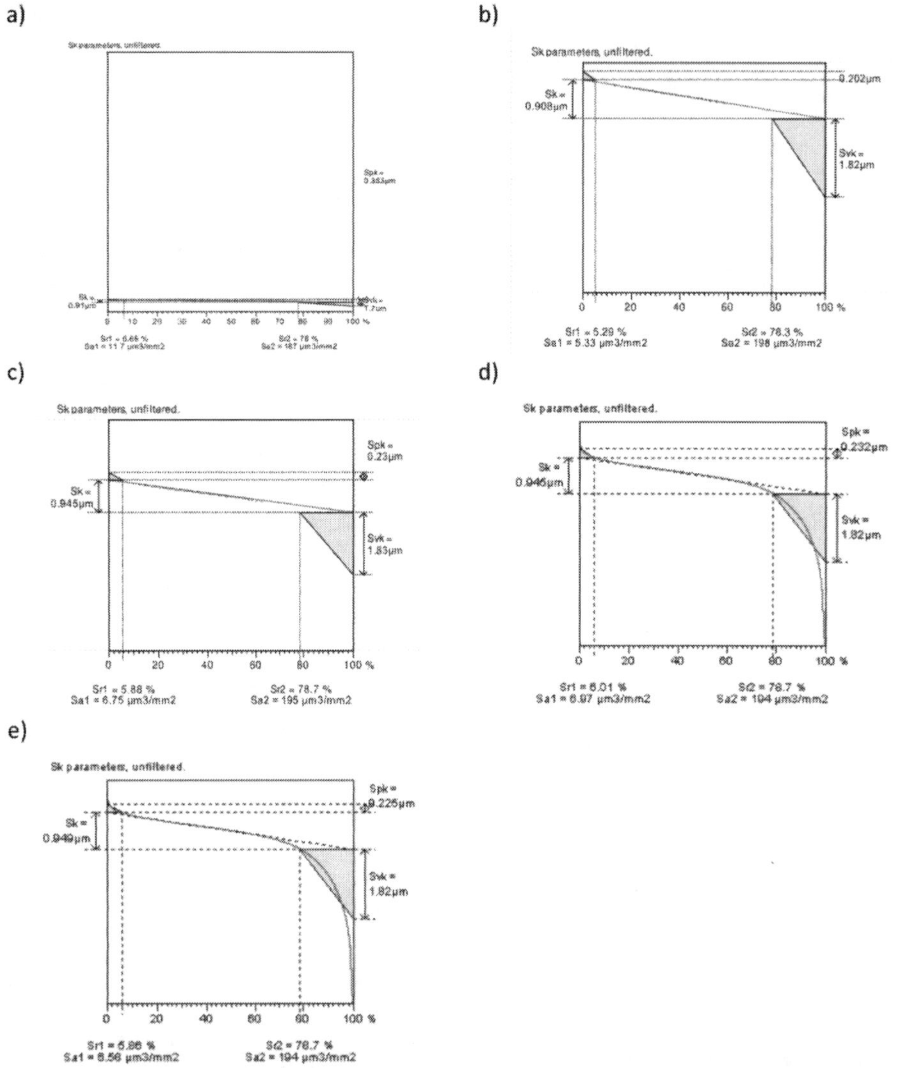

Figure 6. Material ratio of surface containing spike (a), filtered using closing filter of 0.2 mm length of horizontal plane (b), after peak part truncation, corresponding to material ratio of 0.13 (c) and after using algorithms based on standard deviation (d) and rms. slope (e)

5. Conclusions

Surface character should be taken into consideration in selection of parameters in procedures of spikes removal. Elimination of spikes caused decrease of amplitide parameters characterizing maximum height and peak parts; parameter describing valley part are almost constant. Parameters Sq and Sal were found to be stable on surface contrary to Smr and Smc. Change of hybrid parameters as well as feature parameters Spd, Spc, Sda, Sha, Sdv and Shv are also comparatively high.

Applications of algorithms based on slope and standard deviation gave similar results. In these cases hybrid and height parameters are similar to those obtained using surface truncation (thresholding), however application of closing filter with 0.2 mm length of horizontal plane gave the smallest value from all the analysed method.

Use of surface thresholds (truncation) corresponding to material ratio in the range 0.13-99.87% doesn't require selection of any other parameters. This method can be used for one-process and two-process random surface. It can be further modified by replacing ordinates of identified spikes by ordinates calculated from neighboring points, instead of surface truncation.

6. References

[1] Blunt L.: Introduction: the history and current state of 3D surface characterisation. In: Blunt L., Jiang X. (eds): *Advanced Techniques for Assessment Surface Topography*. Kogan Page Science, London and Sterling 2003, 1-13

[2] Schmit J., Olszak A.: High-precision shape measurement by white-light interferometry with real time scanned correction. *Applied Optics*. **33**, 1994, 229-233

[3] Rhee H.-G., Vorburger T., J.Fu, Renegar T.B., Song J.-F.: Comparison of roughness measurements obtained with optical and stylus techniques. *10th International Conference on Metrology and Properties of Engineering Surfaces*, Saint-Etienne, 2005, 39-47

[4] Whitehouse D.: Surface metrology today: complicated, confusing, effective. *Proceedings of the 13th International Conference on Metrology and Properties of Engineering Surfaces*, Twickenham Stadium, UK, 2011, 1-10

[5] Mathia T.G., Pawlus P., Wieczorowski M.: Recent trends in surface metrology. *Wear*, 2011, **271**, 494-508

[6] Jiang X., Whitehouse D.J.: Technological shifts in surface metrology. *CIRP Annals – Manufacturing Technology*, 2012, **61/2**, 815-836

[7] Boedecker S., Rembe C., Schmid H., Hageney T., Kohlein T.: Calibration of the z-axis for Large-scale scanning white-light interferometers. *Proceedings of the 13th International Conference on Metrology and Properties of Engineering Surfaces*, Twickenham Stadium, UK, 2011, 182-188

[8] Lukianowicz C., Tomkowski R.: Analysis of topography of engineering surfaces by white light scanning interferometry. *Proceedings of the 12th International Conference on Metrology and Properties of Engineering Surfaces*, Rzeszow, Poland, 2009, 63-67

[9] Harasaki A., Wyant J.C.: Fringe modulation effect in white-light vertical scanning interferometry. *Applied Optics*, **39**, 2000, 2107-2115

[10] Lou S., Jiang X., Scott P.: A fast algorithm for morphological filters. *Proceedings of the 13th International Conference on Metrology and Properties of Engineering Surfaces*, Twickenham Stadium, UK, 2011, 284

[11] Youinou V., Hacking S.A., Brown C.A.: Measuring the texture of fracture surfaces using confocal microscopy. *Proceedings of the 13th International Conference on Metrology and Properties of Engineering Surfaces*, Twickenham Stadium, UK, 2011, 357

[12] Blunt L., Jiang X.: Numerical parameters for characterization of topography. In L.Blunt, X. Jiang (eds.) *Assessment surface topography*. Kogan Page Science, London and Sterling 2003, 17-42

[13] Pawlus P., Graboń W.: The method of truncation parameters measurement from material ratio curve. *Precision Engineering*, **324**, 2008, 342-347

[14] Pawlus P.: Change of cylinder surface topography in the initial stage of engine life. *Wear* 1997, **209**, 247-254

3D SEM for surface topography quantification - a case study on dental surfaces.

F. Glon[1,2], O. Flys[1,3,4], P.-J. Lööf[1] and B.-G. Rosén[1]
[1]Halmstad University, Box 823, SE-301 18 Halmstad, Sweden
[2]ENISE, 58, rue Jean Parot, 42023, Saint-Etienne, France
[3]SP Technical Research Institute of Sweden, Box 857, SE-501 15 Borås, Sweden

E-mail: olena.flys@hh.se

Abstract. 3D analysis of surface topography is becoming a more used tool for industry and research. New ISO standards are being launched to assist in quantifying engineering surfaces. The traditional optical measuring instrumentation used for 3D surface characterization has been optical interferometers and confocal based instrumentation. However, the resolution here is limited in the lateral dimension to the wavelength of visible light to about 500 nm. The great advantage using the SEM for topography measurements is the high flexibility to zoom from low magnifications and locating interesting areas to high magnification of down to nanometer large surface features within seconds. This paper presents surface characterization of dental implant micro topography. 3D topography data was created from SEM images using commercial photogrammetric software. A coherence scanning interferometer was used for reference measurements to compare with the 3D SEM measurements on relocated areas.
As a result of this study, measurements emphasizes that the correlation between the accepted CSI measurements and the new technology represented by photogrammetry based on SEM images for many areal characterization parameters are around or less than 20%. The importance of selecting sampling and parameter sensitivity to varying sampling is high-lighted. Future work includes a broader study of limitations of the photogrammetry technique on certified micro-geometries and more application surfaces at different scales.

1. Introduction and background

In recent years 3D analysis has become a tool extremely practical and widely used for both industry and research. To assist in measuring of the engineering surfaces new ISO standards were established. Traditionally, the surfaces are characterized by optical interferometers and confocal instrumentation but their resolution are limited in the lateral dimension to the wavelength of the visible light (around 500 nm). Thanks to other instrumentation which uses electromagnetic radiation with smaller scale than the visible light like x-ray-, electron-and neutron irradiation, it is possible to explore smaller features [1].

Today the scanning electron microscope (SEM) is utilized not only in the medical and biological field, but has lot of applications in industry. Unfortunately SEM measurements are just 2D images (in x-y scale) and height details (z-scale) are only represented by color or grey scales. The great advantage in using SEM is that measurements in scale from millimeters to nanometers are possible. Besides investigation of features on surfaces is easy procedure with large magnification options in case of

[4] Corresponding author.

Journal of Physics: Conference Series **483** (2014) 012026 doi:10.1088/1742-6596/483/1/012026

surface measurement. The intermediate between 3D images and SEM measurement is photogrammetry, the technology that involves extraction of 3D geometry information from 2D pictures. The measurements received by this technique promise high accuracy compare to another technique, but still a lot of investigation remains. There are a number of algorithms for reconstruction of SEM stereo images to obtain 3D topographical data, one of them are described by P. Podsiadlo and G.W. Stachowiak [2]. The 3D reconstruction trough stereo-photogrammetry has some critical factors: quality of scanning images and instrument set-up [3].

The aim of this work was to compare the measurement results obtained by photogrammetric technique implemented in SEM and measurement performed using coherence scanning interferometer on a dental implant micro screw.

2. Materials and methods

2.1 Measuring equipment

The coherent scanning interferometer

Surface measurements were effectuated using CSI (MicroXam, ADE Phase shift Technologies)[5]. This is an optical measurement instrument efficient to obtain 3D topographies of surfaces down to nanometer scale. 3D interferometric profiling splits a beam of light, reflecting it off both the test and reference surface, then the light recombines and interference patterns are detected by a high resolution camera. Afterwards this information interprets to provide height information on a pixel basis, building a highly resolved 3D representation of the surface. The resolutions for CSI used in experiment were in the x- and y- axis 0,5 μm and 0,1 nm in z-depth. The accuracy for the instrument is better than 0,1 %.

The Scanning Electron Microscope

The measurements were done using SEM JEOL JSM-6480[6]. The principle of the SEM is to scan the surface of the sample measured by an electron beam in order to explore it by successive lines. A detector picks up the signal generated by the interactions between the beam and the surface. The image obtained by SEM is in 2D. Spatial resolution at 20 kV is 3 nm for secondary electron image, and 4 nm for backscattered electron image.

Photogrammetric technique – principle and algorithm

3D measurement data was created from 2D SEM pictures using photogrammetry technique software (MeX 5.1, Alicona[7]). This type of software provides 3D images reconstructed from 2 or 3 2D pictures taken from different points of view. Thus the method used is based on the principle of the stereoscopy [2]. SEM images have a very high resolution but nevertheless, they are still in 2D, which is incomparable with 3D topographies obtained using the interferometer. The accuracy of the SEM stereoscopy measurement depends on the magnification and the pixel size.

2.2. Sample – Dental implant

In this case study measurements were performed on an anodically oxidized dental implant screw [4]. The roughness for this sample is really important to evaluate because it determines the quality of its anchorage in the jaw bone. Its surface has many pores which are obtained by anodic oxidation. The dental implant has been acquired with the magnifications 30x, 500x, 2000x and 10000x using the SEM and the magnification 100x using the interferometer. The ultrasonic bath with a cleaning solution was used for cleaning of dental implant sample. To compare the images obtained with the SEM and the interferometer, the same area were allocated.

[5] ADE Phase Shift MicroXAM Optical interferometric profiler characteristics [online]. http://www.tcd.ie/CMA/misc/MicroXam.pdf (page accessed in February 2013).

[6] www.jeol.com

[7] http://www.alicona.at

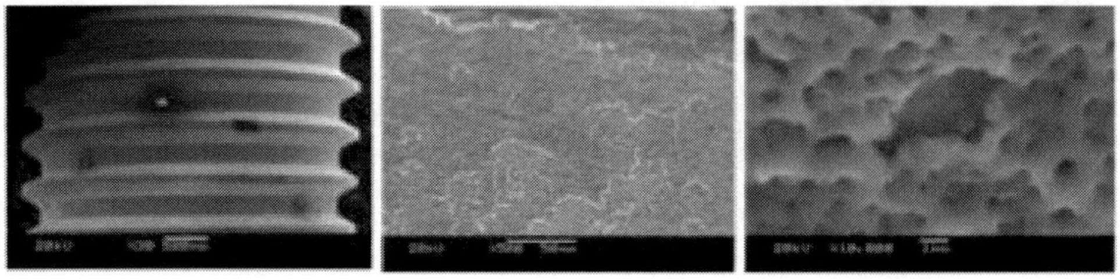

Figure 2.1 Dental implant. SEM pictures at 30x, 500x, and 10 000 magnifications.

2.3 Comparison method

3D surfaces based on SEM images were reconstructed with MeX software, and afterwards exported to the analysis software MountainsMap from DigitalSurf[8]. Three images were used for stereoscopic reconstruction. SEM images were obtained by eucentric tilt of the sample, angle of tilt was 5°.The non-measured points of the surfaces obtained using the interferometer were replaced by a value obtained as compared to their neighboring valid points. As a result, filling-in is perfectly invisible and very efficient. The form of the surface has been removed. In this study the polynomial method of third degree was used.
The relocation area for comparison for the SEM and the interferometer is demonstrated in the (figure 2.2).

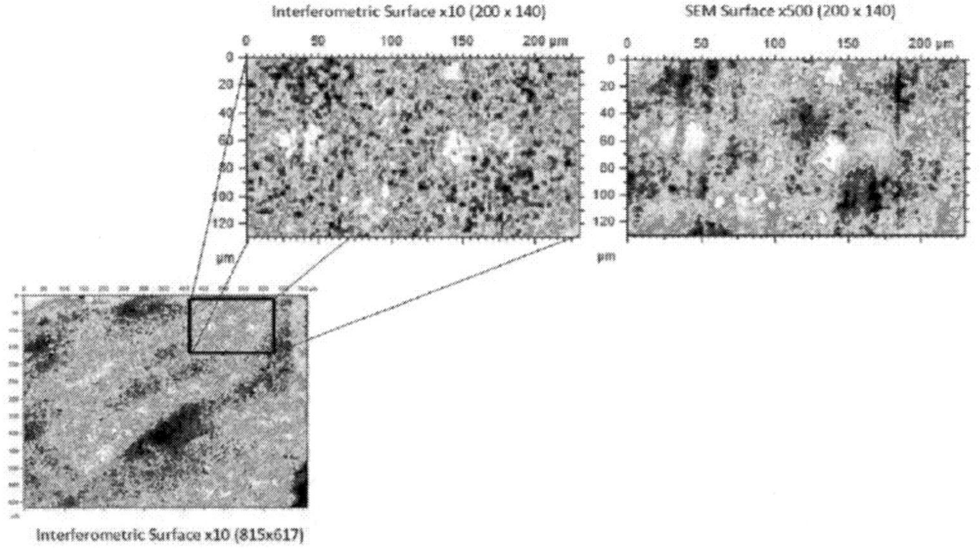

Figure 2.2: Example of comparison using relocated areas for CSI and SEM.

To highlight the differences between the SEM and the interferometer, two comparisons were made. The first compares the parameter S_a of the interferometric surface with that one of the measured surfaces for different wavelengths. To do this, several SL-band pass filters with different wavelengths, -nesting index, were applied to the surfaces. The SEM has a higher resolution than the interferometer and the comparison with the CSI is made with both the SEM surface with the higher resolution and the SEM surface re-sampled to the same resolution as the interferometric surface.

[8]http://www.mountainsmap.com

The second comparison was a global comparison of all of the parameters of the interferometric surface and SEM surface (re-sampled and not). No filter is applied for this second comparison.

3. Result and discussion

Visual and overall Comparison of CSI and SEM measurements

The following results concern the dental implant based on single relocated measurements applying ISO standard Areal parameters [5], [6].

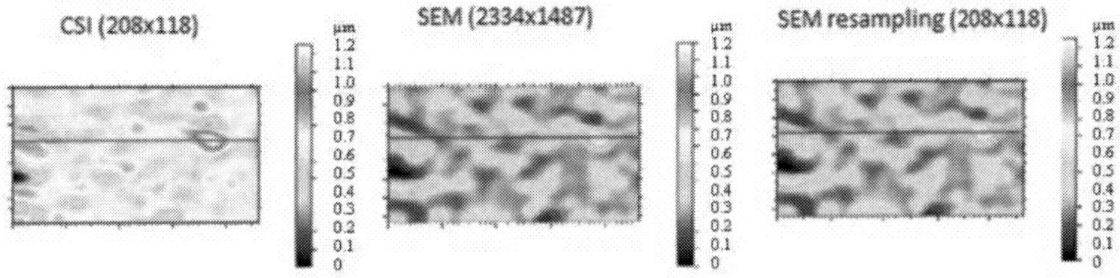

Figure 3.2 Relocated surfaces obtained with the two different measuring instruments. The measurement size is the same for all three measurements (200x140µm) but the sampling numbers differs (208x118, 2334x1487, 208x118). The solid line indicates a profile through the major peak.

The extracted and relocated surfaces show a strong visual correlation in spite the use of two totally different metrology technologies, the one using visible light and the other evaluating the interaction between impacting electrons on the surface.

The amplitude parameters

The second comparison is a comparison of an advanced set of areal roughness parameters obtained from the SEM images, the interferometer and the re-sampled SEM images (fig. 3.3).

The averaging amplitude parameters S_a, S_q, and extreme amplitude parameters Sp (peak height), S_v (valley depth), and Sz (maximum height) give information of the height of the surface profile. The values of the averaging parameters, differs between 18% (S_q) and 21% (S_a) with higher CSI measurement values. Peaks, depths and total height differ all between 5% and 13% and re-sampled SEM measurements seem to underestimate the extreme points (S_q) of the surface with around 5%. However, interferometer and high resolution SEM measurements seems to agree about valley and peak heights on a 5% level indicating a good agreement between the two technologies.

The parameters S_{sk} and S_{ku} (skewness and kurtosis of the amplitude distribution curve) give more information about the distribution of the amplitudes. S_{sk} is the degree of symmetry of the surface heights about the mean plane.

Nominally for the dental surface, S_{sk} is negative (around -1 for all the measurements) indicating the slight predominance of valleys before peaks which describe surfaces with good volume retention.

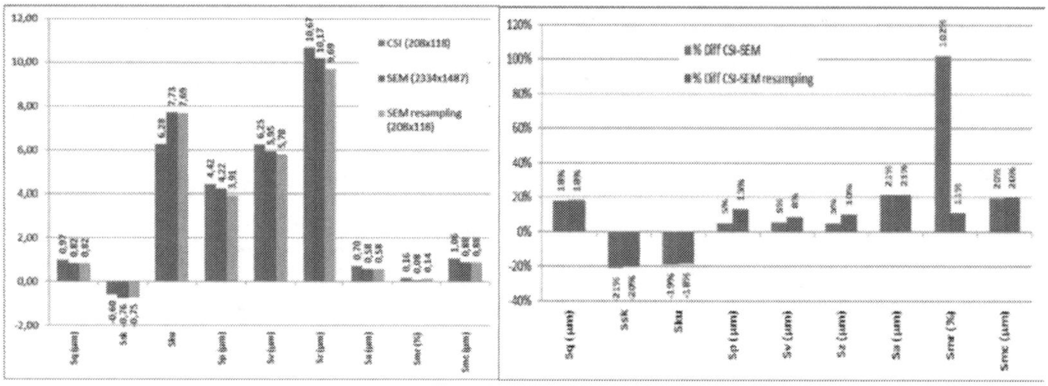

Figure 3.3 Absolute value of differences of the parameters obtained using the SEM, the SEM re-sampling and the interferometer.

The kurtosis, S_{ku} values between 6-8 indicate a more "central" distribution of the amplitudes in comparison the ideal normal distribution of heights. Both S_{sk} and S_{ku} parameters indicate about 20% larger values for CSI measurements (fig. 3.3) hence, stronger negative skewness and higher kurtosis for SEM measurements.

The spatial- and hybride parameters

The spatial parameters (S_{al}, S_{tr}, S_{td}), indicate the wavelength where repeating structures start dominate before random anisotropic structures (S_{al}), degree of anisotropy S_{tr} and S_{td} indicating the texture direction of the global surface profile.

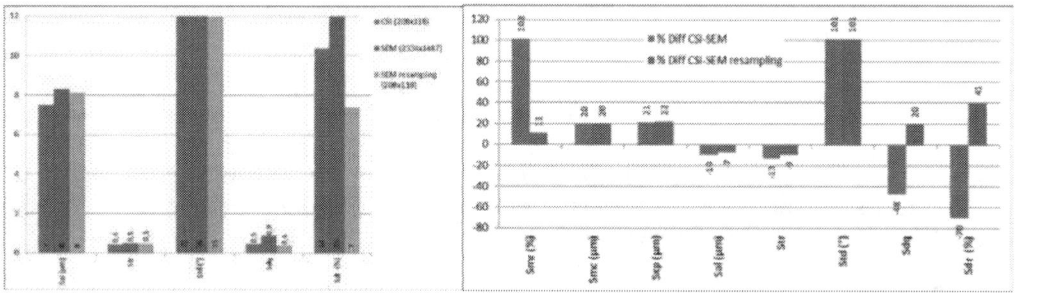

Figure 3.4 Absolute- and relative differences of the parameters obtained using the interferometer (CSI), the SEM, and the SEM re-sampled to the same lateral resolution as the CSI measurements.

S_{al} is around 8um for all measurements indicating that also the interferometer with the lower resolution should be able to pick up dominant repeating features while the sampling distance is well below 8um (fig. 3.4). The S_{tr} is overall close to 0.5 indicating neither clear isotropic nor anisotropic main structures and S_{td} directions are therefore less meaningful for the evaluation. SEM measurements are 7-13% higher for S_{al} and S_{tr} measurements but the 100% difference for S_{td} measurements of texture directions is caused by the ambiguous anisotropy as indicated by the S_{tr} at around 0.5 as discussed above.

S_{dq} and S_{dr} are the hybrids parameters characterizing the rms slope and the developed area ratios where the S_{dr} is expressed as the percentage of additional surface area contributed by the texture as compared to an ideal plane the size of the measurement region. S_{dr} is affected both by texture amplitude and spacing. The SEM resolution (2334x1487 measurement ponts) is much higher than the SEM re-sampled and the CSI interferometer (both 208x118 measurement points), and the increased parameter values for SEM measurements indicate that the dental surface has a structure below the resolution if the CSI.

The volume parameters and named feature parameters

The volume parameters V_m, V_v, V_{mp}, V_{mc}, V_{vp}, and V_{vc} (see fig. 3.5 below) show a minor difference when re-sampling the SEM measurements. The interferometer measurements are between 7% to 28% higher. This increase for interferometer measured parameters could be a result and correlates well to the increased CSI amplitude values observed in fig. 3.3 above.

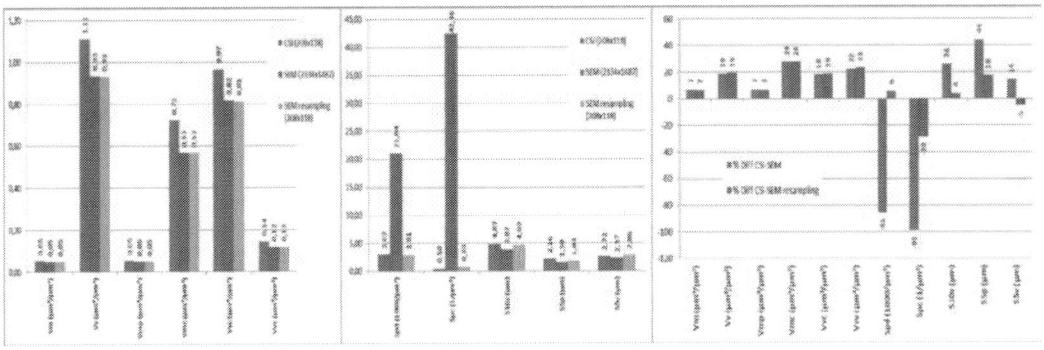

Figure 3.5 Absolute- and relative differences of the volume parameters and named feature parameters obtained using the interferometer (CSI), the SEM, and the SEM re-sampled to the same lateral resolution as the CSI measurements.

The named feature parameters denoting the average 10point height difference, S_{10z}, between the five highest peaks, S_{5p}, and the five deepest valleys, S_{5v}, show both a 20% difference between original measured- and re-sampled SEM measurements due to the sampling sensitivity of the parameters (fig. 3.5). Again, absolute interferometer measurements of the peak- and valley height parameters follow the results from the amplitude height parameters above with higher peak- and valley heights here. 14%-44% smaller heights are measured with the high resolution SEM and as a result of the sampling distance sensitivity of the parameters peaks- and valleys are being redefined resulting in re-sampled SEM measurements are more similar to interferometer measurements (-5%-17% higher CSI parameter values).

The peak density, S_{pd}, and peak curvature, S_{pc}, indicate high differences (85%-98%) between low resolution interferometer and high resolution SEM measurements while a re-sampling of the SEM measurements results in a decreased difference (-7% to 28%) between the CSI- and SEM measurements, again highlighting the sensitivity of lateral sampling distance (fig. 3.5).

Further named feature parameters

Intresting parameters are the named feature parameters dale area, S_{da}, hill area, S_{ha}, dale volume, S_{dv}, and hill volume, S_{hv}, see fig. 3.6 below.

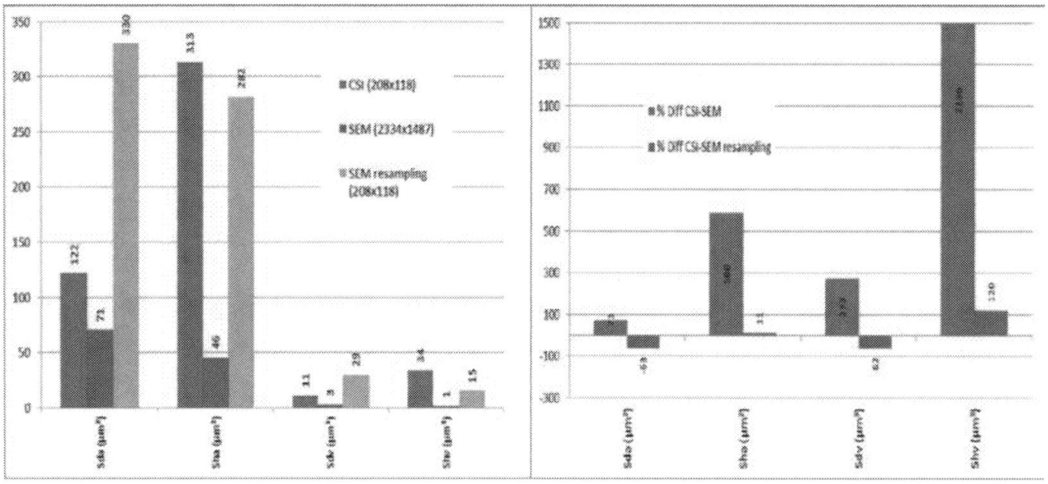

Figure 3.6 Absolute- and relative differences of the feature parameters obtained using the interferometer (CSI), the SEM, and the SEM re-sampled to the same lateral resolution as the CSI measurements.

In absolute values the difference between the instrument types are large. Dale- and hill areas and volumes are following the same trends in absolute differences. The hill areas, dale volumes and hill volumes show a significant sampling distance dependence shown by the 273% to more than 2000% smaller SEM measured parameter values for high resolution measurements. This is also clear through the high difference between parameter values calculated with high resolution SEM compared to the re-sampled values.

4. Conclusions

Several conclusions can be drawn from the dental implant measurement using coherence scanning interferometry (CSI) and 3D SEM technology.

- For measurements using the same sampling distance SEM and Interferometer measurements normally for a majority of characterization parameters not exceed 20% difference.
- SEM average amplitudes parameters S_a, S_q are around 20% smaller than CSI ones.
- Several parameters, S_{mr}, S_{dr}, S_{dq}, S_{pd}, S_{pc} are highly sampling dependent and can vary 56%-100% between the high resolution- and re-sampled SEM measurements.
- Dale- and hill parameters are heavily sampling dependent and for the dental implant huge variations (up to 900%) are found indicating the sensitivity of this set of 3D characterization parameters.
- Dale- and hill parameters show significant differences between SEM- and CSI measurements for both high- and low resolution measurements.

5. Future works

Further comparisons using known certified geometrical features will be made to verify and further explore the results from this study.

More studies at different scales and for other applications will further help the introduction of the new 3D SEM metrology technique and enable limitations and possibilities to be determined.

6. References

[1] Blunt L., Jiang X., "*Advanced Techniques for assessment Surface Topography*", Development of a Basis for 3D Surface Texture Standards "SURFSTAND", Kogan Page Science, ISBN 1 9039 96 11 2, (2003).

[2] Podsiadlo P., Stachowiak G. W.,"*Characterization of surface topography of wear particles by SEM stereoscopy*", Wear, v 206, n 1-2, p 39-52, (1997).

[3] Marinello F., Bariani P., Savio E., Horsewell A., De Chiffre L., "*Critical factors in SEM 3D stereo microscopy*", Measurement Science and Technology, v 19, n 6, June 1, (2008).

[4] Sul Y.-T., Johansson C.B., Roser K.,T, Albrektsson T., "*Qualitative and quantitative observations of bone tissue reactions to anodized implants", Biomaterials 23, 1809–1817, (2002).*

[5] ISO 25178-2:2012, "Geometrical product specification (GPS) – Surface texture: Areal – Part 2: Terms, definitions and surface texture parameters", International Organization for Standardization, (2012).

[6] ISO 25178-3:2012, "*Geometrical product specification (GPS) – Surface texture: Areal – Part 3: Specification operators*", International Organization for Standardization, (2012).

AUTHOR INDEX